L. Yahia

Shape Memory Implants

Springer

Berlin
Heidelberg
New York
Barcelona
Hong Kong
London
Milano
Paris
Singapore
Tokyo

L. Yahia (Ed.)

Shape Memory Implants

With 179 Figures and 67 Tables

Springer

Ph. D. L. Yahia

Ecole Polytechnique de Montreal
Institut de Genie Biomedical
Station „Centre Ville“
Montreal H3C 3A7
Canada

ISBN-13: 978-3-642-64118-3 e-ISBN-13: 978-3-642-59768-8
DOI:10.1007/978-3-642-59768-8

CIP data applied for
Library of Congress Cataloging-in-Publication Data

Shape memory implants / L'Hocine Yahia (ed.). p. cm. Includes bibliographical references and index. 1. Shape memory alloys – Therapeutic use. 2. Medical instruments and apparatus. I. Yahia, Hocine, 1952-R857.S42 S47 2000 610'.28--dc21
00-038819

Springer-Verlag is a company in the BertelmannSpringer publishing group

Softcover reprint of the hardcover 1st edition 2000

Cover design: *design & production* GmbH, Heidelberg
Typesetting: cicero Lasersatz, Dinkelscherben
Printed on acid-free paper – SPIN: 10729915 126/3135 5 4 3 2 1 0

Summary

Shape memory alloys (SMA) are now considered as a maturing technology. This hands-on reference integrates the latest advances in nickel-titanium alloys in orthopaedic, orthodontic and cardiovascular applications.

Emphasizing current and future SMA for clinical and commercial use, "Shape Memory Implants" systematically examines the biocompatibility and biofunctionality of SMA and their highly specific properties qualifiying them as smart biomaterials. It discusses SMA in transarticular fracture, long bone shaft fracture, hand, hip, foot, and spinal surgery. It describes the use of SMA for orthodontic wires as force delivery systems, and for endodontic files as surgical tools. It analyses in detail the use of SMA in stenting of blood vessels as well as extravascular corrections. Finally, it shows how SMA is exploited as a drug delivery system.

With 23 contributions from Germany, Canada, Japan, USA, Belgium, Russia, China, Netherlands, and France, "Shape Memory Implants" is an incomparable reference for biomaterialists, bio-engineers, biomedical researchers, surgeons and medical devices companies, and graduate students in these disciplines.

Dr. L'Hocine Yahia is currently a full professor at the École Polytechnique de Montréal. He received a doctoral degree in Mechanical Engineering (Biomechanics) from the École Polytechnique in 1984.

Since 1991 he has also been the Director of the Biomechanics and Biomaterials Research Group and Codirector of the Orthopedics and Imaging Laboratory of Notre Dame Hospital (CHUM).

About the editor. L'Hocine Yahia is Professor of Biomedical Engineering at the Ecole Polytechnique of Montreal, Quebec, Canada and author or coauthor of over 120 professional papers and book chapters in the fields of advanced biomaterials and biomechanics. Awarded by the International Academy of Shape Memory Materials for Medical Use in 1998 for his contributions in SMA bioperformances, Dr. Yahia received a Doctorat 3ème cycle (1980) in biomechanics from Compiègne University of Technology, France, and a Ph.D degree (1984) in Biomedical Engineering from the Ecole Polytechnique of Montreal, Canada. He is a member of the Society for Biomaterials, International Academy of Shape Memory Materials for Medical Use, International Society for Biomechanics, Canadian Biomaterials Society, and Société Internationale de Recherche Orthopédique et Traumatologique (SIROT).

Preface

This book discusses the distinct and unique mechanical properties of the shape memory alloys: the technology developed to produce the NiTi binary alloys, NiTiMo and porous NiTi used in medecine with special references to quality control; the damping capacity of these alloys; the corrosion resistance and biocompatibility; and their use in medical implant and instrument developments. "Shape Memory Implants" emphazises the solid NiTi that is produced commercially and the porous NiTi whose potential for commercialization is very likely. One of the earliest medical uses of SMA was to manufacture orthopaedic staples which compressed fractures to provide both stability and accelerated healing. The material has also been used to manufacture bone-plates and spine systems, using the memory effect to act as an amplantable actuator. The book covers the current state of orthopaedic implants made from SMA, the clinical experience and futures opportunities.

Orthodontic wires with shape memory and superelastic properties are extensively used during orthodontic treatments, particularly in the initial stages. Many of the characteristics of these wires such as enhanced resilience and low to moderate stiffness, help clinicians improve their efficiency and productivity while patient safety and comfort have improved as well.

SMAs have been utilised in a variety of medical applications, but particularly with intravascular stents and stent grafts. Nitinol stents and stent grafts have proved particularly valuable when their thermal memory allows them to be self-expanding when exposed to body temperature.

During the last decade, endovascular treatment for emergency and elective procedures has emerged as a viable alternative for open surgery with many potential benefits. The increased use of both superelastic and thermally induced shape recovery, SMA has been instrumental in the development of several novel devices which enabled an expansion in the treatment modalities to occur.

Over 23 shape memory alloys experts from academia, industry and clinicians have contributed to this work. Certainly such a distinguished group of authors provides the needed balance and perspective. This book is divided into four parts: the basic properties of shape memory alloys; the orthopaedic applications; the orthodontic applications; and the endovascular applications.

The first part (Chaps. 1–7) of the book highlights the unique properties of the SMA, their corrosion resistance and biocompatibility, their damping capacity, and their mechanical compatibility. The processing and quality control aspects are also covered.

The second part (Chaps. 8–12) deals with the orthopaedic applications of SMA. Studies of different SMA devices conducted in France, Russia, China and Netherlands are presented. Altogether, over 3800 clinical studies with a follow-up ranging from 1–12 years are reported.

The third part (Chaps. 13–17) is devoted to the orthodontic applications. Extensive clinical studies conducted in Germany, Japan and China are presented.

The fourth part (Chaps. 18–23) of the book describes the use of SMA in endovascular surgery. It covers the interventional stenting and a separate chapter is devoted to SMA drug delivery systems.

A few ackowledgements and thanks are in order. First, a special thanks to my wife Djouha Khaouis and our children Rilas Juba, Massine, and Davia for their patience and indulgence: « tanemmirt nwen ! ». Next, I would like to thank all the contributors for making their original work to the benefit of a wider audience; their expertise and perspectives are clearly the backbone of this work.

Finally, a number of individuals at the Ecole Polytechnique of Montreal have contributed to the assembly and production aspects of this work. I offer my special thanks to Irina Chapalo, Rommy Hernandez, and Michel Assad for their assistance and special effort on this important component of the project. My graduate students and researchers involved in our Shape Memory Implants program, Sylvie Lombardi, Michel Assad, Nicola Hagemeister, Franceline Villermaux, Christine Trépanier, Benjamin Thierry, Souad Rhalmi, and Maryam Tabrizian also deserve tremendous credit and accolades for their contributions and insights. This book will be useful for professionals in medical industry, clinics and research, and for all students in biomaterials and medical devices.

L'Hocine Yahia, 15 October 1999

Contents

Basic Properties

Corrosion Resistance and Biocompatibility of Passivated NiTi

Christine Trepanier, Ramakrishna Venugopalan, Alan R. Pelton

The High Damping Capacity of Shape-Memory Alloys

J. Van Humbeeck, Y. Liu

Physical and Biochemical Principles of the Application of TiNi-Based Alloys as Shape-Memory Implants

L.L. Meisner, V.P. Sivokha

Shape-Memory Implants in Spinal Surgery: Long-Term Results (Experimental and Clinical Studies)

Boris M. Silberstein, Victor Gunter

The Use of a Memory-Shape Staple in Cervical Anterior Fusion (about 100 Human Implantations)

Olivier Ricart

The Double Compressive Nickel–Titanium Shape-Memory Staple in Foot Surgery

Louis Samuel Barouk

Orthodontic Applications

Corrosion Behavior of Ni–Ti Alloys in a Physiological Saline Solution

Kazuhiko Endo, Hiroki Ohno

NiTi Alloys in Orthodontics

Andrea Wichelhaus

Other Medical Applications

An Implantable Drug Delivery System Based on Shape-Memory Alloys

Dominiek Reynaerts, Jan Peirs, Hendrik Van Brussel

Authorinfo

R.A. Ayers
BioServe Space Technologies, University of Colorado,
Boulder, Colorado USA

L.-S. Barouk
Polyclinique de Bordeaux, 147 Rue du Tondu, 33000 Bordeaux, France

T.A. Bateman
BioServe Space Technologies, University of Colorado,
Boulder, Colorado, USA

L. Bilodeau
Montreal Heart Institute, 5000 Belanger Street, Montréal,
Québec H1T 1C8, Canada

A.C. Butckevich
Moscow Steel and Alloys Institute, Moscow, Russia

A.P. Chadaev
Moscow Steel and Alloys Institute, Moscow, Russia

K. Dai
Department of Orthopaedics, Shanghai Second Medical University,
Shanghai, People's Republic of China

K. Endo
Department of Dental Materials Science, School of Dentistry,
Health Sciences University of Hokkaido, Hokkaido, Japan

F. Farzin-Nia
Materials Research and Development, ORMCO Corporation,
Indiana, USA

Y. Fengzhi
Beijing General Research Institute for Nonferrous Metals,
Beijing 100088, China

Y. Guansen
Beijing General Research Institute for Nonferrous Metals,
Beijing 100088, China

V. Gunter
Research Institute of Traumatology and Orthopedics,
Republic Center of Vertebrology, Novosibirsk, Russia

W. Hennemann
Intervascular Inc., Florida, USA

N.T. Hyun
Division of Materials Science and Engineering,
Gyeongsang National University, Korea

D. Ibe
Orthodontist Beselerplatz, Hamburg, Germany

G. Jinfang
Beijing General Research Institute for Nonferrous Metals,
Beijing 100088, China

I.Y. Khmelevskaya
Moscow Steel and Alloys Institute, Moscow, Russia

Y. Liu
The Hong Kong Polytechnic University, Department Mechanical
Engineering, Room FG615, Kowloon, Hong Kong

Y. Merhi
Montreal Heart Institute, 5000 Belanger Street, Montréal,
Québec H1T 1C8, Canada

L.L. Meisner
The Institute of Strength Physics and Materials Science, Tomsk, Russia

M. Mertmann
Memory-Metalle GmbH, Nordwalde, Germany

Z. Ming
Beijing General Research Institute for Nonferrous Metals,
Beijing 100088, China

T.-H. Nam
Division of Materials Science and Engineering,
Gyeongsang National University, Gyeongsang, Korea

H. Ohno
Department of Dental Materials Science, School of Dentistry, Health Sciences University of Hokkaido, Hokkaido, Japan

Y. Oshida
Department of Restorative Dentistry, Indiana University School of Dentistry, Indiana, USA

J.D. Pazienza
Intervascular Inc., Florida, USA

J. Peirs
Department of Mechanical Engineering, Katholieke Universiteit Leuven, Heverlee, Belgium

A.R. Pelton
Cordis Corporation - Nitinol Devices and Components, 47533 Westinghouse Drive, Fremont, CA 94539, USA

S.D. Prokoshkin
Moscow Steel and Alloys Institute, Moscow, Russia

I.K. Rabkin
Moscow Steel and Alloys Institute, Moscow, Russia

D. Reynaerts
Department of Mechanical Engineering, Katholieke Universiteit Leuven, Heverlee, Belgium

O. Ricart
Clinique Ambroise Paré, Thionville, France

J. Ryhänen
Department of Surgery, Faculty of Medicine, University of Oulu, Oulu, Finland

E.P. Ryklina
Moscow Steel and Alloys Institute, Moscow, Russia

O. Savadogo
Physics and materials Department, Ecole Polytechnique of Montreal, Montreal, P.O. Box 6079, Station "Centre ville", Montreal, Quebec H3C 3A7, Canada

D. Siegner
Orthodontist Beselerplatz, Hamburg, Germany

B.M. Silberstein
Research Institute of Traumatology and Orthopedics,
Republic Center of Vertebrology, Novosibirsk, Russia

S.J. Simske
BioServe Space Technologies, University of Colorado, Boulder,
Colorado, USA

V.P. Sivokha
The Institute of Strength Physics and Materials Science, Tomsk, Russia

M. Tabrizian
Biomedical Engineering Institute Biomaterial/Biomechanics Research
Group (BBRG), Mechanical Engineering Department,
Ecole Polytechnique of Montréal, Montréal, C.P. 6079, Succ.
"Centre ville", Montréal, Quebec H3C 3A7, Canada

B. Thierry
Biomedical Engineering Institute Biomaterial/Biomechanics Research
Group (BBRG), Mechanical Engineering Department,
Ecole Polytechnique of Montréal, Montréal, C.P. 6079, Succ.
"Centre ville", Montréal, Quebec H3C 3A7, Canada

C. Trepanier
Cordis Corporation – Nitinol Devices and Components,
47533 Westinghouse Drive, Fremont, CA 94539, USA

H. Van Brussel
Department of Mechanical Engineering,
Katholieke Universiteit Leuven, Heverlee, Belgium

J. Van Humbeeck
Department of Metallurgy and Materials Engineering (MTM),
W. de Croylaan 2, B-3001 Leuven (Heverlee), Belgium

A.G. Veldhuizen
Department of Orthopaedics, University Hospital, Groningen,
The Netherlands

R. Venugopalan
Department of Biomedical Engineering, University of Alabama
at Birmingham, 1075 13th Street South, Hoehn 370, Birmingham,
AL 35294-4461, USA

D.J. Wever
Department of Orthopaedics, University Hospital, Groningen,
The Netherlands

A. Wichelhaus
Central Institute for Biomed. Engineering and Department of Orthodontics, University of Ulm, Ulm, Germany

M. Xujun
Beijing General Research Institute for Nonferrous Metals, Beijing 100088, China

L'H. Yahia
Biomedical Engineering Institute Biomaterial/Biomechanics Research Group (BBRG), Mechanical Engineering Department, Ecole Polytechnique of Montréal, Montréal, C.P. 6079, Succ. "Centre ville", Montréal, Quebec H3C 3A7, Canada

C. Youyi
Beijing General Research Institute for Nonferrous Metals, Beijing 100088, China

Basic Properties

Bioperformance of Shape Memory Alloys

L'Hocine Yahia, Jorma Ryhänen

1 Introduction

Materials that allow structures to adapt to their environment are known as actuators. They can change shape, stiffness, position, natural frequency and other mechanical characteristics in response to temperature or electromagnetic fields. The four most common actuator materials being used today are shape-memory alloys, piezoelectric ceramics, magnetostrictive materials and electrorheological and magnetorheological fluids.

Shape-memory alloys (SMAs) are metals that at a certain temperature revert back to their original shape after being strained. In the process of returning to their "remembered" shape, the alloys can generate a large force useful for actuation. SMA can be found in more than fifty alloy systems. Most prominent among them perhaps is the family of the nickel-titanium alloys developed at the Naval Ordnance Laboratory (now the Naval Surface Warfare Center). The material, known as Nitinol (Ni for nickel, Ti for titanium, and NOL for Naval Ordnance Lab), exhibits substantial resistance to corrosion and fatigue and recovers well from large deformations. Strains that elongate up to 8% of the alloy's length can be reversed by heating the alloy, typically with electric current [20].

The basic mechanism governing the properties of SMA is change in crystal structure: a martensitic structure transforms, at a pre-defined temperature, into an austenitic structure during heating and reverts back during cooling. Many materials undergo martensitic transformations; what distinguishes Nitinol from conventional materials is the ability of these particular martensites to de-twin [95]. While other materials deform by slip or dislocation movement, Nitinol responds to stresses by simply changing the orientation of its crystal structure through the movement of twin boundaries.

Japanese engineers are using Nitinol in micromanipulators and micro-robots actuators to mimic the smooth motions of human muscles. The controlled force exerted when the Nitinol recovers its shape allows these devices to grasp delicate paper cups filled with water. Nitinol wires embedded in composite materials have also been used to modify vibrational characteristics [56]. They do so by altering the rigidity or state of stress in the structure, thereby shifting the natural frequency of the composite. Thus, the structure would be unlikely to resonate with any external vibrations, a process known to be powerful enough to bring down a bridge.

Since the discovery of the shape memory effect in nickel-titanium (NiTi) alloys in 1962, there has been great interest on the part of physicians, surgeons and orthodontists in applying these interesting alloys for various clinical procedures. Although the earliest application was as arch wires in orthodontics [2, 3], where biocompatibility was not a problem, the use in devices which would have much longer dwell time in the human body has been hampered by quantitative information on tissue response and the alloy's biocompatibility in various applications.

Extensive in vivo testing and experience indicates that NiTi is highly biocompatible – more so than stainless steel. Implants exist in dentistry, orthopaedics, and in many other branches of medicine, with large numbers of permanent implantations reported in Japan, Germany, China and Russia dating back to the early 1980s [21, 30, 31, 52]. Perhaps the longest and most extensive history pertains to the Dental Implant, in use in Japan since the early 1980s.

2
Medical Applications

The combination of good biocompatibility, good strength and ductility with the specific functional properties of SMA such as the shape memory effect, damping capacity and superelasticity creates a smart material for medical applications. In particular, the damping capacity of NiTi was exploited in many orthopedic and dental implants to solve the mismatching problem leading to loosening. It is known that no hard tissue prosthesis, e.g. total hip prosthesis, lasts a lifetime and there are numerous ideas on how to improve the survival time. Some of the perceived mechanisms of failure have included a lack of good mechanical integration of bone into the metal interface of uncemented stems. This has led to ideas for changing the interface and has also led to a number of modifications. One of the basic problems is that no conventional material, except SMA, is known that has similar mechanical properties to bone. One of the major differences is that bone displays viscoelastic behavior whereas metals do not. This mechanical mismatch may lead to direct cellular reactions through mechano-sensors. The mechanical mismatch will also lead to fretting and several investigations into many aspects of this problem have been carried out [41, 51]. NiTi alloy possesses high damping capacity deriving from high internal friction between the martensitic twins or between the martensitic and the parent phases. The superior damping capacity and quasi-static stress absorbability of NiTi alloy has been previously reported: it was found to transmit less impact stress than titanium or stainless steel [109, 110]. The magnitude of damping in TiNi alloys is at least one order of magnitude greater than in conventional alloys such brass, steel or aluminium where damping, measured as specific damping capacity (SDC), is from 0.5% to 1.5%. Gray cast iron, used for machine tool beds because of its relatively high damping, achieves SDC values of 10–12%. By contrast the SDC of typical SMA is in excess of 40% – values close to those of hard rubber [110]. Damping in these alloys increases with stress but unlike polymers and rubber is relatively insensitive to frequency. This insensitivity is similar to that found in living tissue where the hysteresis loop is almost independent of the strain rate within several decades of the rate variation [28]. Gunter et al. have introduced porous NiTi to further improve the mechanical compatibility (elastic modulus and permeability) of NiTi with spongious bone [30]. The pores of the device are easily filled with growing soft bone tissue, assuring a compatible assimilation of the implant.

The superelastic effect of SMA results in a unique combination of high strength, high stiffness and high pliability. This concept of a metallic material with superelasticity and nearly constant stress levels over a large strain area has found many applications in stenting and in orthodontic wires. The following paragraphs summarize the main developments of applications in the biomedical field.

2.1 Orthopedic Surgery

In late 1960s, Johnson and Alicandri suggested the potential of NiTi as an implant material. After that much further studies have been made to study the ability of the alloy for orthopedic surgery. Some pioneers in this field were Baumgart et al., who examined the NiTi distraction rod in the correction of scoliosis [8]. As early as 1986, Lu et al. implanted NiTi rods in patients with scoliosis with good reported results and no complications [45]. Latter, Matsumoto et al. [46] and Sanders et al. [71] published more in vivo experimental studies.

One of the first orthopedic devices used inside the human body was a NiTi compression staple. They were first introduced in China 1981 [19]. After that, NiTi staples and clamps have been used in several applications like comminuted fractures of the short tubular bone [107], for fixation of mandibular fractures [24], for metatarsal osteotomies [86], for anterior cervical decompression and fusion [48, 63, 78], for fixation of small bone fragments [53], and for some other cursory applications [18, 39, 41].

Mitek G2 suture anchor is the only NiTi-containing orthopedic implant used in the western world. This anchor has superelastic NiTi wings, which prevent the anchor from pulling out of the bone after insertion and secure the tendons or ligaments to the bone [6]. Promising application is also a NiTi hook used to restore the dislocated acromio-clavicular joint [68]. In his book, Gunter has listed over 50 orthopedic and surgical devices using SMA [30].

There are also a number of implants and tools developed from NiTi alloys for dentistry [52, 76]. Superelastic NiTi dental implants are constructed in the manner allowing easy fixation. They have an optimal combination of specific weight, strength, and plasticity, high wear characteristics as well as resistance to fatigue.

2.2 Cardiovascular Surgery

The Simon Nitinol filter (SNF) was the first clinically successful vascular NiTi device. It has been used to treat pulmonary embolism [79]. The filter is inserted as a thin wire via the small bore catheter. When reaching the lumen of the inferior vena cava and sensing body temperature, it reverts to its filter shape and locks into place. Further thromboemboli from the pelvis or the lower limbs will be trapped.

The general trend of stenting seems to be towards self-expandable NiTi-based stents. When thin stent is placed in the narrowed artery, it expands and dilates the artery. The carotid artery stents and the endoluminal polyester-covered NiTi stent-grafts for infrarenal abdominal aortic aneurysms have been shown to be efficient and technically successful, but a careful long-term evaluation is still necessary [11, 100]. Intracoronary [23, 54] and peripheral vascular NiTi stenting [75] also seems to be increasing. Despite the improvements, restenosis and reocclusion still is a problem and the optimal physical and surface properties of an arterial stent have not been defined yet [73]. Endoluminal repair of infrarenal abdominal aortic aneurysms with the use of Dacron-covered NiTi stent-grafts has shown to be safe and clinically effective. Surprisingly, the attempts to advance properties with a heparin-coated Dacron cover have increased inflammatory response [74]. The need to evaluate the biocompatibility of new vascular devices is still evident. Polyurethane stent coating was also associated with an inflammatory tissue response [61]. An occlusion device of atrial septal defect closure has also been recently reported. The implant consists of two umbrellas placed over a long veno-arterial guide-wire. It has been used in a few cases of adults and children [35, 77].

2.3 Gastroenterologic Surgery

The treatment of benign biliary strictures with metallic stents is associated with a low long-term patency rate [10, 65]. They are effective in achieving long-term palliation in-patients with malignant obstructive jaundice. The use of stents reestablishes bile flow in the occluded biliary tree [81].

Esophageal NiTi stents have been found to be easy to implant, provide effective palliation of malignant esophageal obstructions, and have a low risk of severe complications. NiTi stents for esophageal strictures and the palliation of malignomas have been studied by several authors [1, 17, 47]. Some problems were related to incomplete initial stent expansion as well as tumor ingrowth/overgrowth. Covering the NiTi-based stent with a thin Gore-Tex sheath give a possibility to avoid ingrowth and to use the stent also in the case of fistulas. The insertion of NiTi stents in patients with rectosigmoidal carcinoma provide an alternative to repeated palliative laser therapy or palliative surgery in malignant rectosigmoid obstruction [83].

2.4 Urologic Surgery

The use of NiTi prostatic stents has increased since the first reported experiment by Lopatkin et al. [44]. For high-risk patients with prostatic carcinoma or benign prostatic hyperplasia, the insertion of a permanent metal stent system offers a useful alternative to transurethral resection [31–33]. The use of urethral stents was found to decrease the number of repeated dilatations and urethrotomies in recurrent urethral strictures [101]. Despite the good biocompatibility of the material in a long-term study on dogs, no complete covering of the stent by epithelialization was found in a study of Latal et al. [43].

2.5 Other Medical Devices

Orthodontic archwires were the first mass biomedical application [55]. Besides archwires, specific endosseous implants based on the shape memory effect have also been developed [92].

A NiTi-based mesh-expanding prosthesis for laparoscopic hernioplasty significantly shortened the operating time in a study of Himpens [37]. The good holding and atraumatic characteristics of the detachable clamp have been confirmed by use in laparoscopic and thoracoscopic surgery on the gastrointestinal tract [27].

The use of NiTi stents to prevent major airway occlusion was first reported by Rauber et al. [60]. According to the early tests, they seemed to be very useful and effective in inoperable tracheal or bronchial stenosis due to intraluminal tumor invasion [34, 106]. Also, new type of NiTi stapts prosthesis to restore the ossicular fixation after stapedectomy has been reported [40].

2.6 FDA Status of NiTi Medical Devices

Since the first application of SMA in medical application over 20 years ago [14], interest has grown steadily, however not at the extent expected: this is due mainly to the biocompatibility and regulation issues, such as the U.S.A. Food and Drug Administration (FDA). FDA does not regulate the materials of medical devices – but regulates the devices themselves. Most of the NiTi device fall into regulatory class III, which means that human clinical data, is required to support full approval to market the device. The problem of most clinical studies is that they rarely satisfy the quality criteria of scientific study. It is not enough to say that a certain NiTi implant can be used without harm. To be considered fully successful, it must be proved to be better than the existing competitors. At the present, there are no comparative clinical studies and the series have generally been small. Randomized prospective studies are needed to apply new NiTi implant devices for constant clinical use in humans. The only accepted applications include Simon Nitinol Filter (SNF) and recently (1999), FDA released on the market a biliary stent.

Shape memory alloys are used in medical devices in order to take advantage on their highly specific property to provide biofunctionality. With so few alloys displaying the shape memory effect, there is little choice over the composition of the material to be placed in the body and so biocompatibility has to be considered in a different light. The SMAs cannot be selected on the basis of their biocompatibility; they have to be chosen based upon their biofunctionality and then determine whether the biocompatibility characteristics are appropriate. The risk benefit analysis has to take on a different meaning to that associated with conventional metallic biomaterials.

FDA requires significant in-vitro and pre-clinical animal data prior to approving an investigational device exemption (IDE) to permit device usage on human subjects. In fact, the approbation of Simon filter and Mitek means that these companies provided for the FDA enough scientific evidence about safety and effectiveness of NiTi devices and assessment of the risks and benefits of using these NiTi medical devices. However, these data are often kept proprietary and then not available for the public domain. This practice is in contradiction with Association for the advancement of medical Instrumentation (AAMI recommending publishing animal studies to reduce animal sacrifices. The next section will summarize the biocompatibility studies performed in cell cultures, animal models and human subjects.

3 Biocompatibility of NiTi Alloys

3.1 Nickel Issue

As shown above, the NiTi alloy is clinically used in several orthodontic, orthopaedic and cardiovascular applications. No adverse tissue reactions or allergic reactions through these implants have been described so far. In spite of this satisfactory clinical use the biocompatibility of NiTi devices is still being discussed. The apprehension about NiTi's biocompatibility is based on the relatively large nickel content of the alloy. Although, nickel is nutritionally essential; it is well known that Ni is capable of eliciting toxic and allergic responses. Ni is a consistent part of all organs of vertebrae. Lower Ni intake reduces growth, decrease life expectancy of reproducing animals. Ni deficiency is accompanied by biochemical charges, reduced iron resorption, and resulted in anaemia. Ni has a role as a studied component of metallic proteins, etc. However, its excess brings negative effects. It is the most common contact allergen affecting females in USA and Europe. The prevalence of nickel allergy among adult females ranges from 7% to 15% [49]. Alloys with a nickel release exceeding 1 $\mu g/cm^2$/week gave a strong patch test reaction, and those below 0.5 $\mu g/cm^2$/week a weak reaction. The Ni allergy problem can be minimized by the use of alloys with a nickel release rate of less than 0.5 $\mu g/cm^2$/week for metal items designed for prolonged skin contact. The allergic response to nickel is believed by some investigators to be limited to contact with the dermis, but it might also stimulate an allergic response in the deep tissue. Rapid elimination of nickel was found by Merritt et al. by showing that almost all injected nickel is eliminated in the urine in 48 h [50]. Nickel was present only at control levels in the urine, serum, red cells or organs 1 week after injection. Allergic contact dermatitis is caused by Ni^{2+} ions, which binds with a carrier protein. This nickel-protein complex activates Langerhans' cells in the skin, which present an antigen to T lymphocytes. Memory T cells are developed when circulating in the body, these memory cells can start cells motivated immune reactions upon meeting the same allergies again. Control of the allergic reactions also requires inhibitory system, which prevents the immune response from causing systemic damage. To control the reactions, several kinds of suppression T cell are generated at different levels – lack of responsiveness to oral exposure (oral tolerance to Ni is due to the action of these suppression cells).

Titanium, the second compound of the NiTi binary alloys, is not on the list of essential trace elements, which includes V, Cr, Mn, Fe, Co, Ni, Cu, Zn, Mo and W. Its normal tissue concentration in humans is 0.2 ppm. No clinical tissular toxicity has been observed, even at local concentrations higher than 2000 ppm [89]. Titanium is biologically inert and it induces neither toxic nor inflammation reactions in connective or epithelial tissues [72].

The biocompatibility of metallic alloys, amongst other factors, is related to release of ions resulting from the corrosion of these alloys. Past tissue corrosion studies have shown that the surface of NiTi, just like titanium, consists of a chemically stable passive film (TiO_2) [106]. For this reason, the NiTi alloy is believed to approximate the excellent corrosion properties of titanium implants. Finally, despite its high nickel content, the strong intermetallic bond between Ni and Ti is believed to prevent the ion release [29, 30]. NiTi is intermetallic composition, which has covalent coupling. Other alloys like CoNiCrMo (with Ni=33%) or stainless steel (Ni=12%) do not concern to this class.

Addition of third elements opens even more possibilities for adapting the properties of binary NiTi alloys toward more specific applications. Adding a third element implies a relative replacement of Ni and/or Ti. Alloying third elements will influence not only the transformation temperatures, shape memory characteristics, hysteresis, strength and ductility but will also have an effect on the corrosion behavior. Several alloying third elements such as molybdenum (Mo), copper (Cu) and niobium (Nb) have been already explored [93]. Alloying of NiTi with Mo reduces a stable passivation current, bringing the curves of the electrochemical behavior of the NiTiMo alloys closer to the similar curves for conventional titanium alloys [30]. However, active diffusion of NiTiMo begins with somewhat lesser potentials than the investigated titanium alloys. It was found that active corrosion of NiTiMo begins since the moment of appearance of a plastic component of strain, corresponding to the level of 4.5% [30]. Implants made of such superelastic material can be subjected to long alternating strain in the human organism without participation of corrosion processes. Recently, Wen et al. studied the corrosion resistance and biocompatibility of ternary TiNiCu alloys in which Cu substitutes mainly Ni [98]. It was discovered that the addition of Cu raises the repassivation potential of TiNi shape memory alloys and improves their corrosion resistance. After two and three month's implantation, there were no significant differences on tissue reaction parameters between TiNi and TiNiCu. Alloying of NiTi with Nb was also introduced in order to increase the inherent transformation hysteresis of NiTi alloys [93].

The ion release from these ternary alloys, if any, has not yet been established. The toxic potency of different ions or compounds is not clear. Mo is considered to be essential element and humans need a daily amount of 0.1 mg [36]. It maintains the equilibrium of multiple enzyme activities. Ingestion or inhalation of higher amounts of metallic Mo particles may disturb the copper and sulfate metabolism in experimental animals and humans, or induce emaciation and gout symptoms. Cu is also an essential element and is a component of some essential enzyme systems such as tyrosinase and cytochrome oxidase. Cu deficiency induces Menke's disease, a metabolic disease with too rapid Cu clearance. In contrast, Wilson's disease is the accumulation of Cu in the liver and in the nuclei of grey cells [36]. The physiological and biological behavior of other alloying elements such as niobium is not well known.

3.2 In Vitro Biocompatibility (cell cultures)

The results of in vitro studies of cell response to NiTi have been slightly contradictory. This may be due to differences in test protocols, including different observed factors, different cell types, variations in surface treatments, surface roughness, surface area, etc.

Castleman and Motzkin published a preliminary in vitro study [13]. In this study, fibroblasts were used. Stainless steel and Co–Cr alloy did not differ in cell growth from the control cultures, but NiTi and titanium significantly reduced cell growth. Some morphological changes of cells with NiTi were also found.

The effects of increasing dose exposure to NiTi, nickel or titanium in cell cultures were examined in a study by Putters et al. [58]. No significant effects on mitosis in human fibroblasts were found for titanium or NiTi. NiTi was considered biocompatible and comparable to titanium.

Some contradictory results were reported when cytotoxicity assays were performed using Confluent L-929 fibroblasts. Cells were incubated in the presence of NiTi, titanium, Co-Cr-Mo and 316L stainless steel discs. All metals induced a mild biological reaction. The cytotoxicity of NiTi was found to be approximately equal to that of Co-Cr-Mo, both being more than that of pure titanium, Ti-6A1–4V or 316L stainless steel. NiTi samples with plasma surface treatment were found to increase the cytocompatibility of NiTi [4].

Endo et al. reported that human plasma fibronectin (pFN); an adhesive protein can be covalently immobilized onto NiTi substrate [25, 26]. Fibronectin improved fibroblast spreading, suggesting that this chemical modification enabled the controlling of metal/cell interactions.

In the study of Shabalovskaya it was found that the different surface treatments of NiTi could critically affect to the behavior of splenocytes [76]. The hydrogen peroxide surface treatment of NiTi caused a toxic effect. When NiTi was treated by autoclaving in water or steam the reaction was clearly non-toxic. The explanation for this was that the Ni surface concentration may vary from 0.4% to 27%, depending on the specific surface treatments used [76].

Recently, Wever et al. evaluated the short-term biological safety of the NiTi alloy [99]. They performed cytotoxicity, sensitization and genotoxicity tests. The NiTi alloy showed no cytotoxic, allergic or genotoxic activity. The findings were similar to those on AISI 316 LVM stainless steel.

The in vitro genotoxicity of NiTi has also been evaluated using human peripheral blood lymphocytes in the author's laboratory. A comparison was made with commercially pure titanium and 316L stainless steel. Cells were cultured in a semiphysiological medium that had previously been exposed to the biomaterials. An electron microscopy in situ end-labeling assay was performed to provide quantification of in vitro chromatin DNA single stranded breaks. NiTi, titanium and stainless steel induced similar DNA strand breaks of interphase chromatin, but stainless steel induction on metaphase chromatin was more intense than with NiTi or pure titanium. The authors concluded that NiTi genocompatibility is promising in view of its biocompatibility approval [5].

In the in vitro study of Ryhänen et al. NiTi, stainless steel and titanium test discs were compared with human osteoblasts (OB) and fibroblasts (FB) [66]. Cells were incubated for 10 days with test discs of equal size 6x7 mm. The cultures were photographed and the cells counted. Samples from culture media were collected on days 2, 4, 6, 8, and the analysis of metals in the media was done using GF-AAS. The proliferation of FB was 108% (NiTi), 134% (Ti), 107% (Stst) and 48% (C) compared to the control cultures. The proliferation of OB was 101% (Nitinol), 100% (Ti), 105% (Stst) and 54% (C) compared to the controls. Initially Nitinol release more nickel (129–87 μg/l) into the cell culture media than Stst (7 μg/l), but after two days the concentrations were about equal (23–5 μg/l vs 11–1 μg/l). The titanium concentrations from both Nitinol and Ti samples were all <20 μg/l. It was conclude that Nitinol has good in vitro biocompatibility with human osteoblasts and fibroblasts. Despite the higher initial nickel dissolution, Nitinol induced no toxic effects, decrease in cell proliferation or inhibition on the growth of cells in contact with the metal surface.

The fact that SS 316L could induce a higher proportion of DNA damage than NiTi can not be explained by the content of nickel. SS 316L contains 10–14% Ni only compared to 50% in NiTi. The thermodynamically stable Ni can explain this in the intermetallic state. As shown by Barrett et al., a significantly higher Ni was released from SS 316L compared to NiTi after 14 days in the presence of saliva [7]. A second hypothesis regarding the higher genotoxicity of SS 316L compared to NiTi: the presence of other elements such as chromium ions (16–18% Cr in SS 316L). Chromium is known to be a strong genotoxic element, inducing DNA strand breaks, base damage, and DNA protein cross-links.

Rose et al. using Mosmann's MTT test investigated the influence of the corrosion products of different orthodontic wires on the cytotoxicity of a fibroblast culture [64]. NiTi, stainless steel and beta-titanium alloy wires had no effect on the rate of cell proliferation.

To further improve NiTi biocompatibility and corrosion properties, we have developed in our laboratory different surface treatment techniques such as plasma deposition of a tetrafluoroethylene thin film, excimer laser treatment and electropolishing. First, the deposition of polymer thin layers did not affect cores properties (transition temperatures, etc.) and seemed to follow NiTi superelastic properties [96, 103, 104]. Also, using a laser excimer melting treatment we can increase the biocompatible surface oxide layer, while diminishing nickel concentration at the surface [94, 95]. Electropolishing renders the surface oxide layer uniform and increases corrosion resistance [90]. These techniques stabilize NiTi passivity and increase its pitting potential [90, 105]. Preliminary results demonstrate that these treatments improve both general and local corrosion resistances of polished NiTi materials. These treatments also seemed to improve surface conditions of NiTi SMA therefore our cytotoxicity investigation has shown an increase in NiTi compatibility [4, 96]. We are currently studying the effect of different sterilization techniques on corrosion phenomena [88] and on NiTi hemocompatibility [87].

3.3
In Vivo Biocompatibility of NiTi (Animal Models)

3.3.1
Soft-Tissue Response

In the study of Cutright et al., NiTi wire sutures were placed subcutaneously in rats, which were followed for 9 weeks [16]. The tissue reaction was minimal at all check-up points. The reparative process was initiated within 1–2 weeks and generated in a dense, relative avascular fibrous connective tissue capsule by 5–6 weeks, with little change after that. When compared to the tissue reaction to stainless steel seen in earlier experiments, NiTi was similar than stainless steel within similar time periods. It was concluded that 55-NiTi compared favorably with stainless steel and could be used in deep tissues.

The first attempt at a profound biocompatibility evaluation of NiTi was made by Castleman et al. [14]. There were three dogs in the NiTi implant group and one Co-Cr implant and one "sham" as a control at each killing point. The complete NiTi data consisted of 12 beagles examined after exposures of 3, 6, 12 and 17 months. The gross clinical, radiological, and morphological observations of tissue at the implantation sites at autopsy revealed no signs of adverse tissue reactions resulting from the implants. The study supported the conclusion that NiTi had no toxic effects in vivo. The authors concluded that no significant differences were noted between the samples taken from the controls and those taken from the dogs exposed to the implants, and that NiTi alloy is sufficiently compatible with dog tissue to warrant further investigation of its potential as a biomaterial.

In an early study, neutron activation analyses were carried out on a small number of tissue samples from the liver, spleen, brain, and kidneys [14]. The findings of the analysis suggest that there is no metallic contamination in the distant organs due to the implants.

Using NiTi implants in four rabbits, Matsumoto et al. found that the blood Ni concentration after implantation reached a level twice the normal in 6–9 h (28 ± 11 ppb vs 13 ± 5 ppb) [46]. After 4 weeks, the Ni concentration was fourfold in the kidneys (140 ± 43 ppb), twofold in the liver (40 ± 18 ppb), and tenfold in urine (90 ± 35 ppb). The authors concluded that Ni elution from NiTi alloy should be limited by, for example, using some coatings.

In the study of Wen et al. the corrosion resistance and tissue biocompatibility of NiTi and Ti50Ni50-xCux ($x = 1, 2, 4, 6, 8$) alloys were investigated [98]. There were no significant differences in the tissue reaction parameters after two and three months between the alloys. After three months' implantation, no corrosion was observed on the plate surfaces.

In the study of Ryhänen et al., the general soft-tissue response, neural and perineural responses were reported [67]. A comparison was made between Nitinol; stainless steel and Ti-6Al-4V. Test specimens were implanted into paravertebral muscle and near the sciatic nerve of 75 rats. The animals were euthanized at 2, 4, 8, 12 and 26 weeks after implantation. General morphologic and histologic observations were made under light microscopy. Semiautomatic computerized image analysis was used to measure the encapsule membrane thickness around the implants. The muscular tissue response to Nitinol was clearly nontoxic, regardless of the time period. The overall inflammatory response to Nitinol was very similar to that of stainless steel and Ti-6Al-4V alloy. There were no necrosis, granulomas or signs of dystrophic soft tissue calcification. The immune cell response to Nitinol remained low. Only a few foreign body giant cells were present. The detected neural and perineural responses were also clearly nontoxic and nonirritating with Nitinol. No qualitative differences in histology between the different test materials could be seen. At 8 weeks, the encapsule membrane of Nitinol was thicker than that of stainless steel (mean 62 ± 25 µm vs 41 ± 8 µm). At the end of the study the encapsule thickness was equal to all the materials tested. We concluded that Nitinol had good in vivo biocompatibility after intramuscular and perineural implantation in rats in the 26-week follow-up.

Recently, Rhalmi et al. seeded fibroblasts on porous NiTi sheets and found cell ingrowth towards the pores as well as around specimen edges [62]. The results did not demonstrate any cytotoxic effect.

3.3.2
Vascular-Tissue Response

Most of the recent commercial NiTi applications are meant for cardiovascular solutions. The idea behind this is to provide minimally invasive treatment instead of major surgery. Since the first experiments by Cragg et al., several studies have provided further information on the biocompatibility of NiTi as vascular stent material [15].

In the experimental studies of Rabkin et al., endovascular NiTi prostheses were implanted in dogs [59]. Long-term results obtained over a period of 14 months demonstrated good and prolonged permeability of the NiTi prostheses. Morphological investigations showed that the endovascular prosthesis was separated in a ring-like fashion by a thin layer of connective tissue, while inside it was lined with a layer of endothelial cells.

Prince et al. inserted NiTi blood clot filters into the venae cavae of 16 dogs and one sheep [57]. The results were analyzed after periods of one week to 4 years. Cleaned NiTi wire filters remained patent, but some showed venographic filling defects caused by adherent organized thrombi. Surface polishing and filter shape had no observable effect on thrombogenicity. Histologic study revealed patchy chronic inflammation on the surface of uncleaned filters, but only a benign fibrous tissue reaction on cleaned filters. Platelet adhesion and plasma coagulation effects of NiTi wire were tested in vitro in human blood and found to be similar to those of stainless steel. The authors suggest that NiTi may be a promising material for human intravascular prosthetic applications.

Das et al. designed a superelastic NiTi-Dacron atrial septal defect (ASD) closure device and studied its efficacy in a canine model [22]. The defects were created surgically in 20 adult dogs. Percutaneous transcatheter closures were attempted using the new device. The closures were successful in 19 studies and unsuccessful in one. Light microscopy at 8 weeks in three dogs showed the devices to be covered by smooth endocardium enmeshed in mature collagen tissue, with minimal mononuclear cell infiltration. The authors concluded that this new device permits effective and safe ASD closure in a canine model.

Grenadier et al. investigated patency rates and the histologic responses of coronary arteries to NiTi coil stent [33]. Twenty-two stents were implanted in sixteen dogs, which were monitored for 1–2 weeks, 1 month, 3 months, 6 months and 1 year. The animals underwent angiography and histopathologic examination. A histologic study showed outward stent pressure compressing the internal elastic membrane and the media in most cases. Intimal hyperplasia started at 2 weeks and was most apparent at 3 months and 6 months. Therefore, the NiTi self-expandable stent provokes a mild cellular proliferative response that reaches its maximum in 3–6 months without further progression.

Restenosis after intracoronary stenting still remains a significant problem. In the present study by Carter et al., the vascular response of a NiTi stent was compared to a balloon-expandable stent in porcine coronary arteries [12]. Eleven NiTi and eleven stainless steel stents were implanted. On histology at 3 days, the stainless steel stents had more inflammatory cells adjacent to the stent than did NiTi. After 28 days, the vessel reaction was similar for the NiTi and stainless steel designs. The mean neointimal area and the percentage of stenosis were significantly lower in the NiTi than in the stainless steel group. It seems that a NiTi stent has a more favorable response on vascular remodeling with less neointimal formation, than a balloon-expandable design. Progressive intrinsic stent expansion after the implantation does not appear to stimulate neointimal formation and may therefore prevent in-stent restenosis.

Recently, Trépanier et al. reported on the improvement of the NiTi stent's corrosion resistance by using different surface treatments (electropolishing, heat treatment, and nitric acid passivation) [90]. A cell proliferation test was performed to evaluate the cytotoxicity of surface treated NiTi using human fibroblasts. A stent implantation was performed in rabbit paramuscular muscle to study the inflammatory response. A gradual overall reduction with time of the fibrocellular capsule thickness surrounding the implants was demonstrated [91]. After a 12-week implantation period, the fibrous capsules surrounding the different implants tended toward the same value of 0.07 mm, which suggested that all surface treatments produced a similar biological response. This low value of the fibrocellular capsule indicated that the NiTi surface treated implants were relatively inert [91].

In a second study performed in our laboratory, Thierry et al. studied the effects of surface modification induced by sterilization processes (steam autoclave, ethylene oxide, Sterrad plasma, and liquid peracetic acid) on the thrombogenicity of NiTi stents [87]. An ex-vivo AV shunt porcine model was used to compare the thrombogenicity of NiTi stents to that of stainless steel stents of identical design. No significant modification was found for the NiTi stents. Conversely, stainless steel stents enhanced thrombus formation after sterilization in comparison to NiTi stents.

Based on the above studies, the histopathological changes caused by vascular NiTi stents are associated with a mild inflammatory response, some atrophy of vessel media, acceptable fibrocellular tissue growth and endothelization. The biocompatibility of NiTi stents seems to be equal or better compared to stainless steel stents.

3.3.3
Bone-Tissue Response

NiTi is one of the most innovative concepts introduced in the field of metallic biomaterials in the recent years, but its biocompatibility remains controversial, especially in bone. The effects of NiTi on fracture healing, new bone formation, remodelation and osteointegration have been without an answer.

The first efforts to study NiTi as a bone implant were made by Castleman et al. [14]. NiTi bone plates were implanted into the femurs of 12 beagles. Commercial cobalt-chromium (Co-Cr) alloy bone plates served as reference controls. The plates were removed from the animals and examined after exposure for 3, 6, 12 and 17 months. There was no evidence of either localized or general corrosion on the surfaces of the bone plates. No sign of harmful tissue reactions resulting from the NiTi implants were seen. Histological samples showed no evidence of bone resorption in specimens adjacent to the plate. Nor were any significant differences noted in the sham-operated controls. The data used in neutron activation analyses suggested that there is no nickel contamination in bone due to the implants. However, in the NiTi group, some high nickel concentrations were also observed, but these were attributed to contamination.

Yang et al. used NiTi internal fixing device to fractured femoral shafts of dogs [108]. Comparison was done with a 316L stainless steel plate-screw system. Osteotomy on both sides of the femoral diaphyses was performed in 15 dogs. One side was plated with a bone plate and the other with a NiTi device. Dogs were killed at 4, 8, 12 weeks after operation. Radiographic examination, light microscopy and transmission electron microscopy methods were used. The fracture healing and callus remodeling were similar in these two groups. However, the cortical bone remodeling underneath the fixator near the osteotomized area was significantly different. The authors suggested that since the elastic modulus of the NiTi shape memory alloy is lower, the stress-shielding effect in the bone underneath the NiTi device is less. The axial compression stress of the fracture line is kept greater and the contact of NiTi implant with the bone was not so close. This might be beneficial for the recovery of blood supply and bone remodeling.

Simske and Sachdeva published a preliminary report on the use of porous NiTi [80]. This porous material has a certain open structure that provides a possibility for the ingrowth of bony tissue into the body of the implant, resulting in firm fixation to bone. NiTi implants were placed to either side of the frontal bone of rabbits. A hydroxyapatite implant of was used as a control. No adjacent macrophage cells were seen for either implant type. Both materials made bone contact with the surrounding cranial bone, and the percentage of ingrowth increased with the time. Porous NiTi implants appear to allow for significant cranial bone ingrowth after as few as 12 weeks. Compared to HA, the NiTi implants showed a trend for less total apposition and more total ingrowth after 6 weeks and 12 weeks of implantation. The authors concluded that porous NiTi appears to be suitable for craniofacial applications.

Bone reaction to porous NiTi was studied in our Laboratory in rabbit tibias after three, six and twelve weeks of implantation [62]. Bone tissue demonstrated good healing of the osteotomy without any adverse reaction. There was bone remodeling characterized by osteoclastic and osteoblastic activity in the cortex. The results confirmed the good biocompatibility acceptance of porous NiTi reported by Gunter et al. [30].

Ryhänen et al. evaluated the new bone formation, modeling and cell material interface responses induced by nickel titanium shape memory alloy after periosteal implantation [69]. We used a regional acceleratory phenomenon (RAP) model, in which a periosteal contact stimulus provokes an adaptive modeling response. NiTi was compared to stainless steel and Ti-6Al-4V. The test implant was placed in contact with the intact femur periosteum, but it was not fixed inside the bone. Histomorphometry was used to determine the bone formation and resorption parameters. The ultrastructural features of cell-material adhesion were analysed with scanning electron microscopy (FESEM). A typical peri-implant bone wall modelation was seen due to the normal RAP. The maximum new woven bone formation started earlier (2 weeks) in the Ti-6Al-4V group than in the NiTi ($P<0.01$) group, but also decreased earlier, and at 8 weeks, the NiTi ($P<0.05$) and stst ($P<0.005$) groups had greater cortical bone width. At 12 weeks and 26 weeks, no statistical differences were seen in the histomorphometric values. The histological response of the soft tissues around the NiTi implant was also clearly non-toxic and non-irritating. Cell adhesion and focal contacts were similar between the materials studied by FESEM. We conclude that NiTi had no negative effect on total new bone formation or normal RAP.

Kasano and Morimitsu established a new type of ear stapes prosthesis made of nickel-titanium shape memory alloy wire [40]. Its biocompatibility was examined in ears of cats. The prosthesis was implanted at the long crus of the incus and the incus was examined 27–355 days after operation. There was no progressive bone resorption, which was prosthesis-induced. The authors concluded that the biocompatibility of the nickel-titanium alloy stapes prosthesis with the long crus of the incus was hereby proven.

The above studies suggested that NiTi is quite well tolerated into bone. However, there are some conflicting studies, in which NiTi has been found to have inferior properties compared to the other implant materials.

In the author's laboratory, Berger-Gorbet et al. evaluated the biocompatibility of NiTi screws using immunohistochemical methods [9]. The distribution of bone proteins during the bone remodeling process around a NiTi implant was observed. The control materials were screws made of Vitallium, c.p. titanium, Duplex austenitic-ferritic stainless steel (SAF), and stainless steel 316L. The test materials were implanted in rabbit tibias for 3 ($n=2$), 6 ($n=2$) and 12 ($n=2$) weeks. The biocompatibility results of the NiTi screws compared with the other screws showed a slower osteogenesis process characterized by no close contacts between the implant and bone, disorganized migration of osteoblasts around the implant, and a lower activity of osteonectin synthesis.

The bone reaction to NiTi implants inserted transcortically to rat tibiae was quantitatively assessed by Takeshita et al. [84]. The control materials were composed of pure titanium, anodic oxidized titanium (AO-Ti), and Ti-6Al-4V alloy.

Three rats were killed 7, 14, 28, 84 and 168 days after operation (n=3). Essentially the same histological findings were made for NiTi, Ti, Ti-6Al-4V and AO-Ti implants. NiTi was progressively encapsulated with bone tissues. Histometric analysis revealed no significant differences between the tissue reactions to Ti, AO-Ti and Ti-6Al-4V, but NiTi implants showed a significantly lower bone contact than any of the other titanium or titanium alloy materials. In terms of bone contact thickness, there were no significant differences between NiTi and the other three materials (Ti, AO-Ti and Ti-6Al-4V).

In the latest study of Ryhänen et al. femoral osteotomies of 40 rats were fixed with either NiTi or stainless steel (StSt) intramedullary nails [70]. The rats were killed at 2, 4, 8, 12, 26 and 60 weeks. Bone healing was examined with radiographs, peripheral quantitative computed tomography and histologically. The corrosion of the implants was studied by electron microscopy. Analysis of trace metals from several organs was done by GF-AAS or ICP-AES. More healed bone unions were found in the NiTi than the StSt group at early (4 weeks and 8 weeks) time points. Callus size was similar between the groups. The total and cortical bone mineral densities did not differ between the NiTi and StSt groups. Mineral density in both groups was lower in the osteotomy area than in the other areas along the nail. Density in the nail area was lower than in the proximal part of the operated femur or the contralateral femur. Bone contact to NiTi was close. A peri-implant lamellar bone sheet formed in the metaphyseal area after 8 weeks, indicating good tissue tolerance. The FESEM assessment showed surface corrosion changes to be more evident in the StSt implants. There were no statistically significant differences in nickel concentration between the NiTi and StSt groups in any of the organs. Based on this study, NiTi appears to be an appropriate material for further intramedullary use.

3.4 Clinical Studies of NiTi Orthopedic Devices

NiTi has also been used as a bone implant material in humans. There are reports that NiTi material has been successfully used in bone-related human applications in Russia and China in a large number of patients [18, 41, 76, 108]. Worldwide medical applications have been hindered for a long time because of the lack of knowledge of the biocompatibility of NiTi. Very few well-monitored studies have been published in peer-reviewed journals up until now. Also, no controlled or randomized studies have been published so far. FDA has lately approved a bone anchor (MITEK G2) which includes a small piece of superelastic NiTi wire. In the USA, FDA limits the marketing of long-term implanted NiTi devices because their biocompatibility has not been proved.

Drugacz et al. tested the clinical application of $Ti_{50}Ni_{48.7}Co_{1.3}$ alloy shape-memory clamps for the fixation of mandibular fractures using transoral access [24]. The clamps were used to treat all types of fractures occurring between the mandibular angles. The clamps were removed after a period of at least 6 weeks, and tissue samples were taken for microscopic examination. Seventy-seven patients with mandibular fractures were treated using the clamps. In 72 patients, the treatment progressed satisfactorily, while in five cases, infections occurred.

Tissue samples for histologic examination was taken from 58 patients after removal of the clamps. There were no pathologic or atypical tissue reactions or signs of disturbed cell maturation. The authors concluded that the application of shape memory clamps for the surgical treatment of mandibular fractures facilitates treatment while ensuring stable fixation of the bone fragments.

In the studies of Sysolyatin et al. [82] and Itro et al. [38], NiTi implants were used in the surgical correction of maxillo-facial fractures. The results showed a good stability and rapid bone healing. Also the time needed for operative procedures and rehabilitation was reduced.

Von Salis-Soglio reported the results of ventral intercorporeal lumbar spondylodesis with a NiTi implant [97]. The primary stabilization of the unstable segment was achieved using implant that was inserted intercorporeally following ventral removal of the intervertebral disc. The results included 51 cases of bony fusion within an average postoperative period of 9 months, one case of pseudoarthrosis and 11 cases of delayed bony fusion. The author concluded that, in view of the easier operative technique, the earlier mobilization of the patients and the good fusion rate, the memory spondylodesis seems to have important advantages over the transplantation of bone chips only. The use of a NiTi staple to lock a tricortical iliac bone graft in cervical anterior fusion was used by Ricart [63]. Fifty patients with several clinical diagnoses were operated. Good and very good results were reported in 80% of the cases. The average bone fusion rate was fast.

Silberstein reported a clinical study where 84 patients with fractures, tumors or intervertebral disc disease of the cervical and lumbar spine were treated with anterior fusion and porous NiTi implant grafts [78]. They concluded that porous NiTi implants can be successfully used and the material itself shows a high degree of biocompatibility.

Thirty-six metatarsal osteotomies fixed with NiTi compression staple for hallux valgus were performed in a study by Tang et al. [86]. The recovery period to normal work and walking were resumed an average of 41 days postoperatively. Twenty patients (35 feet) experienced complete pain relief. Radiographic analysis of the feet showed that all the osteotomies healed. No external fixation by plaster splintage was needed. According to the authors, the benefits of this internal fixator were that the period of bone healing was shortened and the patients were able to bear weight earlier.

Musialek et al. reported the fixation of small bone fragments with NiTi clamps in 64 patients [53]. Clamps were used for compressive stabilization in several kinds of fractures. Bone union, wound healing problems and histology were studied. Non-union occurred in four patients treated with only one fixative. Two clamps implanted in non-parallel planes seem to be advisable to exclude the need for longer immobilization. Histological evaluation of the tissue covering the implants in 22 patients did not reveal any adverse reactions. The study suggests that by using NiTi clamps in an appropriate way, satisfactory outcomes could be achieved with respect to both biofunctionality and biocompatibility. In conclusion, on the basis of a few studies, it seems the NiTi material in itself has no deleterious effects in human use.

References

1. Acunas B, Rozanes I, Akpinar S, Tunaci A, Tunaci M, Acunas G (1996) Palliation of malignant esophageal strictures with self-expanding nitinol stents: drawbacks and complications. Radiology 199:648–652
2. Andreasen GF, Hilleman TB (1971) An evaluation of 55 cobalt substituted Nitinol wire for use in orthodontics. J Am Dent Assoc 82:1373–1375
3. Andreasen GF, Fahl JL (1987) Alloys, Shape Memory. In: Webster JG (ed) Encyclopedia of medical devices and instrumentation (vol 2). Wiley, New York, pp 15–20
4. Assad M, Lombardi S, Berneche S, Desrosiers EA, Yahia LH, Rivard CH (1994) Essais de cytotoxicite sur l'alliage a memoire de forme Nickel-Titane. Ann Chir 48:731–736
5. Assad M, Yahia LH, Rivard CH, Lemieux N (1998) In vitro biocompatibility assessment of a nickel-titanium alloy using electron microscopy in situ end-labeling (EM-ISEL). J Biomed Mater Res 41:154–161
6. Barber FA, Herbert MA, Click JN (1996) Suture anchor strength revisited. Arthroscopy 12:32–38
7. Barrett RD, Bishara SE, Quinn JK (1993) Biodegradation of orthodontic appliances. Part I. Biodegradation of nickel and chromium in vitro. Am J Orthod Dentofacial Orthop 103:8–14
8. Baumgart F, Bensmann G, Haasters J, Nolker A, Schlegel KF (1978) Zur Dwyerschen Skoliosenoperation mittels Drahten aus Memory- Legierungen. Eine experimentelle Studie. Arch Orthop Trauma Surg 91:67–75
9. Berger-Gorbet M, Broxup B, Rivard C, Yahia LH (1996) Biocompatibility testing of NiTi screws using immunohistochemistry on sections containing metallic implants. J Biomed Mater Res 32:243–248
10. Bezzi M, Orsi F, Salvatori FM, Maccioni F, Rossi P (1994) Self-expandable nitinol stent for the management of biliary obstruction: long-term clinical results. J Vasc Interv Radiol 5:287–293
11. Blum U, Voshage G, Lammer J, Beyersdorf F, Tollner D, Kretschmer G, Spillner G, Polterauer P, Nagel G, Holzenbein T (1997) Endoluminal stent-grafts for infrarenal abdominal aortic aneurysms. N Engl J Med 336:13–20
12. Carter AJ, Scott D, Laird JR, Bailey L, Kovach JA, Hoopes TG, Pierce K, Heath K, Hess K, Farb A, Virmani R (1998) Progressive vascular remodeling and reduced neointimal formation after placement of a thermoelastic self-expanding nitinol stent in an experimental model. Cathet Cardiovasc Diagn 44:193–201
13. Castleman LS, Motzkin SM (1981) The biocompatibility of Nitinol. In: Williams DF (ed) Biocompatibility of clinical implant materials (vol 1). CRC, Boca Raton, pp 129–154
14. Castleman LS, Motzkin SM, Alicandri FP, Bonawit VL (1976) Biocompatibility of nitinol alloy as an implant material. J Biomed Mater Res 10:695–731
15. Cragg A, Lund G, Rysavy J, Castaneda F, Castaneda-Zuniga W, Amplatz K (1983) Nonsurgical placement of arterial endoprostheses: a new technique using nitinol wire. Radiology 147:261–263
16. Cutright DE, Bhaskar SN, Perez B, Johnson RM, Cowan GSJ (1973) Tissue reaction to nitinol wire alloy. Oral Surg Oral Med Oral Pathol Oral Radiol Endod 35:578–584
17. Cwikiel W, Willen R, Stridbeck H, Lillo-Gil R, Von HC (1993) Self-expanding stent in the treatment of benign esophageal strictures: experimental study in pigs and presentation of clinical cases. Radiology 187:667–671
18. Dai K, Chu Y (1996) Studies and applications of NiTi shape memory alloys in the medical field in China. Biomed Mater Eng 6:233–240
19. Dai KR (1983) Orthopedic application of a Ni-Ti shape-memory alloy compression staple. Chung Hua Wai Ko Tsa Chih 21:343–345
20. Duerig TW, Pelton AR, Stockel D (1996) The utility of superelasticity in medicine. Biomed Mater Eng 6:255–266
21. Dambaev GZ, Gjunter VE, Radionchenko AA, Itin VI, Kouzenko IG, Hodorenko VN, Bazilevich LR, Goural KA (1996) Porous permeable superelastic implants in surgery. Tomsk University, Tomsk
22. Das GS, Voss G, Jarvis G, Wyche K, Gunther R, Wilson RF (1993) Experimental atrial septal defect closure with a new, transcatheter, self-centering device. Circulation 88:1754–1764
23. De Jaegere PP, Eefting FD, Popma JJ, Serruys PW (1996) Clinical trials on intracoronary stenting. Semin Interv Cardiol 1:233–245
24. Drugacz J, Lekston Z, Morawiec H, Januszewski K (1995) Use of TiNiCo shape-memory clamps in the surgical treatment of mandibular fractures. J Oral Maxillofac Surg 53:665–671
25. Endo K (1995a) Chemical modification of metallic implant surfaces with biofunctional proteins (part 1). Molecular structure and biological activity of a modified NiTi alloy surface. Dent Mater J 14:185–198
26. Endo K (1995b) Chemical modification of metallic implant surfaces with biofunctional proteins (part 2). Corrosion resistance of a chemically modified NiTi alloy. Dent Mater J 14:199–210

27. Frank T, Willetts GJ, Cuschieri A (1995) Detachable clamps for minimal access surgery. Proc Inst Mech Eng [H] 209:117–120
28. Fung YC (1981) Biomechanics, mechanical properties of living tissues. Springer, Berlin Heidelberg New York
29. Gjunter VE, Itin VI, Monassevich LA, Yu IP, et al. (1992) Shape memory effects and their applications in medicine. Nauka, Novosibirsk
30. Gjunter VE (1995) Superelastic shape memory implants in maxillofacial surgery, traumatology, orthopaedics and neurosurgery. Tomsk University, Tomsk
31. Gottfried HW, Gnann R, Brandle E, Bachor R, Gschwend JE, Kleinschmidt K (1997) Treatment of high-risk patients with subvesical obstruction from advanced prostatic carcinoma using a thermosensitive mesh stent. Br J Urol 80:623–627
32. Gottfried HW, Schimers HP, Gschwend J, Brandle E, Hautmann R (1995) Erste erfahrungen mit dem Memotherm-stent in der behandlung der BPH. Urologe A 34:110–118
33. Grenadier E, Shofti R, Beyar M, Lichtig H, Mordechowitz D, Globerman O, Markiewicz W, Beyar R (1994) Self-expandable and highly flexible nitinol stent: immediate and long- term results in dogs. Am Heart J 128:870–878
34. Hauck RW, Lembeck RM, Emslander HP, Schomig A (1997) Implantation of Accuflex and Strecker stents in malignant bronchial stenoses by flexible bronchoscopy. Chest 112:134–144
35. Hausdorf G, Schneider M, Franzbach B, Kampmann C, Kargus K, Goeldner B (1996) Transcatheter closure of secundum atrial septal defects with the atrial septal defect occlusion system (ASDOS): initial experience in children. Heart 75:83–88
36. Hilbebrand HF, Hornez J-C (1998) Biological response and biocompatibility. In: Helsen JA, Breme HJ (eds) Metals as biomaterials. Wiley, New York, pp 265–290
37. Himpens JM (1993) Laparoscopic inguinal hernioplasty. Repair with a conventional vs a new self-expandable mesh. Surg Endosc 7:315–318
38. Itro A, Garau V, Tartaro GP, Colella G (1997) La nostra esperienza su di una metodica di fissazione rigida in chirurgia maxillo-facciale mediante clips a memoria di forma. Minerva Stomatol 46:381–389
39. Iwabuchi T, Suzuki S, Ebina K, Honma T (1975) Memory clip for intracranial aneurysm surgery. Technical note. J Neurosurg 42:733–735
40. Kasano F, Morimitsu T (1997) Utilization of nickel-titanium shape memory alloy for stapes prosthesis. Auris Nasus Larynx 24:137–142
41. Kuo PP, Yang PJ, Zhang YF, Yang HB, Yu YF, Dai KR, Hong WQ, Ke MZ, Cai TD, Tao JC (1989) The use of nickel-titanium alloy in orthopedic surgery in China. Orthopedics 12:111–116
42. Lacy SA, Merrit K, Brown SA, Puryear A (1996) Distribution of nickel and cobalt following dermal and systematic administration with in vitro and in vivo studies, J Biomed Mater Res 32:279–283
43. Latal D, Mraz J, Zerhau P, Susani M, Marberger M (1994) Nitinol urethral stents: long-term results in dogs. Urol Res 22:295–300
44. Lopatkin NA, Afanas Z, Zakhmatov IM, Varentsov GI, Chepurov AK (1989) Endourethral drainage of the bladder in patients with prostatic adenoma. Urol Nefrol (Mosk) 3:5–7
45. Lu SB, Wang JF, Guo JF (1986) Treatment of scoliosis with a shape-memory alloy rod. Chung Hua Wai Ko Tsa Chih 24:129–32, 187
46. Matsumoto K, Tajima N, Kuwahara S (1993) Correction of scoliosis with shape-memory alloy. Nippon Seikeigeka Gakkai Zasshi 67:267–274
47. May A, Selmaier M, Hochberger J, Gossner L, Muhldorfer S, Hahn EG, Ell C (1995) Memory metal stents for palliation of malignant obstruction of the oesophagus and cardia. Gut 37:309–313
48. Mei F, Ren X, Wang W (1997) The biomechanical effect and clinical application of a Ni-Ti shape memory expansion clamp. Spine 22:2083–2088
49. Menné T, Brandrup F, Thestrup-Pedersen K, Veien NK, Andersen JR, Yding F, Valeur G (1987) Patch test reactivity to nickel alloys. Contact Dermatitis 16:255–259
50. Merrit K, Crowe TD, Brown SA (1989) Elimination of nickel, cobalt, and chromium following repeated injections of high dose metal salts. J Biomed Mater Res 23:845–862
51. Merrit K, Brown SA (1995) Released of hexavalent chromium from corrosion of stainless-steel and cobalt-chromium alloys J Biomed Mater Res 29:627–633
52. Mirgazizov MZ, Gjunter VE, Itin VI, et al. (1993) Superelastic implants and devices from shape memory alloys in dentistry. Quintessenz Verlag Berlin
53. Musialek J, Filip P, Nieslanik J (1998) Titanium-nickel shape memory clamps in small bone surgery. Arch Orthop Trauma Surg 117:341–344
54. Oesterle SN, Whitbourn R, Fitzgerald PJ, Yeung AC, Stertzer SH, Dake MD, Yock PG, Virmani R (1998) The stent decade: 1987 to 1997. Stanford stent summit faculty. Am Heart J 136:578–599
55. Oshida Y, Sachdeva RC, Miyazaki S (1992) Microanalytical characterization and surface modification of TiNi orthodontic archwires. Biomed Mater Eng 2:51–69

56. Pomerleau E, Gauvin R, Trochu F, Lours T (1994) Caractérisation de l'amortissement de matériaux composites renforcés par un alliage à mémoire de forme. In: Proceedings of the conference on industrial applications of SMA, Montreal, Canada. pp 375–393
57. Prince MR, Salzman EW, Schoen FJ, Palestrant AM, Simon M (1988) Local intravascular effects of the nitinol wire blood clot filter. Invest Radiol 23:294–300
58. Putters JL, Kaulesar SD, De ZG, Bijma A, Besselink PA (1992) Comparative cell culture effects of shape memory metal (Nitinol), nickel and titanium: a biocompatibility estimation. Eur Surg Res 24:378–382
59. Rabkin DI, Minkina SM, Kadnikov AA, Khasenov BP (1986) Experimental-morphological study of the roentgeno-endovascular prosthesis. Med Radiol (Mosk) 31:55–63
60. Rauber K, Franke C, Rau WS, Syed AS, Bensmann G (1990) Peroral einfuhrbare endotracheale stutzgeruste aus der memory-legierung NiTi–Tierexperimentelle studie. Rofo Fortschr Geb Rontgenstr Neuen Bildgeb Verfahr 152:698–701
61. Rechavia E, Litvack F, Fishbien MC, Nakamura M, Eigler N (1998) Biocompatibility of polyurethane-coated stents: tissue and vascular aspects. Cathet Cardiovasc Diagn 45:202–207
62. Rhalmi S, Odin M, Assad M, Tabrizian M, Rivard C-H, Yahia L'H (1999) Hard, soft tissue and in vitro cell response to porous nickel-titanium: a biocompatibility evaluation. Biomed Mater Eng 9:151–162
63. Ricart O (1997) The use of memory shape staple in cervical anterior fusion. In: Pelton AR, Hodgson D, Russell SM, Duerig TW (eds) Proceedings of SMST 1997. Shape Memory and Superelastic Technologies, Pacific Grove, pp 623–626
64. Rose EC, Jonas IE, Kappert HF (1998) In vitro investigation into the biological assessment of orthodontic wires. J Orofac Orthop 59:253–264
65. Rossi P, Bezzi M, Rossi M, Adam A, Chetty N, Roddie ME, Iacari V, Cwikiel W, Zollikofer CL, Antonucci F (1994) Metallic stents in malignant biliary obstruction: results of a multicenter European study of 240 patients. J Vasc Interv Radiol 5:279–285
66. Ryhänen J, Niemi E, Serlo W, Niemelä E, Sandvik P, Pernu H, Salo T (1997) Biocompatibility of nickel-titanium shape memory metal and its corrosion behavior in human cell cultures. J Biomed Mater Res 35:451–457
67. Ryhänen J, Kallioinen M, Tuukkanen J, Junila J, Niemelä E, Sandvik P, Serlo W (1998) In vivo biocompatibility evaluation of nickel-titanium shape memory metal alloy: muscle and perineural tissue responses and encapsule membrane thickness. J Biomed Mater Res 41:481–488
68. Ryhänen J, Raatikainen T, Kaarela O (1998) Complete acromioclavicular dislocation repair with a new shape memory AC-hook implant: an operative technique and prospective pilot study. In: Proceedings of 7th international congress on surgery of the shoulder, Sydney, Australia, p 292
69. Ryhänen J, Kallioinen M, Tuukkanen J, Lehenkari P, Junila J, Niemelä E, Sandvik P, Serlo W (1999a) Bone modeling and cell-material interface responses induced by nickel-titanium shape memory alloy after periosteal implantation. Biomaterials 20:1309–1317
70. Ryhänen J, Kallioinen M, Serlo W, Perämäki P, Junila J, Sandvik P, Niemelä E, Tuukkanen J (1999b) Bone healing and mineralization, implant corrosion and trace metals after nickel-titanium shape memory metal intramedullary fixation. J Biomed Mater Res 47:472–480
71. Sanders JO, Sanders AE, More R, Ashman RB (1993) A preliminary investigation of shape memory alloys in the surgical correction of scoliosis. Spine 18:1640–1646
72. Schroeder A, Van der Zypen E, Stoch H, Sutter F (1981) The reaction of bone connective tissue and epithelium to endosteal implants with titanium sprayed surfaces. J Maxillofac Surg 9:15–25
73. Schurmann K, Vorwerk D, Kulisch A, Stroehmer-Kulisch E, Biesterfeld S, Stopinski T, Gunther RW (1995) Experimental arterial stent placement. Comparison of a new nitinol stent and wallstent. Invest Radiol 30:412–420
74. Schurmann K, Vorwerk D, Uppenkamp R, Klosterhalfen B, Bucker A, Gunther RW (1997) Iliac arteries: plain and heparin-coated Dacron-covered stent-grafts compared with noncovered metal stents – an experimental study. Radiology 203:55–63
75. Schwarzenberg H, Muller-Hulsbeck S, Gluer CC, Wesner F, Heller M (1998) Restenosis of peripheral stents and stent grafts as revealed by intravascular sonography: in vivo comparison with angiography. AJR Am J Roentgenol 170:1181–1185
76. Shabalovskaya SA (1996) On the nature of the biocompatibility and on medical applications of NiTi shape memory and superelastic alloys. Biomed Mater Eng 6:267–289
77. Sievert H, Babic UU, Ensslen R, Scherer D, Spies H, Wiederspahn T, Zeplin HE (1995) Transcatheter closure of large atrial septal defects with the Babic system. Cathet Cardiovasc Diagn 36:232–240
78. Silberstein B (1997) Subtotal and total vertebral body replacement and interbody fusion with porous Ti–Ni implants. In: Pelton AR, Hodgson D, Russell SM, Duerig TW (eds) Proceedings of SMST 1997. Shape Memory and Superelastic Technologies, Pacific Grove, pp 617–621
79. Simon M, Kaplow R, Salzman E, Freiman D (1977) A vena cava filter using thermal shape memory alloy. Experimental aspects. Radiology 125:87–94

80. Simske SJ, Sachdeva R (1995) Cranial bone apposition and ingrowth in a porous nickel–titanium implant. J Biomed Mater Res 29:527–533
81. Smits M, Huibregtse K, Tytgat G (1995) Results of the new nitinol self-expandable stents for distal biliary structures. Endoscopy 27:505–508
82. Sysolyatin PG, Gyunter VE, Starokha AV, Makarova IA, Sysolyatin SP, Denisov VN, Rahman BQ (1994) The use of Ni–Ti implants in maxillofacial surgery. In: Pelton AR, Hodgson D, Duerig TW (eds) Proceedings of SMST 1994. Shape Memory and Superelastic Technologies, Pacific Grove, pp 470–475
83. Tack J, Gevers AM, Rutgeerts P (1998) Self-expandable metallic stents in the palliation of recto-sigmoidal carcinoma: a follow-up study. Gastrointest Endosc 48:267–271
84. Takeshita F, Takata H, Ayukawa Y, Suetsugu T (1997) Histomorphometric analysis of the response of rat tibiae to shape memory alloy (nitinol). Biomaterials 18:21–25
85. Tang L, Ugarova TP, Plow EF, Eaton JW (1996) Molecular determinants of acute inflammatory responses to biomaterials. J Clin Invest 97:1329–1334
86. Tang RG, Dai KR, Chen YQ (1996) Application of a NiTi staple in the metatarsal osteotomy. Biomed Mater Eng 6:307–312
87. Thierry B, Tabrizian M, Savadogo O, Yahia L'H (2000) Effects of sterilization processes on NiTi alloy: surface characterization. J Biomed Mater Res 49:88–98
88. Thierry B, Tabrizian M, Yahia L'H, Savadogo O (1999) Effect of sterilization processes on the ions: release of NiTi Alloy. In: Trochu F, Brailovski V, Galibois A (eds) Shape memory alloys: fundamentals, modeling and industrial applications. MET SOC, pp 373–381
89. Toth RW, Parr GR, Gardner LK (1985) Soft tissue response to endosseaous titanium oral implants. J Prosthet Dent 54:546–567
90. Trepanier C, Tabrizian M, Yahia LH, Bilodeau L, Piron DL (1998) Effect of modification of oxide layer on NiTi stent corrosion resistance. J Biomed Mater Res 43:433–440
91. Trepanier C, Leung TK, Tabrizian M, Yahia L'H, Bienvenu D, Tanguay J-F, Piron DL, Bilodeau L (1999) A preliminary investigation of biological response to shape memory NiTi stents. J Appl Biomater 48:165–171
92. Van Humbeeck J, Stalmans R (1998) Characteristics of shape memory alloys. In: Otsuka K, Wayman CM (eds) Shape memory materials, Cambridge University, Cambridge, pp 149–183
93. Van Humbeeck J (1997) Shape memory materials: state of the art and requirements for future applications. J Phys IV France 7:3–12
94. Villermaux F, Tabrizian M, Yahia L'H, Meunier M, Piron DL (1997) Excimer laser treatment of NiTi shape memory alloy biomaterials. Appl Surface Sci 62:109–115
95. Villermaux F, Nakatsugawa I, Tabrizian M, Piron DL, Meunier M, Yahia L'H (1997) Corrosion kinetics of laser treated NiTi shape-memory alloy biomaterials. In: George EP, Gottherdt R, Otsuka K, Trolier-McKinstry S, Wun-Fogle M (eds) Materials for smart systems II (Materials Research Society, vol 459). Materials Research Society, Pittsburgh, pp 477–482
96. Villermaux F, Tabrizian M, Yahia L'H, Czeremuszkin G, Piron D (1996) Corrosion resistance improvement of NiTi osteosynthesis staples by plasma polymerized tetrafluoroethylene coating. Biomed Mater Eng 6:241–254
97. Von Salis-Soglio GF (1989) Die memory-spondylodese an der lendenwirbelsaule – Ergebnisse nach 76 operationen. Z Orthop Ihre Grenzgeb 127:191–196
98. Wen X, Zhang N, Li X, Cao Z (1997) Electrochemical and histomorphometric evaluation of the TiNiCu shape memory alloy. Biomed Mater Eng 7:1–11
99. Wever DJ, Veldhuizen AG, Sanders MM, Schakenraad JM, Van Horn JR (1997) Cytotoxic, allergic and genotoxic activity of a nickel–titanium alloy. Biomaterials 18:1115–1120
100. Wholey MH, Wholey M, Bergeron P, Diethrich EB, Henry M, Laborde JC, Mathias K, Myla S, Yahia LH, Lombardi S, Piron D, Klemberg-Sapieha JE, Wertheimer MR (1996) NiTi shape memory alloys treated by plasma-polymerized tetrafluoroethylene. A physicochemical and electrochemical characterization. Med Prog Technol 21:187–193
101. Yachia D (1993) The use of urethral stents for the treatment of urethral strictures. Ann Urol (Paris) 27:245–250
102. Yahia L'H, Lombardi S, Piron D, Okuyama M (1999) Corrosion protection of nickel-titanium shape memory alloy for implants. fundamentals, modeling and industrial applications. In: Trochu F, Brailovski V, Galibois A (eds) Shape memory alloys: fundamentals, modeling and industrial applications. MET SOC, pp 383–392
103. Yahia L'H, Lombardi S, Piron D, Klemberg-Sapieha JE, Wertheimer MR (1997) NiTi shape memory alloys treated by plasma-polymerized tetrafluoroethylene. A physicochemical and electrochemical characterization. Med Prog Technol 21:187–193
104. Yahia L'H, Lombardi S, Hagemeister N, Drouin G, Rivard CH, Assad M, Okuyama M (1995) Improvement of cytocompatibility and biomechanical compatibility of NiTi shape memory alloys. J Appl Biomech 10:19–24

105. Okuyama M, Yahia L'H, Piron D, Lombardi S (1995) Biocompatibility of NiTi shape memory alloy: an in vitro electrochemical study. Biomater Living Syst Interact 3:145–160
106. Yanagihara K, Mizuno H, Wada H, Hitomi S (1997) Tracheal stenosis treated with self-expanding nitinol stent. Ann Thorac Surg 63:1786–1789
107. Yang PJ, Tao JC, Ge MZ, Yang QM, Yang HB, Sun Q (1992) Ni-Ti memory alloy clamp plate for fracture of short tubular bone. Chin Med J (Engl) 105:312–315
108. Yang PJ, Zhang YF, Ge MZ, Cai TD, Tao JC, Yang HP (1987) Internal fixation with Ni-Ti shape memory alloy compressive staples in orthopedic surgery. A review of 51 cases. Chin Med J (Engl) 100:712–714
109. Yoneyama T, Tanabe Y, Doi H, Kobayashi E, Hamanaka H, Bonfield W (1996) Stress reduction by NiTi alloy in impact compression test. In: Proceedings of the Fifth World Biomaterials Congress, p 88
110. Yoneyama T (1988) Studies on NiTi alloys for dental casting. Part 2: damping capacity and shock absorptive characteristics. J Jpn Soc Dent Mater Devices 7:262–269

Processing and Quality Control of Binary NiTi Shape Memory Alloys

Matthias Mertmann

1 Introduction

Binary alloys based on the intermetallic system NiTi are at present the commercially most important shape memory materials on the market. Especially in the field of medical technology the NiTi alloys (NiTiNOLA) had their breakthrough during the past years as implant or instrument e.g. for minimal invasive surgery. The improved understanding of the metallurgical principles was certainly the basis for this success. But probably more important were the achievements related to the reliable fabrication of NiTi with tight tolerances and high reproducibility. As the temperature of the human body varies only within a very narrow window the characteristics temperatures of a shape memory alloy have to be adjusted and controlled very carefully to meet the unique requirements of the medical product. On the other hand, the constant temperature of the human body is a very good prerequisite for the application of shape memory alloys in general as all effects related to the martensitic phase transformation are very sensitive to changes of environmental temperature (Fig. 1). Therefore, most pro-

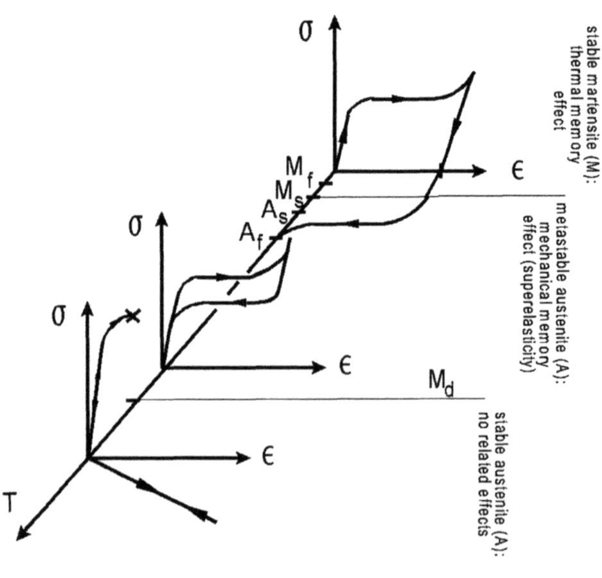

Fig. 1. Temperature dependence of the mechanical properties (stress-strain behavior) of NiTi shape memory alloys. With increasing temperature the behavior change from "one-way effect" (thermal memory effect) over "Superelasticity" (mechanical memory effect) to the stress-strain characteristic of conventional metals. The position of the human body temperature on T-axis can be adjusted sensitively by the chemical composition and the thermomechanical treatment of the material

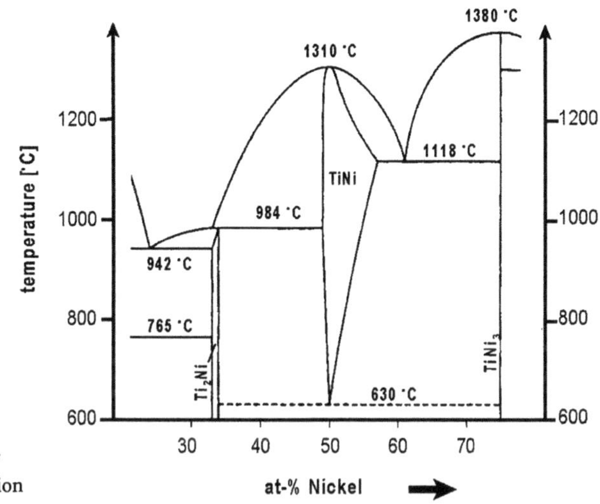

Fig. 2. Relevant section of the phase diagram of Ni and Ti showing especially the area of near stoichiometric composition

perties of NiTiNOL depend strongly on the temperature [1]. These facts show clearly the importance of an advanced quality control concept for NiTi alloys because once the tight tolerances were achieved, the market situation for these unique materials in the field of medical implants and instruments became very interesting and promising.

The high sensitivity of the NiTi alloy system with nearly stoichiometric composition can be explained by analyzing the binary phase diagram (Fig. 2), which shows that the intermetallic NiTi phase has only a very small stability range at temperatures below 630 °C. This stabilty range widens towards the Ni_3Ti if the temperature increases above 630 °C. This means that Ni-rich precipitations are possible in general and affect significantly the chemical composition and therefore the characteristic temperatures of the martensitic transformation [2]. Precipitation hardening is not possible on the Ti-rich side of the phase diagram. In general the decreasing Ti- and respectively the increasing Ni-content lead to decreasing transformation temperatures: Ni-rich alloys show austenite transformation temperatures (A_s or A_f) at below body or room temperatures whereas Ti-rich alloys achieve A_{peak}-temperatures of 110 °C. For medical products the latter materials are less important and will receive less attention in the following.

The mentioned physical prerequisites given by the phase diagram are combined with new and uncommon functional properties. Together with the thermomechanical history of a NiTi alloy, shape or component, this gives us the answer for the obvious high sensitivity of the binary NiTi alloys towards changes in the fabrication process, which is true for the Ni-rich alloys in particular. NiTi alloys owe part of their success to the implementation of quality control systems of the ISO 9000 ff standard which allow the control of several important properties under industrial conditions. The difficulties to make advantage of such a quality control system in order to achieve improved control of the properties of NiTiNOL is to be explained in the following.

2
Production and Processing of NiTiNOL

The fabrication of NiTiNOL products starts today with conventional melting techniques like "vacuum induction melting" (VIM) or "vacuum arc melting" (VAM) using elementary components of highest purity especially in terms of oxygen and carbon content. Ingots produced by VIM are melted in graphite crucibles or water cooled copper crucibles. Major advantages of the VIM process are the high homogeneity of the ingots, which is caused by the convection in the melt originated by the induction current together with the possibility of analyzing the melt during production by extraction of specimen and subsequent alloying of Ni or Ti for fine adjustment of the transformation temperatures. The VAM process allows the fine adjustment of transformation temperatures by adding further alloying elements during the re-melting procedure, which has to be carried out to achieve the necessary homogeneity of the ingot. Due to the absence of any crucible and the possibility of re-melting the purity of the VAM ingots in terms of non-metallic impurities tends to be higher compared to the VIM ingots.

The most important impurities appearing in NiTi ingots are carbides and oxides, and those have to be controlled carefully. Carbides usually appear as very brittle and large particles of 10 μm and more in size. They initiate cracks and affect the mechanical properties causing serious problems during the fabrication of shapes like thin wire or thin-walled tubes. Oxides are much smaller in size (1–5 μm) and are not as brittle as carbides. The appearance of oxides can usually not be avoided with the standard production processes. But this is not such important because the effect of oxides on the mechanical properties is less significant compared with the effect of carbides. On the other hand, oxides and carbides can affect the chemical balance of the alloy because they take away the Ti (more than Ni) from the matrix and cause a decrease of transformation temperatures due to the change in chemical composition.

These examples show the importance of controlling the content of the non-metallic inclusions together with the chemical balance of the alloy. Standard ingot qualities contain less than 500–600 ppm of oxygen and carbon. Higher ingot qualities, which are adequate for the production of advanced shapes like capillary tubing or very thin sheet, contain less than 200 ppm of carbon and less than 300 ppm of oxygen. Moreover, this gives an idea for the urge of finding and controlling the chemical balance of the alloy to produce an ingot of desired chemical composition. This necessity is even more obvious if one takes the relevant composition range of NiTiNOL alloys into account and compares it with the related transformation temperature window, demonstrating the very high sensitivity of NiTiNOL towards smallest deviations of the chemical composition: a change of 0.1 at -% of the Ni:Ti ratio leads to a change of ingot transformation temperatures of about 5–10 °C. In order to meet the requirements of the human body it is sometimes necessary to produce alloys with transformation temperature tolerances as tight as ±2 °C and even less.

The as-cast structure of a NiTiNOL ingot has to be broken by hot deformation, e.g. rolling or forging. This is carried out at temperatures between 850 °C and

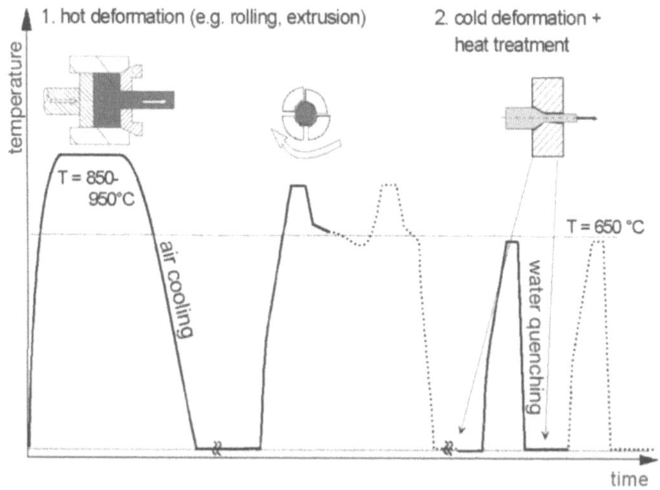

Fig. 3. Schematic depiction of the fabrication process of a binary NiTi ingot to a final product like e.g. a fine wire

950 °C and followed by further cold deformation with deformation degrees up to 45% (Fig. 3). The very important cold deformation during the last deformation step is what the metallurgists call the "thermomechanical treatment" if combined with a specific heat treatment.

3 Thermomechanical Treatment (TMT) and Functional Properties

From the metallurgical viewpoint the TMT of NiTi is besides the chemical composition the most important factor to control the functional properties. The TMT consists of the cold deformation during cold rolling or wire drawing, and a subsequent specific heat treatment. Further steps are possible, for instance a pre-straining or a two-way shape memory training procedure. This implements a large number of parameters for the characterization of the TMT process:

1. Deformation degree during cold work
2. Heat treatment temperature and duration
3. Initial condition of the alloy before final deformation step (solution heat treated microstructure or precipitates, etc.)
4. Sequence of final heat treatments and deformation step (during training procedure)

For a medical implant or instrument the required functional properties may be highly versatile and complex. It is very useful to reduce the total of possible functional properties to the following ones, which have shown to be the most important for a high number of medical applications:

1. Transformation temperatures (mostly A_{peak} or A_f)
2. Plateau stress of the superelastic effect at either body or room temperature
3. Plateau strain at either body or room temperature
4. Permanent set (plastic deformation) after a certain amount of total strain

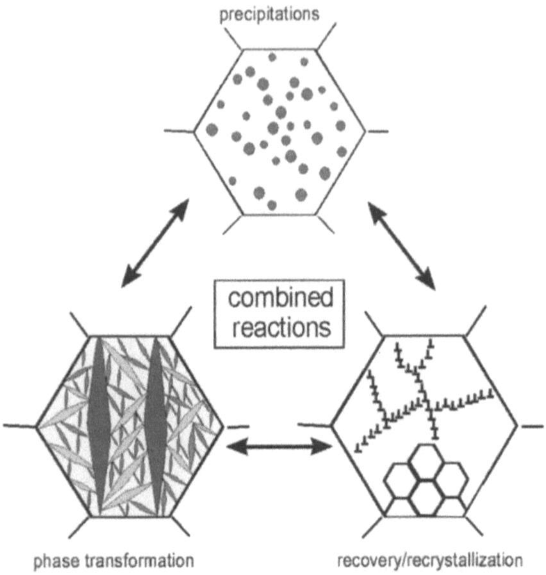

Fig. 4. Combinations of different microstructural reactions affecting the functional and structural properties of NiTiNOL. The different reactions (precipitation hardening, phase transformation, recrystallization) may occur simultaneously or sequentially

It is important to point out, that the applied TMT affects strongly the structural as well as the functional properties of a NiTiNOL shape or component. As the number of interesting functional properties may be increased up to above ten, it is important to reduce the requirements to what is needed for the respective application. This is especially true for the Ni-rich alloys showing precipitation hardening and though much more complex microstructural reactions (Fig. 4). Those are initiated by the TMT sequence, which is of importance because it is possible to identify different types of combined reactions A based on the existence or absence of [3]:

1. Dislocations as nucleation sites for precipitations
2. Supersaturated solid solutions
3. Activation energy for recovery, recrystallization or grain growth
4. And finally the occurrence of a phase transformation

All of these (and even some more) parameters affect each other strongly. The respective microstructural effects may appear simultaneously or successively, each of them leading to different functional properties.

In the literature there are countless publications about the effect of TMT on selected functional or structural properties [4, 5]. The following figures (Fig. 5; Table 1) can only give a rough idea of what is possible in terms fine-tuning of transformation temperatures and pseudoelastic strain/reversibility as a function of the applied TMT and the resulting microstructural reactions. From this example, it becomes clear that the comprehensive knowledge and control of the TMT together with the qualified characterization of the initial condition of the alloy is the key to the understanding of the resulting functional properties and basis for the application of a quality control system on an industrial level.

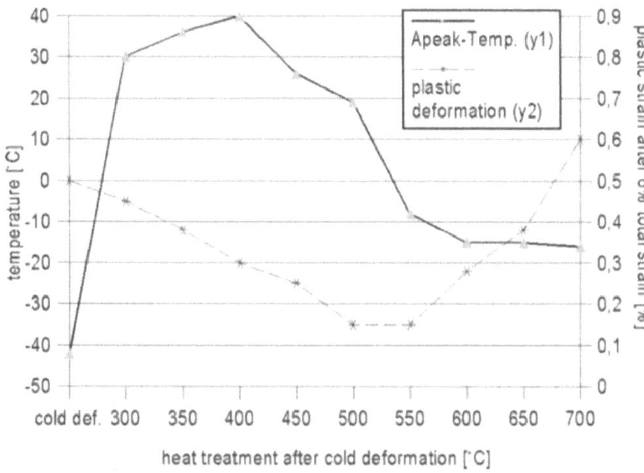

Fig. 5. Effect of TMT on two different important functional properties of NiTiNOL (alloy with approximately 50.7 atomic % Ni, condition before deformation: homogenized, 850°C, 15 min) after cold rolling (degree of deformation: ε≈20%)

Table 1. Effect of different annealing temperatures on microstructural reactions of a Ni (50.8 atomic %) – Ti alloy after different thermomechanical treatments (TMT) and resulting functional and structural properties

TMT (homogenization 850° C, cold rolling 20%) annealing	Microstructural reactions	Pseudoelastic properties (average)	Structural properties (average)
400° C/5–10 in.	Little recovery, very small precipitations	Poor	High UTS, poor strain
450° C/5–10 in.	Some subgrains, growing precipitations	Poor	High UTS, poor strain
500° C/5–10 in.	Subgrains, aging of precipitations	Good	Decreasing UTS, but still good mechanical properties
550° C/5–10 in.	Beginning of recrystallization, over-aged precipitates	Still fairly good	Poor UTS, plateau stress too low

UTS, ultimate tensile strength

4 Quality Control of NiTiNOL Semi-Finished Shapes

The main reasons for the introduction of a certified quality control system of the ISO 9000 ff standard are the reduction of the product liability risk and the possibility to increase the economic efficiency of the production. Basis for the certification is the existence of industrial technical standards. Due to missing standards for shape memory alloys the certification of products and processes based on NiTiNOL is only possible according to the "state of technique" which is sometimes difficult to define. In Germany some manufacturers try to reject the product reliability for their NiTiNOL products. Hence, in Europe there is only a French national standard which gives some definition of terms in a kind of glossary and defines rules for the measurement of transformation temperatures and the interpretation of the results. The most promising attempt to define a technical standard is actually reported from the United States where scientists, manufacturers and producers of medical products are working jointly together on a standard for shapes and products made of NiTiNOL especially for medical industry.

The problems arising with the quest for a technical norm shall be explained by giving some examples for the difficulties to find international standards. Therefore, the definition of all relevant technical terms is as essential as the characterization of fundamental test and interpretation methods. Together with the lack of appropriate tools for simulation and exact calculations, the situation in R&D is until now, that most innovations and developments are made on a "trial-and-error" basis. The existing finite element models do not represent reality good enough because the laws of linear elastomechanics are invalid and the parameters affecting the material's functional and structural properties are versatile and highly complex.

Additional difficulties arise with the new European medical device directive in which a CE marking is required for each medical product or device. CE marks can only be conferred if the products are based on valid standards or single component part tests [5]. But such tests are very expensive and the test methods are oftenly not very clear or defined. Thus, the manufacturer bears a certain risk if using NiTiNOL for his products. The fact that most engineers do not know enough about the special properties and problems in understanding of the material makes the situation even worse, leading sometimes to a complete rejection of a projected application of NiTiNOL due to the virtually increased risk with the material.

The following examples demonstrate the obstacles on the way to a international standard but will also give some ideas of how to overcome or solve the obvious problems.

4.1 Definition of Terms

Shape memory alloys are applied in very different areas of industrial products, such as medical industry, high-tech areas like micro-systems, satellites or computers but also in less pretentious area like toys or consumer goods. Together with the people from R&D and universities working with NiTiNOL, this leads often to a confusion with the definition of terms because the communication between these very different technical groups is difficult. Many effects are not named by just one unique meaning and many material properties can be misinterpreted. For instance the Young's modulus of a superelastic NiTiNOL material cannot be defined by just making a tensile test with the material (Fig. 6). Moreover, the Young's modulus is a figure with a completely different meaning for NiTiNOL compared to conventional materials. From Figure 6a, it becomes obvious that there are a couple different possible interpretations and all lead to figures for Young's modulus but none of the numbers does really make sense if applied in equations used for the calculation of a NiTiNOL component. Figure 6b shows the quantity of parameters, which have to be measured in order to entirely

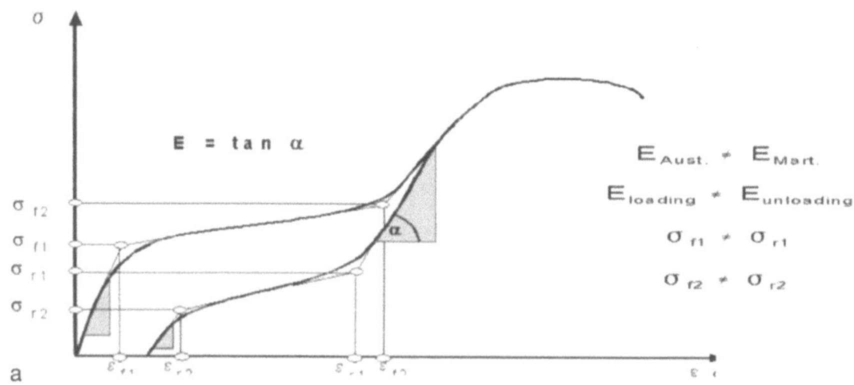

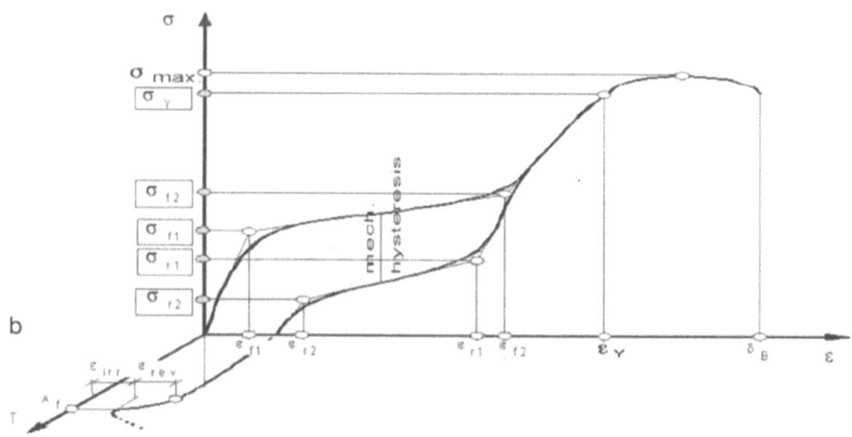

Fig. 6a,b. a Due to the physical properties of NiTiNOL the definition of terms is different to conventional materials. b Moreover, the number of parameters to define the structural and physical behavior entirely is much higher compared to conventional materials [5]

characterize the material in a stress-strain-temperature diagram [5]. The quite easy access yield stress characterizing the beginning of the plastic deformation in conventional materials has about five pendants if measured in NiTiNOL specimen. These two examples demonstrate the difficulties of defining the relevant terms for shape memory alloys, which is even worse with respect to the diversity of the different disciplines working in this field.

4.2 Measurement of Relevant Functional Properties

The measurement of important functional properties, such as the transformation temperatures or the transformation strain has to be standardized in order to guarantee reproducible measurements and to find an industrial norm. A couple of different tests are applied for measuring the transformation temperatures of

Table 2. Choice of applied test methods for measuring of the relevant austenite transformation temperatures

Test method	Measurable variables	Registered parameters	Pros	Cons
DSC, ingot material	All transformation temperatures (including R phase)	Usually A_{peak} temperature	Easy and cost-effective method; very accurate	No characterization of the final condition
DSC, final condition	All transformation temperatures (including R phase)	Usually either A_s or A_f temperature	More expensive, but precise method; very accurate	No measurement of the "active" austenite temperatures
Bending test	A temperatures (M temperatures are possible)	Either active A_s or A_f temperature	Identifies for the customer important active A temperatures	Not very precise; depends on the person doing the test; depends on amount of pre-straining (bending)
Dilatometric test	A-temperatures (M temperatures are possible)	Either active A_s or A_f temperature	Precise method; identifies for the customer important active A temperatures	Depends on amount of pre-straining

DSC, differential scanning calorimetry

NiTiNOL. Besides the calorimetric methods (differential scanning calorimetry; DSC) we can identify dilatometric methods, resistivity measurements and even bending tests are applied. Each of these tests leads to different results due to the nature of the test method. In the field of medical industry it has been widely accepted to measure on the austenite temperatures to characterize the thermal properties. This reduces the number of parameters drastically. Even though, we find in the field of medical applications different test methods for the transformation temperatures (Table 2). It has shown during the past that the most practicable and reliable method is the DSC measurement (Fig. 7), where the NiTiNOL specimen and a reference specimen are heated with a constant temperature gradient. During the phase transformation the NiTiNOL specimen changes its temperature compared to the reference, which is due to the endothermic character of the transformation (Fig. 7). Bending tests [6] are also applied on the final condition of the material and are usually carried out in a temporized water bath. The specimen is bent at temperatures below M_f and heated with a constant temperature gradient. Upon reaching the "active A_s temperature" the shape recovery occurs, which is completed until "active A_f" is reached.

But even DSC measurements of one and the same specimen carried out in different laboratories can cause serious differences between the results, which are normally unacceptable (Fig. 8). Therefore, it has shown to be useful to measure both, the "active transformation temperatures" using a bending or dilatometric measurement method and to additionally characterize the A-temperatures using a standard DSC test on the final condition of the material.

These examples demonstrate some of the problems arising with the attempt to introduce a quality control system of the ISO 9000 ff standard. Nevertheless it is possible to achieve certification for the products if the supplier is able to prove that he uses reliable methods for measurement that are state-of-the-art, that he

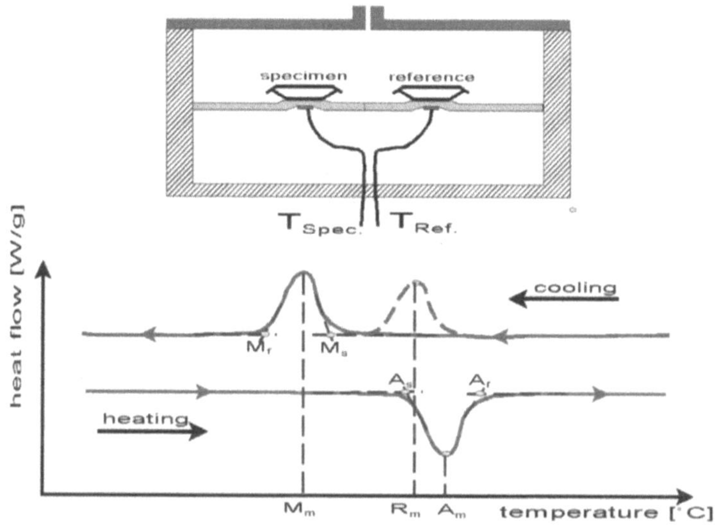

Fig. 7. DSC method for determination of transformation temperatures by measuring of the heat flow from a specimen and a reference. The result is a plot where beginning and end of the martensitic transformation can be identified by tangent methods

calibrates his devices for production on a regular basis and that he has introduced a closed and logical documentation system for the complete fabrication route of his products. All people working in production relevant areas have to be instructed in the measurement methods and have to be aware of any problem during production affecting the quality.

Thus, different test methods are today accepted and applied by most suppliers of NiTiNOL shapes and components. Besides DSC measurements on the ingot and the final condition, bending or dilatometric tests are only carried out on special

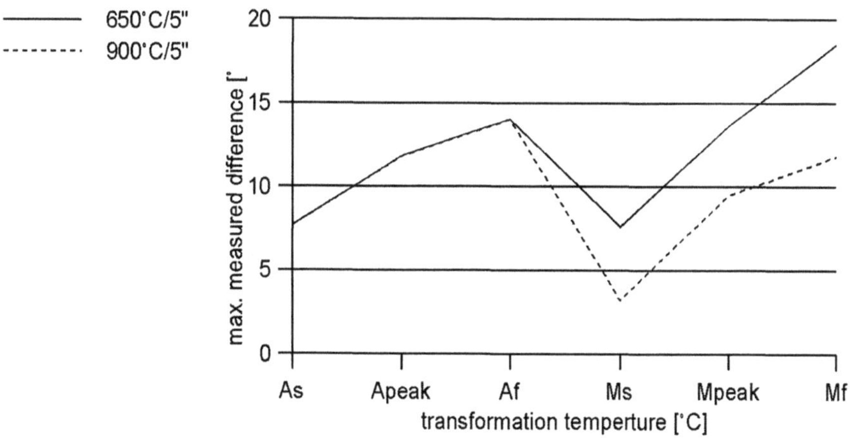

Fig. 8. Maximum difference appearing different labs using all the same samples and parameters

request of the customer. Also, the environmental temperature can be adjusted to the later application situation. This is especially useful, if a superelastic material is for instance used as an implant in the human body, where the temperature is very well defined. The active A-temperatures are very important figures because they have to be at least a couple of degrees below the human body temperature. Further tests on the structural and functional mechanical properties (stress-strain measurements) are possible and carried out upon customer's request. Critical products and implants like lasers cut stents are often characterized by a couple of more attributes like e.g. geometrical size, concentricity or straightness. Fulfilling these requirements is usually very difficult because some properties affect each other and upon adjusting e.g. the transformation temperature to the required figure, the plateau stress might run out of the specified range.

5 Conclusions

NiTiNOL alloys are very sensitive towards even smallest changes in chemical composition, so that the measurement of the composition does not make sense for the full characterization of the material. Therefore other methods have to be chosen. The material characterization starts with the complete documentation of the production route of the alloy, which is especially true for the important final steps, the so called thermomechanical treatment (TMT), which sets the very special properties of the material. The defined TMT reproducible tests and a complete documentation gives the manufacturer the possibility to supply a high quality NiTiNOL product and to achieve a ISO 9000 ff certification for his production even without an existing international technical standard.

However, the lack of standardization has to be overcome during the next years by a joint effort of manufacturers, end users and scientists in order to find widely acceptable and practicable standards for NiTiNOL, in which the relevance of functional and structural properties for each field of application have to be defined anew.

References

1. Van Humbeeck J, Cederstrom J (1994) The present state of shape memory materials and barriers still to be overcome. In: Pelton AR, Hodgson D, Duerig TW (eds) Proceedings of SMST 1994. Shape Memory and Superelastic Technologies, Pacific Grove, pp 1–6
2. Wasilewski RJ, Butler SR, Hanlon JE, Worden D (1971) Homogenity range and the martensitic transformation in NiTi. Metals Trans 2:229–238
3. Treppmann D, Hornbogen E, Wurzel D (1995) The effect of combined recrystallization and precipitation processes on the functional and structural properties in NiTi alloys. J Phys IV France 5:569–574
4. Todoroki T, Tamura H (1987) Effect of heat treatment after cold working on the phase transformation in TiNi alloys. Transactions of the japanese institute of Metals 28:83–94
5. Treppmann D (1997) Thermomechanische Behandlung von NiTi (mit Lösungsansätzen für Qualitätssicherung und Normung von Formgedächtnislegierungen). Fortschrittberichte VDI 5 No. 298. Duesseldorf
6. Wick A, Vöhringer O, Pelton AR (1995) The bending behaviour of superelastic NiTi. J Phys IV France 5:789–794

Corrosion Resistance and Biocompatibility of Passivated NiTi

Christine Trepanier, Ramakrishna Venugopalan, Alan R. Pelton

1 Introduction

Equiatomic nickel-titanium (NiTi) or Nitinol possess a unique combination of properties, including superelasticity and shape memory, which are very attractive for biomedical applications. NiTi has been used in orthopedic and orthodontic implants for several decades and has contributed to significant improvements in these fields [1, 2]. This alloy is rapidly becoming the material of choice for self-expanding stents, graft support systems, filters, baskets and various other devices for minimally invasive interventional procedures (Fig. 1) [1, 3]. While the superior performance of NiTi over conventional engineering materials for implants is well documented [1, 4, 5], the high nickel content of the alloy (55 weight % Ni) and its

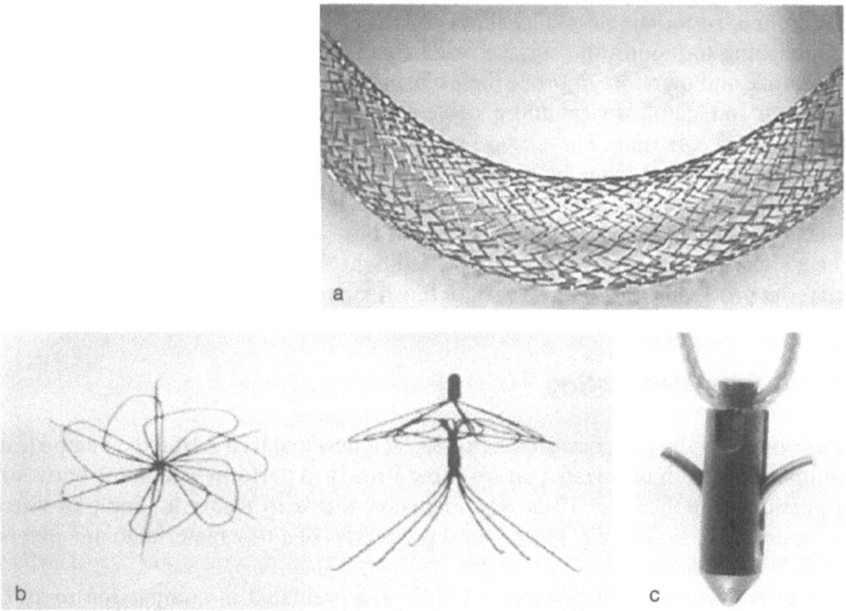

Fig. 1. Various types of minimally invasive interventional NiTi devices. **a** Cordis SMART stent. **b** Simon filter. **c** Mitek bone anchor

possible influence on biocompatibility continues to be an issue of concern. This concern is further complicated by the conflicting literature on corrosion resistance.

Human tissue contains approximately 0.1 ppm of nickel, which is essential in nutrition for biological functionality of the human body [6, 7]. The potential for higher nickel concentration release from implant material may generate harmful allergic, toxic or carcinogenic reactions [7–9]. Besides NiTi and 316L stainless steel, another high nickel containing alloy, MP35N (35 weight % Ni), exhibits good biocompatibility and is used for implants in orthodontics, orthopedics and cardiovascular applications [10–12]. Furthermore, the atomic bonding forces between Ni and Ti in intermetallic NiTi are considerably higher than in a Ti alloy with a small amount of Ni [13], and will not produce the same reactions as pure metals. Thus, it is important to recognize the synergistic effect of alloying elements when evaluating the biocompatibility of any alloy.

The limitation in the use of NiTi for medical implants is due to literature references that report moderate corrosion resistance and cell culture compatibility. However, a careful study of these references reveal that NiTi material processing history and surface conditions are generally not well documented. It is now well known that NiTi requires controlled processing to achieve optimal mechanical and thermal properties. Optimization of the thermo-mechanical processing provides good fatigue life and general mechanical properties to meet the stringent structural requirements of medical implants. Similarly, surface processing is required in order to promote optimal corrosion resistance and biocompatibility of the material. In fact, ASTM F86 [14] standard recommends an appropriate chemical treatment of metallic implants to ensure passive surface condition. The treatments recommended for stainless steel alloys consist of a nitric acid passivation or electropolishing to modify the surface oxide characteristics, increase their corrosion resistance and therefore improve their biocompatibility. NiTi is a passive alloy like titanium and stainless steel and a stable surface oxide protects the base material from general corrosion. The surface is predominantly composed of titanium oxide and thus its passivity may be further enhanced by modifying the thickness, topography and chemical composition of the surface by selective treatments [15–17]. The purpose of this chapter is to focus on the bio-corrosion properties of NiTi, and their effect on biocompatibility and to highlight the importance of documenting material processing history and surface finish for such evaluations.

2 Active Corrosion Testing

Our understanding of corrosion behavior of a new material is based on empirical comparisons with materials that we know already to perform well in the body for a particular application. Thus, any corrosion test is in reality a simple in vitro comparison of limited electrochemical properties of a new material to one that is already in clinical use.

The active corrosion behavior of NiTi was evaluated in comparison to 316L stainless steel discs passivated and sterilized according to ASTM F86 standard practices for metallic implants. Potentiodynamic polarization testing was conducted per ASTM G5 [18] in de-aerated Hank's physiological solution at 37°C. Tafel extrapolation and Stern-Geary currents were used to calculate the corrosion current density (I_{corr}) in ampere/cm² at the corrosion potential (E_{corr}). The breakdown potential (E_{bd}) was determined from the y-axis co-ordinate corresponding to the intersection of a line fit extrapolation of the passive and transpassive regions. The protection potential (E_{prot}) was the y-axis co-ordinate of the point where the reverse polarizations scan crossed over the forward scan.

Overlaid polarization plots of NiTi and 316L stainless steel are presented in Figure 2. The E_{corr} values for NiTi were more active compared to 316L stainless steel. The I_{corr} values were in the nA/cm² range for both NiTi and 316L stainless steel. The E_{bd} values for NiTi were almost three times greater than 316L stainless steel. The NiTi samples exhibited instantaneous repassivation (no hysteresis) on scan reversal at the vertex potential compared to the significant hysteresis exhibited by the 316L stainless steel. The approximately 150 mV region between the E_{prot} and E_{bd} of the 316L stainless steel makes it more susceptible to propagation of existing surface damage than NiTi. It should be noted that these results are valid only for conditions where no surface damage is involved. Similar results were presented by Venugopalan et al. [19] in their testing on small diameter NiTi and 316L stainless steel stents. Nevertheless, most minimally invasive devices may be susceptible to scratch damage due to the nature of their deployment. Thus, it is imperative that the corrosion behavior of scratched NiTi be fully characterized and understood.

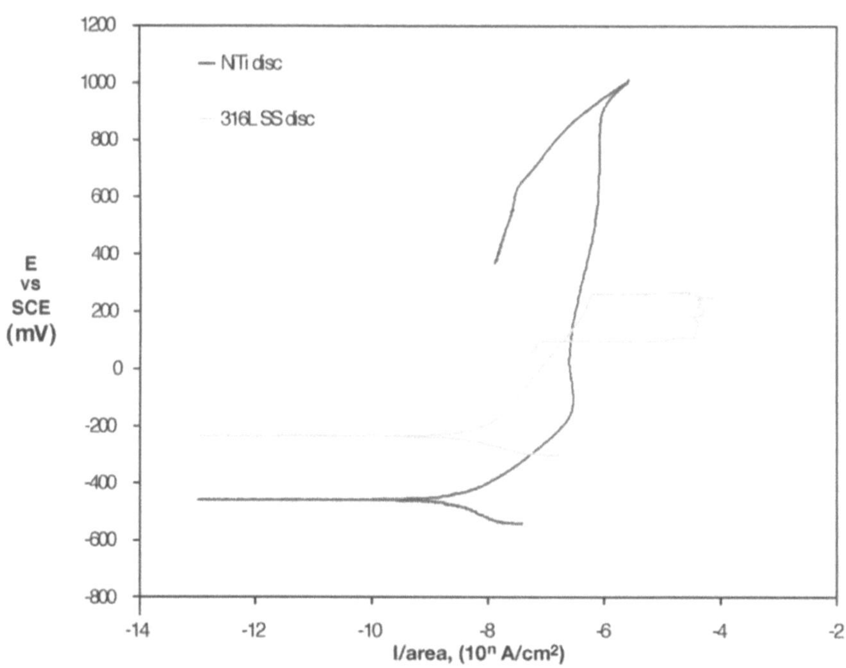

Fig. 2. Potentiodynamic polarization curves for NiTi and 316L stainless steel in de-aerated Hank's physiological solution at 37°C

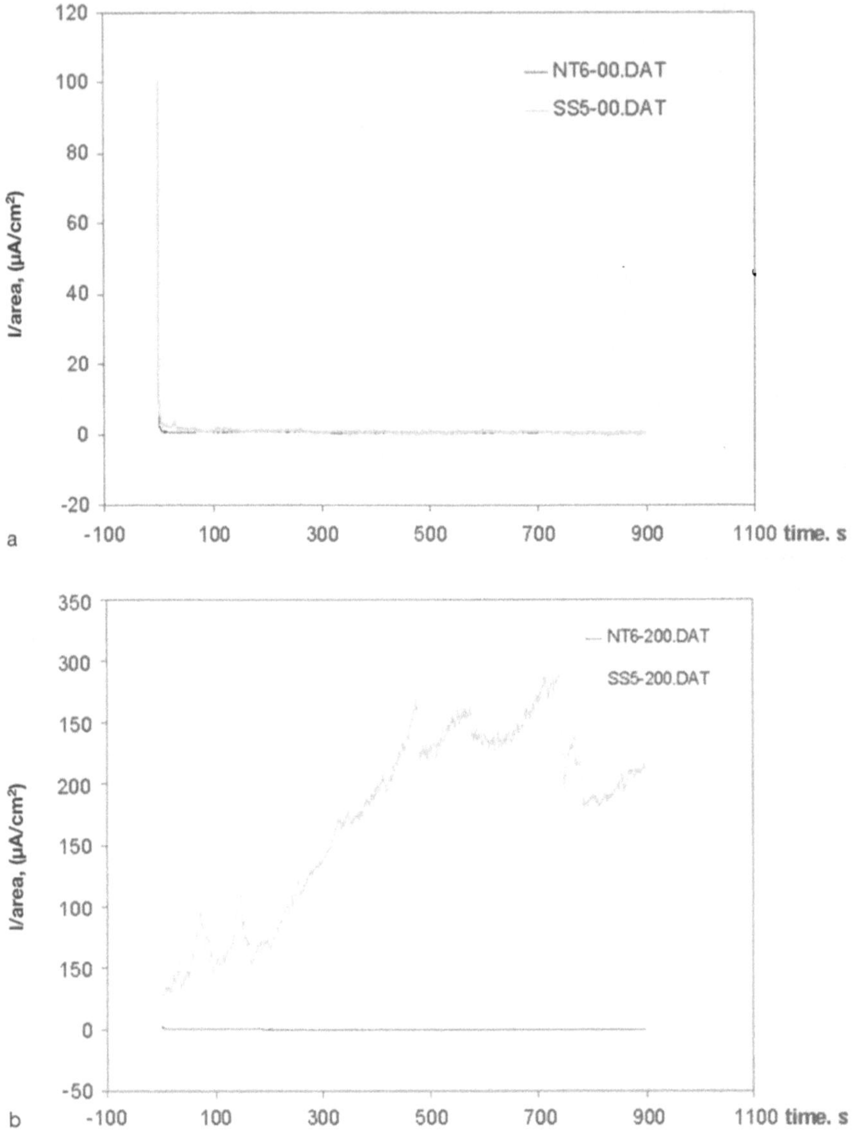

Fig. 3. Current density profiles after scratch tests at specific potentiostatic holds in de-aerated Hank's physiological solution at 37°C: **a** 0 mV, **b** 200 mV, **c** 400 mV and **d** 600 mV. All potentials are expressed with reference to a standard calomel electrode

NiTi and 316L stainless steel samples were subject to step polarization experiments in de-aerated Hank's physiological solution at 37°C [20]. The samples underwent physical scratch damage using a diamond stylus as opposed to the potentiostatic surface rupture method described in ASTM F746 [21]. The samples

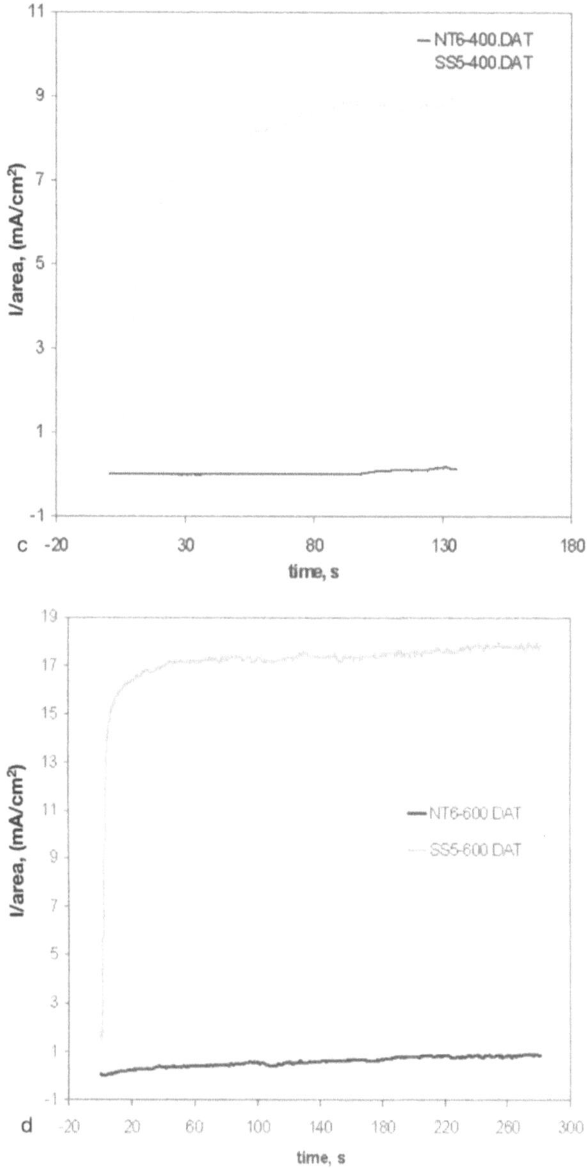

were scratched before the potentiostatic holds (P-Hold) at 0 mV, 200 mV, 400 mV and 600 mV with reference to a saturated calomel electrode (SCE) and the current density profiles were monitored. A decreasing current density trend indicated that the material was able to repassivate the surface damage while an increasing current density trend indicated that the sample was not able to repassivate the damage at that P-hold. Current density >500 nA/cm² was used as a threshold to define total loss of ability to repassivate scratch damage.

Representative overlaid current density plots at each potentiostatic hold are presented in Figure 3. The NiTi discs and the 316L stainless steel samples exhibited decreasing current densities and hence complete repassivation after scratch damage at the 0 mV potentiostatic hold. At the 200 mV potentiostatic hold, the NiTi samples exhibited decreasing current densities compared to the increasing current densities exhibited by the 316L stainless steel samples, indicating the 316L stainless steels alloy's inability to achieve total repassivation. However, the 316L stainless steel current densities did not exceed 500 nA/cm^2, a threshold value for total lack of repassivation ability. At 400 mV and 600 mV potentiostatic holds, the current densities for NiTi and the 316L stainless steel samples exceeded 500 nA/cm^2. It should be noted that the 316L stainless steel samples exhibited a faster current density transient to the 500 nA/cm^2 current density benchmark value. In conclusion, the region of repassivation capability after scratch damage for the NiTi was approximately 200 mV potential range greater than the 316L stainless steel.

3 Passive Corrosion Behavior

Passive dissolution studies in simulated physiological environments allow us to track a corrosion process through its initiation and propagation and to discriminate between the two. The effect of the environment on the device can be ascertained by visual inspection or scanning electron microscopy of the device removed from the environment at predetermined time segments of the study. The effect of the device on the environment can be determined by analyzing the media for ionic by-products using inductively coupled plasma (ICP) or atomic absorption spectroscopy (AAS).

NiTi, MP35N, 316L stainless steel alloy, and commercially pure nickel were obtained in the form of discs (surface area approximately 4.5 cm^2) and polished to 1200 grit surface finish. The alloys were passivated based on ASTM F86 standard and all samples were sterilized under UV light. The samples were placed in tissue culture plates and the plates were filled with 4 mL of Hank's physiological solution using aseptic procedure. The samples were then placed in a water-jacketed incubator with humidity, temperature and multiplex gas control. The tests were conducted under a mixed-gas environment (20.9% O_2, 5.0% CO_2, and air) at 37°C. The samples were removed 1, 6, 12, 24, 48, 72, 96 and 120 h, and 7, 14 and 21 days after placing in the incubator. Three samples of each group were removed per time period. The media was extracted, made up to 4 ml to normalize concentration effects resulting from evaporation, and analyzed using AAS to determine the ionic content of Ni.

A semi-log plot format was used to overlay the results (Fig. 4) as they spanned over multiple orders of magnitude. The 316L SS and NiTi alloy exhibited the least (tens of ppb) Ni ion release into the media. The MP35N exhibited an order of magnitude increase (hundreds of ppb) in Ni ion release into the media compared to the NiTi and 316L SS alloys. The negative control, commercially pure nickel, exhibited the highest amount (thousands of ppb) of Ni ion release into the media (as expected).

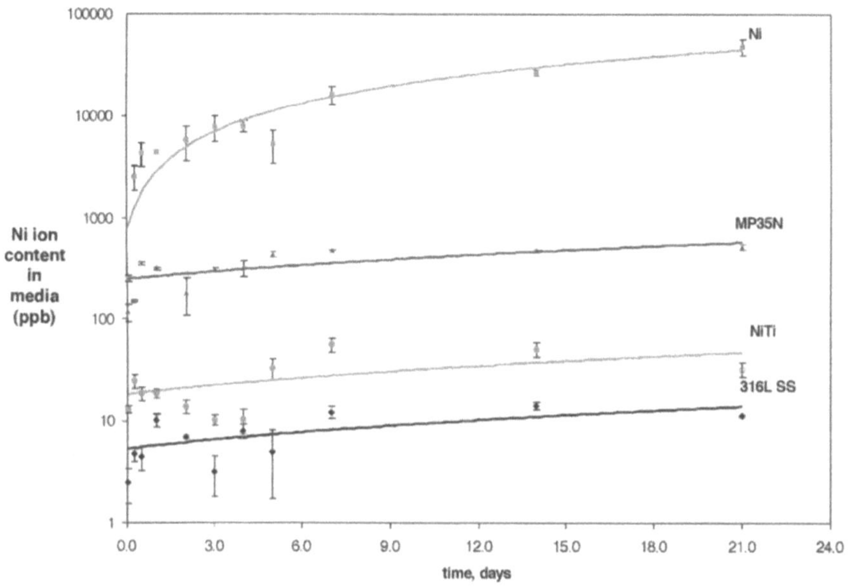

Fig. 4. Ni-ion concentrations released by Ni, MP35N, 316L stainless steel, and NiTi in Hank's physiological solution at various time periods of removal during dissolution study. The dissolution study was conducted in mixed-gas atmosphere at 37°C

4
Effect of Surface Layer on Corrosion Resistance

Several studies have demonstrated that passivated NiTi surface layers consist predominantly of a titanium oxide layer (TiO_2) [15–17, 22] similar to that found on Ti alloys [23]. This is in agreement with theoretical thermodynamics, which specify that the free energy of formation of TiO_2 is favored over other nickel or other titanium oxides [22]. This oxide layer serves two purposes:

1. Increases the stability of the surface layers by protecting the bulk material from corrosion
2. Creates a physical and chemical barrier against Ni oxidation by modifying the oxidation pathways of Ni [24]

The stability of the surface layer on NiTi and its ability to protect the material from corrosion have been investigated in several studies by electrochemical experiments.

Early studies performed by Kimura and Sohmura [25] showed that passivation promotes the growth of an oxide layer on NiTi and resulted in its improved corrosion resistance in 1% saline solution at 37°C. More recently, Trepanier et al. [16] investigated the effects of electropolishing and heat treatments of NiTi stents on their corrosion resistance in Hank's physiological solution at 37°C. These results indicated a significant improvement in the corrosion resistance of NiTi stents that was attributed to the formation of a thin and very uniform Ti-based oxide layer. The authors concluded that uniformity rather than thickness of the oxide was most important to the improved corrosion resistance for this kind of devices. Furthermore, as was shown by Kimura and Sohmura [25], a thin oxide layer is preferable to maintain the integrity of the surface layer to sustain the large deformation induced by the shape memory effect.

A comparative study of the corrosion resistance of passivated Ti-6Al-4V, 316L stainless steel and NiTi was performed in Hank's physiological solution by Wever et al. [15]. Their results show that while Ti-6Al-4V was the most corrosion resistant, NiTi samples were more resistant to chemical breakdown of their passive film than 316L stainless steel samples. Our results are in agreement with Wever et al. regarding the corrosion behavior of NiTi in comparison to stainless steel. These results highlight the importance of a well-controlled and optimal surface preparation process to achieve good and reproducible corrosion resistance for both materials. Furthermore, our scratch test investigations demonstrated that both NiTi and stainless steel exhibit a decreased resistance to pitting once their surface is severely damaged. Nevertheless, in the event of a similar surface damage, NiTi is still characterized by a higher resistance to localized corrosion compared to stainless steel.

5 Nickel Release and Biocompatibility

Since nickel release during the bio-degradation of NiTi is an important concern for its use as an implant, several studies have been undertaken to measure this value. For example, Barret et al. [26] and Bishara et al. [27] investigated nickel release from NiTi arch wires (processed by the manufacturer) in saliva. During an in vitro dissolution study, they found that NiTi and stainless steel appliances released a similar total amount of Ni around 18 ppm after a 28 days dissolution study. In a second study, orthodontic patients with NiTi appliances had Ni-concentration in their blood measured for a period of 5 months. Results show no significant increase in the nickel blood level throughout this study.

A comparative in vitro cell culture study was undertaken by Ryhanen et al. [28] in which they measured Ni released from NiTi and 316L stainless steel in a fibroblast and osteoblast cell culture media. In both media, Ni levels were higher in the NiTi group the first day and decreased rapidly as a function of time to achieve similar levels as 316L after 8 days. It is important to note that even though Ni release was higher in the NiTi group, it did not reach toxic values and cell proliferation or cell growth near the implant surface was not affected. Furthermore, NiTi was only mechanically polished without additional passivation treatments, whereas the stainless steel was electropolished according to the guidelines of the manufacturer. Ryhanen et al. [28] hypothesized a further decrease in Ni release if additional passivation treatments, such as electropolishing, are performed on NiTi. Wever et al. [15] conducted a similar comparative study with passivated NiTi and 316L stainless steel in Hank's solution. Ni release from NiTi was maximum the first day (14.5×10^{-7} $\mu g/cm^2/s$) and reached undetectable levels similar to 316L stainless steel after 10 days.

More recently, Jia et al. published their results on Ni release from NiTi and stainless steel orthodontic appliances [29]. Their study showed that NiTi released more Ni (maximum of 4.1 ppb) than stainless steel arch wires in a period of 24 hours. Furthermore, in agreement with several studies, they have shown that a threshold value of 30 ppm is needed to trigger a cytotoxic response during in vitro experiments. Our results on electropolished NiTi and 316L show that this Ni release threshold is far from being reached, even after 21 days of immersion in Hank's physiological solution.

Biocompatibility of a material may be simply defined as its ability to be well accepted by the body. Since every material will generate a "foreign body reaction" when implanted in the body, the degree of biocompatibility is related to the extent of this reaction. In order to study this phenomenon, in vitro testing with cell cultures allows isolation of the reaction from each cell and physiological media, whereas, in vivo testing provides a more complete response involving the biological environment and immune system. Both types of tests have been undertaken to better understand the biological response to NiTi.

A recent in vitro study revealed no significant differences between the cell growth behavior near the surfaces of different implant materials (mechanically polished Ti and NiTi, electropolished 316L stainless steel) [28]. A microscopy analyses also showed that the cells had grown very near to Ti and NiTi alloys while they were less close to the stainless steel samples. The authors concluded that NiTi showed very good biocompatibility and that it had an excellent potential for clinical applications. Also, passivated NiTi showed no cytotoxic, allergic or genotoxic activity based on a MEM extract cytotoxicity test, a guinea-pig sensitization test and genotoxicity testing, respectively [30]. Similar results were obtained for the control group composed of passivated 316L stainless steel samples. In a different study that addressed only the genocompatibility of the material, NiTi exhibited a good biocompatibility behavior similar to Ti and 316L stainless steel on cellular chromatin [31].

Cutright et al. [32] have studied the tissue response to subcutaneous implantation of NiTi wire sutures in rats for a period of 9 weeks. The inflammatory response was minimal starting 3 days after implantation and the healing process initiated after 1–2 weeks consisted of a fibrous capsule formation around the implant. This reaction was similar to the one generated by similar stainless steel wires. In addition, Castleman et al. [33] evaluated the biocompatibility of chemically passivated NiTi by inserting plates into beagle femurs for periods ranging from 3 months to 17 months. The histological analysis of muscular tissue surrounding the implantation site showed no significant difference between NiTi and Cr–Co plates. Neutron activation analyses near the NiTi implants have indicated that there was no significant presence of metallic Ni in the muscle. Based on their observations, they concluded that the material was safe to conduct further testing.

More recently, Trepanier et al. [34] performed an in vivo study on passivated NiTi stents. Implantation of the material in rabbit paravertebral muscles and study of the inflammatory reaction for periods ranging from 3 weeks to 12 weeks demonstrated good biological response to NiTi. Analysis of the fibrous capsule surrounding NiTi stents revealed a decrease in thickness with time. A comparative 26-week follow-up study was conducted on rats to assess the effect of different materials on soft tissues [35]. In this study, short-term biocompatibility of polished NiTi was similar to polished Ti-6Al-4V and electropolished stainless steel when in contact with muscle and perineural tissue. These results indicate promising soft tissue compatibility of NiTi.

6 Conclusions

Based on the abundance of literature reports, passivated NiTi has improved corrosion resistance compared to stainless steel. NiTi is protected from corrosion by a highly stable and biocompatible Ti-based oxide layer. This good corrosion behavior will prevent degradation of the material in the physiological environment and therefore will promote biocompatibility. Ni release from NiTi has been shown to be minimal in every study. The Ni dissolution rapidly decreases from a maximum (well below cytotoxic levels) to nearly non-detectable levels few days following NiTi immersion in a physiological media. Corrosion resistance of NiTi can be further enhanced by different surface treatments such as electropolishing which promote a very uniform oxide layer. In vitro and in vivo studies show that NiTi exhibits good biocompatibility and does not promote toxic or genotoxic reactions when in contact with a physiological environment. Therefore, passivated or properly treated NiTi can be considered a biologically safe implant material with unique mechanical properties.

References

1. Duerig TW, Pelton AR, Stockel D (1996) The utility of superelasticity in medicine. Biomed Mater Eng 6:255–266
2. Haasters J, Salis-Solio G, Bonsmann G (1990) The use of Ni-Ti as an implant material in orthopedics. In: Duerig TW, Melton KN, Stockel D, Wayman CM (eds) Engineering aspects of shape memory alloys. Butterworth-Heinemann, Boston, pp 426–444
3. Frank TG, Xu W, Cuschieri A (1997) Shape memory applications in minimal access surgery – the Dundee experience. In: Pelton AR, Hodgson D, Russell SM, Duerig TW (eds) Proceedings of SMST 1997. Shape Memory and Superelastic Technologies, Pacific Grove, pp 509–514
4. Shabalovskaya SA (1996) On the nature of the biocompatibility and on medical applications of NiTi shape memory and superelastic alloys. Biomed Mater Eng 6:267–289
5. Lu S (1990) Medical applications of Ni-Ti in China. In: Duerig TW, Melton KN, Stockel D, Wayman CM (eds) Engineering aspects of shape memory alloys. Butterworth-Heinemann, pp 445–451
6. Anke M, Groppel B, Kronemann H, Grun M (1984) Nickel – an essential element. In: Saunderman FW (ed) Nickel in the human environment. International Agency for Research on Cancer, Lyon, pp 339–366
7. Williams DF (1981) Toxicology of implanted metals. (Fundamental aspects of biocompatibility, vol 2) CRC, Boca Raton, pp 45–61
8. Bass JK, Fine H, Cisneros GJ (1993) Nickel hypersensitivity in the orthodontic patent. Am J Orthod Dentofacial Orthop 103:280–285
9. Takamura K, Hayashi K, Ishinishi N, Sugioka Y (1994) Evaluation of carcinogenecity and chronic toxicity associated with orthopedic implants in mice. J Biomed Mater Res 28:583–589
10. Liotta D (1998) Assisted circulation for end-stage chronic heart failure. Artif Organs 22:230–236
11. Brown SA, Hughes PJ, Merritt K (1988) In vitro studies of fretting corrosion of orthopaedic materials. J Orthop Res 6:572–579
12. Speck KM, Fraker AC (1980) Anodic polarization behavior of Ti-Ni and Ti-6Al-4V in simulated physiological solutions. J Dent Res 59:1590–1595

13. Hultgren R, Desai PD, Hawkins DT, Gleiser M, Kelley KK (1973) Selected values of the thermodynamic properties of binary alloys. American Society for Metals, Materials Park, pp 1244–1246
14. American Society for Testing and Materials F86 (1995) Standard practice for surface preparation and marking of metallic surgical implants. In: ASTM (ed) Annual book of ASTM standards. (Medical devices and services, vol 13.01) American Society for Testing and Materials, Philadelphia, pp 6–8
15. Wever DJ, Veldhuizen AG, De Vries J, Busscher HJ, Uges DRA, van Horn JR (1998) Electrochemical and surface characterization of a nickel-titanium alloy. Biomaterials 19:761–769
16. Trépanier C, Tabrizian M, Yahia L'H, Bilodeau L, Piron DL (1998) Effect of the modification of the oxide layer on NiTi stent corrosion resistance. J Biomed Mater Res 43:433–440
17. Trigwell S, Selvaduray G (1997) Effects of surface finish on the corrosion of NiTi alloy for biomedical applications. In: Pelton AR, Hodgson D, Russell SM, Duerig TW (eds) Proceedings of SMST 1997. Shape Memory and Superelastic Technologies, Pacific Grove, pp 383–388
18. American Society for Testing and Materials G5 (1995) Standard reference test method for making potentiostatic and potentiodynamic anodic polarization measurements. In: ASTM (ed) Annual book of ASTM standards. (Medical devices and services, vol 03.02) American Society for Testing and Materials, Philadelphia, pp 48–58
19. Venugopalan R (1999) Corrosion testing of stents: A novel fixture to hold entire device in deployed form and finish. J Biomed Mater Res 48:829–832
20. Venugopalan R, Trepanier C, Pelton AR, Lucas LC (1999) Comparative electrochemical behavior of NiTi and 316L stainless steel. In: Society for Biomaterials (ed) Proceedings of the 25th annual meeting of the Society for Biomaterials and the 31st International Biomaterials Symposium. Society for Biomaterials, Providence, p 144
21. American Society for Testing and Materials F746 (1995) Standard test method for pitting or crevice corrosion of metallic surgical implant materials. In: ASTM (ed) Annual book of ASTM standards. (Medical devices and services, vol 13.01) American Society for Testing and Materials, Philadelphia, pp 192–197
22. Chan CM, Trigwell S, Duerig T (1990) Oxidation of a NiTi alloy. Surface Interface Anal 15:349–354
23. Lausmaa J, Mattsson L, Rolander U, Kasemo B (1986) Chemical composition and morphology of titanium surface oxides. (Materials Research Society Symposium Proceedings, vol 55) Materials Research Society, Pittsburgh, pp 351–359
24. Espinos JP, Fernandez A, Gonzalez-Elipe AR (1993) Oxidation and diffusion processes in nickel-titanium oxide systems. Surface Sci 295:420–410
25. Kimura H, Sohmura T (1987) Surface coating on TiNi Shape memory implant alloys. J Osaka Univ Dent Sch 27:211–223
26. Barrett RD, Bishara SE, Quinn JK (1993) Biodegradation of orthodontic appliances. Part I. Biodegradation of nickel and chromium in vitro. Am J Orthod Dentofacial Orthop 103:8–14
27. Bishara SE, Barrett RD, Selim MI (1993) Biodegradation of orthodontic appliances. Part II. Changes in the blood level of nickel. Am J Orthod Dentofacial Orthop 103:115–119
28. Ryhanen J, Niemi E, Serlo W, Niemela E, Sandvik P, Pernu H, Salo T (1997) Biocompatibility of nickel-titanium shape memory metal and its corrosion behavior in human cell cultures. J Biomed Mater Res 35:451–457
29. Jia W, Beatty MW, Reinhardt RA, Petro TM, Cohen DM, Maze CR, Strom EA, Hoffman M (1999) Nickel release from orthodontic arch wires and cellular immune response to various nickel concentrations. J Biomed Mater Res 48:488–495
30. Wever DJ, Veldhuizen AG, Sanders MM, Schakenraas JM, van Horn JR (1997) Cytotoxic, allergic and genotoxic activity of a nickel-titanium alloy. Biomaterials 18:1115–1120
31. Assad M, Yahia L'H, Rivard CH, Lemieux N (1998) *In vitro* biocompatibility assessment of a nickel-titanium alloy using electron microscopy in situ end labeling (EM-ISEL). J Biomed Mater Res 44:154–161
32. Cutright DE, Bashker SN, Perez B, Johnson RM, Cowan Jr GSM (1973) Tissue reaction to nitinol wire alloy. Oral Surg Oral Med Oral Pathol Oral Radiol Endod 35:578–584
33. Castleman LS, Motzkin SM, Alicandri SM, Bonawit VL (1976) Biocompatibility of nitinol alloy as an implant material. J Biomed Mater Res 10:695–731
34. Trépanier C, Leung TK, Tabrizian M, Yahia L'H, Bienvenu J-G, Tanguay J-F, Piron DL, Bilodeau L (1999) Preliminary investigation of the effects of surface treatments on the biological response to shape memory NiTi stents. J Biomed Mater Res 48:165–171
35. Ryhanen J, Kallioinen M, Tuukkanen J, Junila J, Niemela E, Sandvik P, Serlo W (1998) In vivo biocompatibility evaluation of nickel-titanium shape memory metal alloy: muscle and perineural tissue responses and encapsule membrane thickness. J Biomed Mater Res 41:481–488

The High Damping Capacity of Shape Memory Alloys

J. Van Humbeeck, Y. Liu

1 Introduction

The ability of damping out rapidly mechanical vibrations or noise created by impact loading is considered as an important and useful material property. Materials that can fulfil this condition are qualified as high damping materials. Especially metallic materials belonging to this group are classified as "hidamets" (**hi**gh **dam**ping **met**als).

Shape memory alloys are gaining an increased interest as passive as well as active damping materials. This damping ability when applied in structural elements can lead to a better noise control, improved lifetime and even better performance of the envisaged tools.

By passive damping, it is understood that the material converts a significant part of unwanted mechanical energy into heat. This mechanical energy can be a (resonance) vibration, impact loading or shock waves.

This high damping capacity finds its origin in the thermoelastic martensitic phase due to the hysteretic mobility of martensite-variants or different phase interfaces. The damping capacity increases with increasing amplitude of the applied vibration or impact and is almost frequency independent. Special interest exists moreover for damping extreme large displacements by applying the mechanical hysteresis performed during pseudoelastic loading. This aspect is nowadays very strongly studied as a tool for protecting buildings against earthquakes in seismic active regions and could also be useful in some orthopaedic devices.

Active damping can be obtained in hybrid composites by controlling the recovery stresses or strains of embedded shape memory alloy wires. This controls the internal energy of a structure which allows controlled modal modification and tuning of the dynamical properties of structural elements. But also impact damage, acoustic radiation, dynamic shape control can be actively controlled. As a consequence improved fatigue-resistance, better performance and a longer lifetime of the structural elements can be obtained.

The quality and quantity of a high damping level are a matter of a subjective choice, but a consideration of the experimental data available suggests that a lower bound of $Q^{-1}=10^{-2}$ is not unreasonable [1]. Q^{-1} is called the loss factor (many times also given the symbol η) and is defined as:

$$Q^{-1}=\frac{1}{2\pi}\cdot\frac{\Delta W}{W} \tag{1}$$

where ΔW is the energy (generally converted into heat) absorbed after loading and unloading and W is the applied energy during loading.

The origin of the damping capacity of a metal is related with the internal friction occurring during the hysteretic movement of defects and the interaction of (different kinds of) defects, mainly dislocations and interfaces (interfaces between different phases, magnetic domain walls, interfaces between martensite variants) [2].

The internal friction occurring in shape memory alloys during transformation and in the martensitic state has been studied quite in detail. Moreover, the study of the internal friction behaviour in all types of shape memory alloys has revealed a lot of information on the structure of the martensite and β phase and on the dynamic process of the martensitic transformation itself. Several reviews have already been dedicated to the particular damping behaviour of materials exhibiting a thermoelastic or non-thermoelastic martensitic transformation [2–5]. The most relevant part of the present state of knowledge, mainly related to Ni-Ti shape memory alloys, will be explained further.

2 Internal Friction Behaviour of Shape-Memory Alloys

Two main temperature regions have to be considered separately: the temperature region in which the material is completely martensitic and the one in which the material transforms and thus the β phase and the martensite coexist. The β phase will not be discussed here, since no high damping has been observed although it can be mentioned that interesting relaxation phenomena have been observed in this phase [6–8].

2.1 Internal Friction during Martensitic Transformation

During the martensitic transformation, an internal friction peak is observed concurrently with a strong modulus minimum [9]. As pointed out by Bidaux et al. [10], in materials in which the two phases during the phase transformation can coexist over a limited temperature range, one should consider three separate contributions to the total internal friction, illustrated in Figure 1.

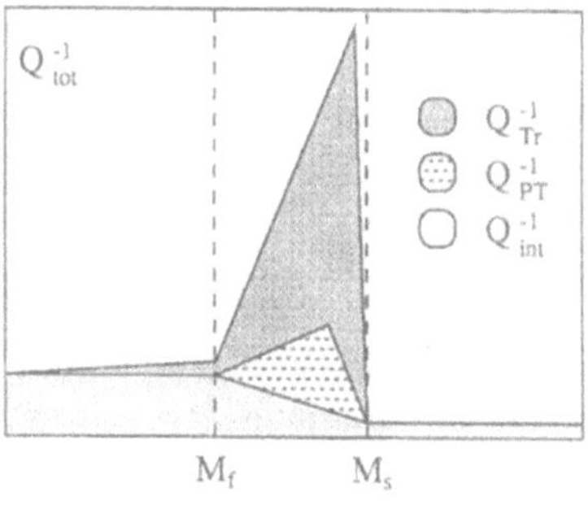

Fig. 1. Schematic of Q^{-1}_{tot} and its contributions Q^{-1}_{Tr}, Q^{-1}_{PT} and Q^{-1}_{int} during the martensitic phase transformations

$$Q^{-1}_{tot} = Q^{-1}_{Tr} + Q^{-1}_{PT} + Q^{-1}_{int} \quad (2)$$

Q^{-1}_{Tr} is the transient part of Q^{-1}_{tot}, and it exists only during cooling or heating ($\dot{T} \neq 0$). It depends on external parameters like temperature rate ($\dot{T}$), resonance frequency (f) and oscillation amplitude (σ_0). Q^{-1}_{Tr} depends on the transformation kinetics and is therefore proportional to the volume fraction which is transformed per unit of time. Q^{-1}_{PT} is related to mechanisms of the phase transformation (PT), which are independent of the transformation rate, such as the movement of parent martensite and martensite/martensite interfaces. Q^{-1}_{PT} exhibits a small peak when the interface mobility is maximum. Q^{-1}_{int} is composed of the IF contributions of each phase and is strongly dependent on microstructural properties (interface density, vacancies), especially in the martensitic phase.

Several models have been developed to describe and analyse those different contributions, taken into account the parameters: temperature, temperature rate, frequency, amplitude, and the transformed volume fraction. Those models are summerized in ref. [11, 12]. It is very important to notice that those models describe an inherent relation between $\dot{T}$ and Q^{-1}_{Tr}. Generally the internal friction or damping capacity of shape memory alloys are measured at a constant heating or cooling rate and thus a significant damping peak appears during transformation. However at $\dot{T}=0$, thus at constant temperature, Q^{-1}_{Tr} becomes zero. This is also actually seen during the experiment and was first described in NiTi by Mercier et al. [13]. The remaining damping capacity remains a summation of Q^{-1}_{int} and Q^{-1}_{PT}. As a consequence, the damping capacity becomes a function of the volume fraction of the martensite, and the very high damping capacity during transformation is lost. For applications where a continuous vibration is applied at constant temperature a 100% martensite condition offers therefore a more stable damping. In the case of impact loadings at very low frequency, the two-phase region can be interesting since the martensite will be now stress-induced concomitant with an exothermic heat effect.

2.2 Internal Friction in the Martensitic Phase

The high damping capacity of the martensitic phase is related to the hysteretic movement of interfaces (martensite variant interfaces, twin boundaries). Many publications have appeared related to this subject and the most important references can be found in some review papers [2, 4, 11, 14].

It has also been established that the global internal friction is also controlled by dislocations and their interactions with other lattice defects. This has been recently described by S. Kustov et al. [15] who found a high damping capacity at low temperatures (4–200 K) in Cu–Al–Ni alloys. This particular behaviour has been related to a decrease in the concentration of obstacles, pinning the dislocations, with decreasing temperature. The influence of vibration frequency, heating/cooling rate (temperature rate) and amplitude is shortly reviewed in the following paragraphs.

2.2.1
Frequency

From the published data on the internal friction values, exclusive of some relaxation peaks, no frequency dependence was detected. This parameter was explicitly studied by Vandeurzen in Cu–Zn–Al, Ni–Ti and Cu–Mn alloys in the range between 10 Hz and 150 Hz [16]. This result seems reasonable for a hysteretic type of damping.

2.2.2
The Temperature Rate

Although no explicit experiments where carried out to determine the dependence of this parameter, some authors report different results obtained with $\dot{T} \neq 0$ and $\dot{T} = 0$ [13, 16–18]. When the internal friction was measured at $\dot{T}=0$ a decrease was noticed. However, the internal friction was restored by interruption of the vibration [17, 18], by an amplitude discontinuity [13, 18] or by starting heating or cooling [13, 18]. This means that the loss of damping has no permanent character but is probably influenced by pinning defects which can easily vanish or can be displaced or annihilated. The point defects can be disordered atom pairs formed by the motion of the interface [18]. When the vibration is interrupted, a restoration occurs, while the concentration of disordered atom pairs is lowered. It is concluded that these atom pairs reduce the mobility of the interfaces. The other hypothesis proposes an increasing concentration of vacancies along the interface dislocations due to the vibration. Interrupting the vibration would allow that these vacancies can be redistributed again in the matrix [17].

2.2.3
The Amplitude

From experimental observations in Cu–Zn–Al and Ni–Ti, Koshimizu [19] found that three amplitude domains can be distinguished (Fig. 2). The amplitude-dependence observed in region C was dedicated to the mechanism proposed by Granato and Lücke [20]: an amplitude-dependent dislocation damping due to unpinning from the weak pinning points. The amplitude should be limited to $10^{-7} - 10^{-6}$. A similar result was obtained by Zhu Jin-Song et al. in Au–Cd [21]. This behaviour was found in the martensitic region as well as in the transformation region and the parent phase. The highest value for the IF was obtained in the

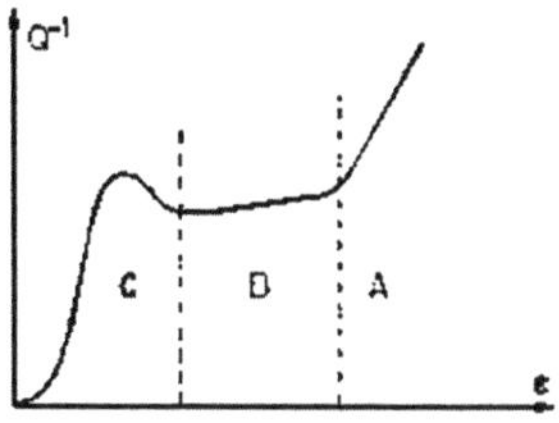

Fig. 2. The dependence of the IF on the oscillation amplitude (ε_0) measured in Cu–Zn–Al martensite according to Koshimizu. Domain C: $\varepsilon<10^{-6}$, domain B: $\varepsilon<10^{-5}$, domain A: $\varepsilon>10^{-5}$

martensitic state. A further analysis of the IF in this region such as the dependence on the temperature would be very interesting since using the Granato-Lücke (G-L) model more information on the dislocations, such as their average length and concentration or even their localisation can be obtained. Region B between 10^{-6} and about 5×10^{-6} to 10^{-5} shows almost no amplitude dependence. The IF should be due to the interaction between dislocations and weak pinning defects. More information in the literature can be found on the amplitude at which the IF becomes again amplitude dependent. This critical amplitude is found to be between 5×10^{-6} and 2×10^{-5} depending on the system. For a Cu-Zn-Al alloy Koshimizu et al. [19] find a value of 5×10^{-6}. For Ni-Ti Mercier et al. [22] give a value of 2×10^{-5}, a value that is also given by Tirbonod and Koshimizu for Ni46Ti7Cu [23]. For Cu-Al-Ni one can derive the value of 1×10^{-5} from the given "break-away stress" measured by Sugimoto et al. [24]. Another value 2×10^{-5} can be found in [25]. It is generally proposed that this critical amplitude is the amplitude where the interfaces starts to "break-away", analogous to the G-L model for dislocations. Some authors therefore have applied the G-L model in this amplitude region (A) and indeed a linear relationship is found between ln $Q^{-1}\varepsilon$ and ε^{-1} [19, 26]. It should be noticed thereby that two lines with different slopes are obtained. For amplitudes larger than 10^{-4} the slope is much higher than for amplitudes lower than 10^{-4}, which seems to indicate another critical amplitude [17, 19]. Of course one should be prudent by applying models, derived for different conditions. Other damping models like the one of Takahashi ($Q^{-1}=B\varepsilon_0^{\alpha}$ [27]) seems also to fit very good with the experimental results [17].

This model seems to work better where concentrated Cottrell clouds of point defects are present along the dislocations. A third interesting model is the one of Peguin et al. [28]. This model extends the G-L model. Once the dislocations are broken away, (micro-) plastic phenomena can occur with increasing amplitude. This means that the observed IF behaviour in a broad amplitude range is the summation of a G-L behaviour Q_H^{-} and a plastic contribution Q_P^{-1}. Though no experimental evidence is given so far for it, other observations make the model quite acceptable. Indeed, Kajiwara and Kikuchi [29] demonstrated that thermal cycling through the transformation zone increases the dislocation density due to the movement of the interfaces through the lattice. Morin et al. [30] found that thermal cycling indeed influences the amplitude dependent behaviour of the IF.

2.2.4
Time Dependence

This effect has been investigated in detail by Mercier et al. for Ni-Ti alloys [13] and Van Humbeeck et al. and Morin et al. for CuZnAl alloys [18, 31–36]. The experimental results can be summarised as follows:

- At constant amplitude, the damping follows either a monotonously decreasing trend or goes through a maximum (at frequencies between 10 Hz and 100 Hz). Complementary, the modulus increases monotonously or goes through a minimum (concurrently with the maximum in damping).
- The time dependence is also controlled by frequency, strain amplitude and temperature. The effect seems to be thermally activated.

- The loss of damping under constant vibrating amplitude can be restored by a discontinuity in the amplitude (Sect. 2.2.1).

Two different types of maxima in the damping behaviour have been observed [31, 33]. Since these maxima occur as a function of time, they are called the peaking effect (PE; PE1: peaking effect of the first kind; PE2: peaking effect of the second kind) in analogy with the peaking effect observed during irradiation of metals [37]. In this case the peaking effect is ascribed to the interaction between oscillating dislocations and irradiation produced point defects. The PE1 in the martensite is interpreted in terms of an interaction between the martensite variant boundaries (or dislocations constituting these boundaries) and the excess vacancies induced as a result of the quenching procedure and of the phase transformation. This peaking effect has been observed during vibration in the frequency range of the order of 10 Hz [31, 35] and occurs at room temperature after a few minutes, depending on the amplitude.

The PE2 describes the fact that the overall damping capacity of the martensite goes through a maximum as function of the ageing time in the martensite without applied vibration mode [33]. It seems to be related with the process of stabilisation of martensite. The maximum damping is now obtained after a few hours to several days of ageing depending on the temperature of ageing. In both cases, PE1 and PE2, the increase of damping is of the order of 10–50% relative to the initial value.

2.2.5
Relaxation Peaks in Ni–Ti and Cu-Based Martensites

Several types of relaxation peaks have been reported. For Ni-Ti, Hasiguti and Iwasaki [38] and Postnikow et al. [39], report a peak at 203 K with an activation energy of 0.38 eV and a frequency factor of $6 \times 10^9\ s^{-1}$. This peak might be due to a movement of dislocations or a movement of point-defect-pinned dislocations.

Tirbonod and Koshimizu [23] report a Bordoni type peak in an equiatomic Ni–Ti alloy. The presence of this peak should be due to the motion of imperfect dislocations with a Burgers vector smaller than the lattice parameter. These can be twinning dislocations but the dislocations situated at the interface between the different variants cannot be ruled out. Experiments on monovariant specimens could make this matter clearer.

Cu–Al–Ni alloys with γ'-martensite exhibit a twin relaxation peak as observed by K. Suzuki et al. [25]. The intensity of this peak is even higher than the one of the transformation peak. It was also strain-amplitude-dependent in a range from 5×10^{-5} to 3×10^{-3}. From the peak shift method, an activation energy of 7.8×10^4 J/mol was calculated.

2.2.6
How Large is the Damping Capacity?

The previous paragraphs should have made clear that there is no unique damping value for one material. First of all, the external parameters temperature, time, frequency and most important the amplitude can change the damping capacity. The type of material, grain size, martensite interface density, defect structure are important internal variables.

In any case one may state that the martensite of Cu-based alloys and Ni–Ti show a damping capacity of at least an order of magnitude higher than classic materials such as steels. For high amplitudes ($\varepsilon \approx 10^{-4}$), the loss factor in martensite can be of the order of 6–8%. During impact loading 10% and more can be obtained. This loss factor decreases to about 2–4% for amplitudes in the order of 10^{-5}.

2.3 Specific Results on Ni–Ti Shape-Memory Alloys

Recently it was shown experimentally that the damping capacity in NiTi is directly related to the stress for martensite reorientation, also called the barrier stress [40]. The internal friction of the material at low strain amplitude (3×10^{-5}) was measured by means of DMA (Dynamical Mechanical Analyzer), the IF appeared to be a function of annealing temperature as shown in Figure 3a. It increases systematically with increasing annealing temperature below 550°C and decreases after annealing above 550°C. Annealing at 550°C gives the highest damping of the martensite phase. The barrier stresses (σ_{mr}) for martensite reorientation are also found to be a function of annealing temperature as shown in Figure 3b. With increasing annealing temperature, σ_{mr} passes through a minimum at about 550°C, suggesting that annealing above the recrystallisation temperature has a drastic effect on σ_{mr}.

Plotting the barrier stresses for martensite reorientation as a function of the martensite damping capacity in Figure 3c, shows that the stress for martensite reorientation and the corresponding values of tan delta (Q^{-1}) obey generally a linear relationship. A higher barrier stress corresponds to a lower damping of martensite and vice versa.

A recent renewed interest for the damping capacity of NiTi alloys has extended the investigated amplitude range up to 6×10^{-2}. The interest was to investigate the stability of the closed stress-strain hysteresis loop when the sample was cyclically stressed and compressed in order to investigate the reversibility of martensite reorientation, even when some plasticity occurs [41]. The results are summarised in Figure 4 [41].

A typical stress-strain curve of a NiTi bar under tension-compression cyclic deformation for large amplitudes is shown in Figure 5. Further experimental details concerning the cyclic tests can be found in ref. [33, 34]. The cyclic test started with tensioning to +1%/+2%/+4% strain proceeded with compression to −1%/−2%/−4% strain. The process was continued to 50 cycles for each strain level.

The area (ΔW) of the stress-strain loops and the corresponding value of internal friction ($\Delta W/2\pi W$) are both calculated from the tension-compression curves. The stresses at maximum strains, 4% and −4 % strains in the present cases, which increase with increasing number of cycles, tend to stabilize with further cycling. The stresses at 0% strain under both tension and compression,

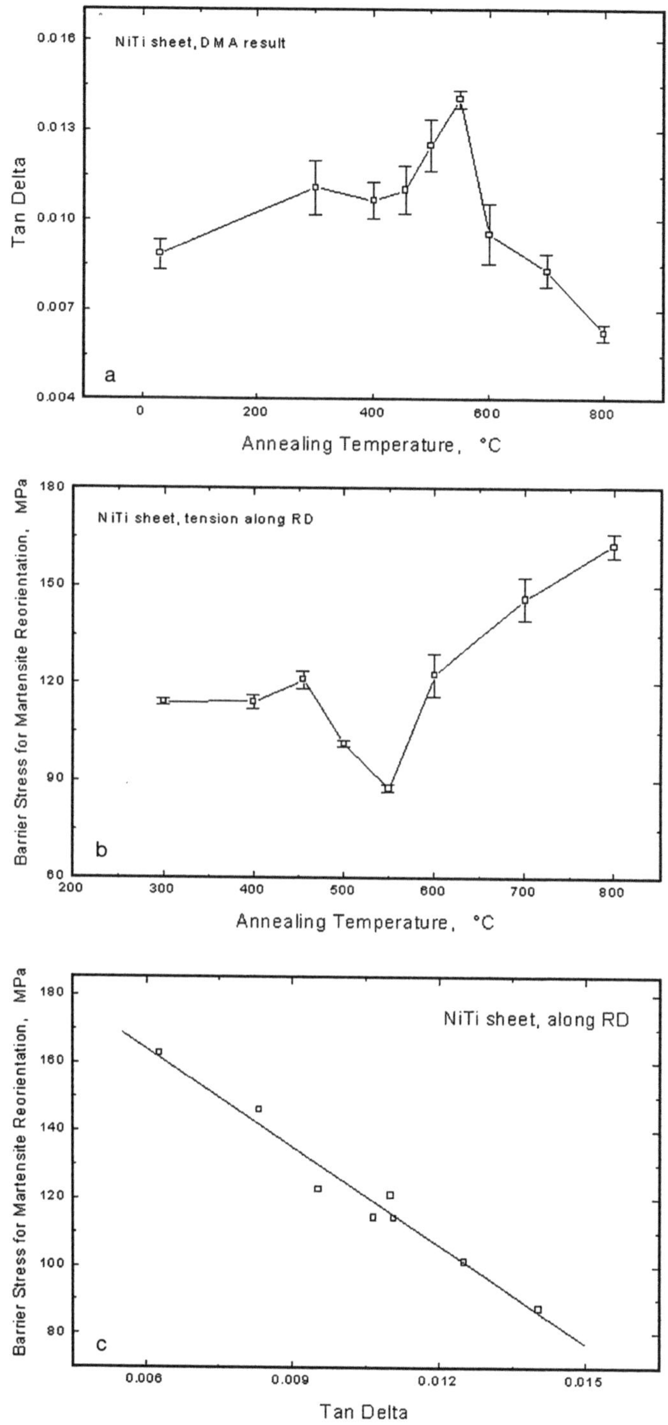

Fig. 3a–c. **a** Barrier stress for martensite reorientation as a function of tan delta for specimens annealed at different temperatures. Tensile testing temperature was 27°C. Tan delta values were taken from the DMA testing data at 27°C during heating at vibration strain amplitude of 3×10^{-5}. **b** Internal friction of NiTi SMA as a function of strain amplitudes for several annealing treatments. **c** The linear relation between the barrier stress and tan delta (Q^{-1})

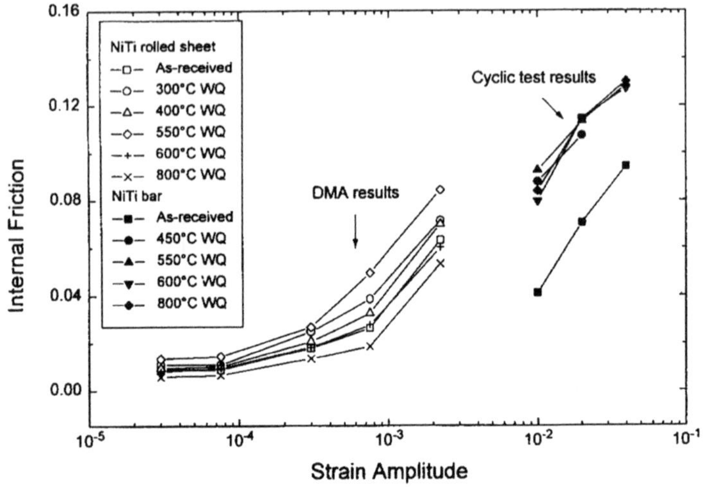

Fig. 4. Internal friction of (a) cold rolled NiTi sheet and (b) cold drawn NiTi bar as a function of annealing temperature

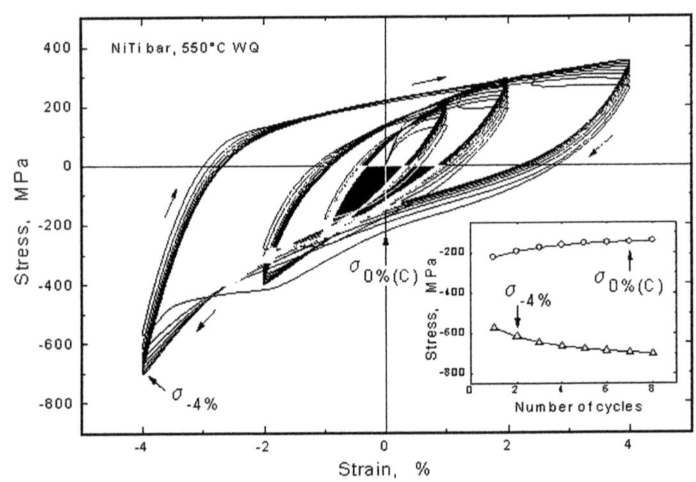

Fig. 5. Stress-strain curves of a NiTi bar during tension-compression cyclic deformation at a strain rate of 1.6 x 10^{-2} s^{-1}

which decrease with increasing number of cycles, also tend to stabilize with further cycling. In the first 10 cycles, the characteristic stresses of the compression partial loops (lower part of the stress-strain curves), $\sigma_{-4\%}$ and $\sigma_{0\%(C)}$, respectively increase and decrease more significant than that of the tension partial loops (upper part of the stress-strain curves), $\sigma_{4\%}$ and $\sigma_{0\%(T)}$, suggesting a more significant cyclic hardening/softening process occurring during compression.

The internal friction as a function of number of cycles for both partial loops and full loops is shown in Figure 6. Internal friction of the upper loop (under

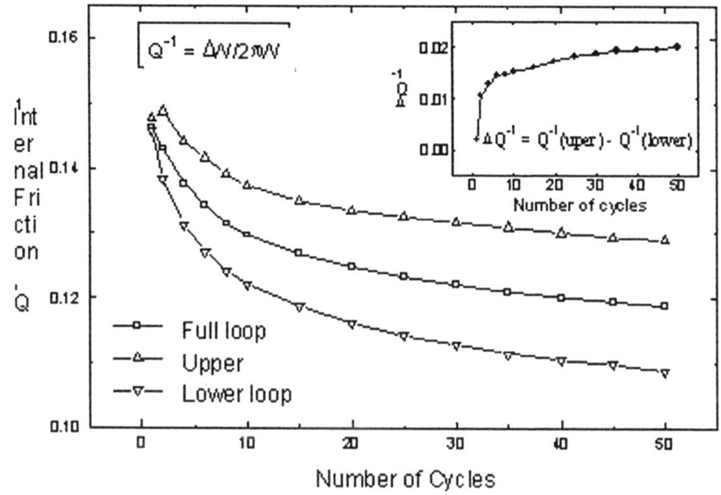

Fig. 6. The internal friction of martensitic NiTi SMA during tension-compression cyclic deformation. The IF of the upper loop (under tension) is much higher than that of the lower loop (under compression)

tension) is much higher than that of the lower loop (under compression), and the difference between these two internal friction values increases quickly at the beginning of the cycling and it slowly increases with further cycling. The martensite damping capacity decreases with increasing number of testing cycles. However, the decreasing tendency slows down when the test exceeds about ten cycles.

The martensite internal friction of the NiTi bar as a function of annealing temperature and amplitude is also plotted in Figure 4. Plotting the internal friction data of both the NiTi sheet and the NiTi bar in Figure 4 as a function of strain amplitude clearly shows that, although the testing methods and the materials processing histories were different, the overall tendency is somehow similar, and the internal friction magnitude from both types of tests falls into a comparable value. This can be understood by the fact that both the mechanisms of internal friction are related to the same type of microstructural defects, i.e., martensite twins. At lower strain amplitude ($\sim 10^{-4}$), about 6% of the mechanical energy can be consumed during one cycle of loading, while at higher strains ($\sim 4 \times 10^{-2}$), nearly 80% of the mechanical energy can be dissipated.

The martensite microstructure developed during tension-compression cycling has been systematically studied by the present authors [42]. Before deformation, the strain associated with martensitic transformation in the undeformed samples is minimized by the self-accommodation of martensite plates. Several types of twinning mechanisms co-exist as a result of martensitic transformation, while the <011> type-II twinning is the main lattice invariant shear.

After tension-compression cyclic deformation within ±4% strain, the martensite variants are still self-accommodating. A high density of dislocations have been generated inside the <011> type II twins. The $(11\bar{1})$ type-I twins have been more frequently observed in the cyclic tested specimens, suggesting that under the tension-compression cyclic deformation, a stress-induced re-configuration of martensite twins has taken place. In the cyclically tested specimens, most observed $(11\bar{1})$ type-I martensite twins were extending through the whole martensite plates. A high density of dislocations have been formed along the $(11\bar{1})$ twin plane and (110) shear plane. In addition to dislocations, shear bands along the (110) plane appear regularly in the twin bands. Between the shear bands, a high density of stacking faults on the basal plane (001) have been formed. The junction plane area between two martensite variants is highly strained and largely deformed [42]. These results show that the deformation mechanisms involved in the cyclic test are not a simple martensite reorientation mechanism. Further research shows that the martensite deformation mechanism is different for tension and for compression [43]. During tension to 6% strain, a reorientation of martensite twins occurs. Several neighboring <011> type-II twins have been reoriented to one orientation most favorable to the applied stress through migration of the variant interfaces. Under compression, however, no significant martensite reorientation has been observed. Stead, a high density of dislocations has been generated both inside the martensite twins and along the junction plane areas [43]. The generation of new lattice defects may explain the martensite strain hardening process occurring during tension-compression cycling, while the reconfiguration of martensite twins might be responsible for the cyclic softening effect.

2.4 Energy Loss during Pseudoelastic Loading

The pseudoelastic stress-strain curve as schematically shown in Figure 7, exhibits an important hysteresis during loading and unloading. The released energy

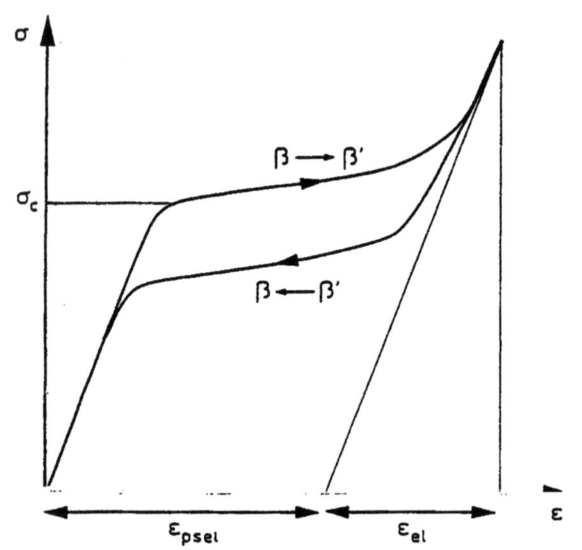

Fig. 7. Pseudoelastic deformation behaviour observed during stress-induced transformation in shape memory alloys (σ_c is the critical stress, ε_{psel} and ε_{el} are the pseudoelastic and elastic strains)

during unloading is therefore significantly lower than the applied energy to deform the material and the total energy loss is quite high.

This particular behaviour becomes very interesting when very large strain amplitude (>10^{-2}) vibrations have to be damped. Especially civil engineering applications are nowadays developed especially to protect buildings and bridges against earthquake damage. The shape restoration capacity in combination with the high damping capacity and the stiffness of the material offers interesting prospects for base isolation and energy absorbers within the construction [44–51].

However in applying the pseudoelastic loops for damping purposes one should take into account that both stress-plateau's shift to higher stress-values with increasing temperature, thereby reducing the relative damping capacity. Amplitude and frequency of the applied vibration will also significantly influences the damping capacity as pointed out in [52].

At the other side, the thermomechanical processing and composition can be selected in such a way that the hysteretic loop becomes optimized. Parameters such as degree of cold deformation, temperature and time of the post-annealing treatment can be selected in such a way that maximum internal friction occurs during the strain-induced transformation. But also parameters such as grain size, texture and amount of cycles can change the hysteretic behaviour [53]. Regarding the composition, Ni-rich compositions might be preferred in order to control better the microstructure leading to an optimized damping. But alloying third elements might also influence significantly the hysteresis. Cu will decrease the hysteresis and thus the damping while Nb will increase it [54, 55].

2.5 Some Remarks on the Fatigue Life of SMA Devices

Since the damping capacity of Ni–Ti SMA becomes especially interesting at rather high strain-amplitudes (>10^{-4}), attention should be given to the fatigue properties especially when high amount of loading cycles are expected. Fatigue should also be interpreted in a large sense. Thermal and/or mechanical cycling will both influence the lattice defect structure leading to changes in hysteresis, transformation temperatures, functional properties [54, 55].

When more focussed on mechanical stability of NiTi, Dauskardt et al. [56] came to important conclusions:

1. Fatigue-crack growth rates in TiNi are significantly faster and fatigue treshold values (ΔK_{TH}) significantly lower, compared to other metallic engineering alloys of similar strength.
2. Contrary to first-order expectations, fatigue-crack growth rates are slowest in the stable (non-transforming) microstructures, particularly the stable austenite and fastest in the unstable (transforming) microstuctures, particularly involving a reversible transformation to martensite.

From those results, it should thus be concluded that crack-initiation should be postponed as much as possible. This can be reached by proper surface conditioning or probably also surface treatment, but few attention has been given to those aspects so far. Apart from the surface, the presence of stress-induced R-phase might also promote high brittleness as observed by Brachet et al. [57].

3 Conclusions

In the past, there have been many attempts to apply martensitic CuZnAl or NiTi in several devices such as tennis rackets, saw blades, bullet proof materials, but so far none of them have become a commercial success. One of the main reasons is that SMA is still too expensive to be used as a structural material and also the implementation of SMA into the design has its price. Moreover SMA are difficult to machine and even almost impossible to weld. Also, it appeared that the total noise reduction obtained was rather limited due to the restricted amount of mechanical vibration energy that can be transmitted to the SMA-part. This was for example the case in tennis rackets and saw blades. Nevertheless, recently a new interest appeared for the damping capacity of SMA. Stöckli-Ski, a Swiss ski producer, is testing composite skis in which laminated Cu–Zn–Al strips are embedded. Those strips have the martensitic transformation temperatures slightly above 0°C once in contact with snow, the skis will cool down while the Cu–Zn–Al elements will transform into martensite. This way, vibrations will be damped significantly, giving the skis a much better performance [58].

There is also a large military interest for shock wave absorption in armor material but also in other bullet proof materials, since the high strain amplitude in combination with a single impact allows the optimal conditions of SMA for high damping properties. Exploration of martensite reorientation and superelasticity is going on for applications in this field [45]. Similarly, also civil engineering applications of SMA are recently attracting more attention, especially for protection of civil constructions, such as buildings and bridges, against earthquake vibration damage [44, 46–51].

The damping capacity of orthopaedic devices might also become an important selection criterion. By the earlier and recent interests, it became clear that, when the damping capacity is the envisaged functional property of the device, the design should take this into account, taking care that maximum mechanical energy is transmitted to the SMA part. Expected amplitude-range, frequency of vibration, impacts, temperature range should be taken into account in order to process the SMA towards optimal damping and fatigue properties.

References

1. De Batist R (1983) High damping materials: mechanisms and applications. J Phys France 44(Suppl. 12):39
2. Van Humbeeck J (1985) The high damping capacity of martensitic copper–zinc–aluminium alloys. In: Rath BB, Misra MS (eds) Proceedings of the International Symposium on the Role of Interfaces in Material Damping. American Society for Metals, Materials Park, p 59
3. Van Humbeeck J (1984) Internal friction in shape memory alloys showing a thermo-elastic martensitic transformation. In: Garczyca S, Magalas LB (eds) Proceedings of the Summer School on Internal Friction in Solids, Wydawnictwo AGH, Krakow, p 131
4. Van Humbeeck J (1989) Internal friction in shape memory alloys. In: Ke TS (ed) Proceedings of ICIFUAS-9, Beijing, June 17–20, 1989. Pergamon, New York, p 337

5. De Batist R (1992) Mechanical energy dissipation related with martensitic transformation processes. In: Kinra VK, Wolfenden A (eds) Proceedings on the Mechanics and Mechanisms of Material Damping. American Society for Testing and Materials, Philadelphia, p 45
6. Clarebrough LM (1957) Internal friction of β-brass. Acta Metallica 5:413
7. Ghilarducci A, Ahlers M (1980) Internal friction in quenched â-phase Cu-Zn and Cu-Zn-Al alloys. Scripta Metallurgica 14:1341
8. Ghilarducci A, Ahlers M (1983) Internal friction and point defects in ordered β Cu–Zn and β Cu–Zn–Al. J Phys France Metal Phys. 13:1757
9. Van Humbeeck J, Delaey L (1983) The internal friction behaviour of martensitic Cu–Zn–Al alloys. J Phys France 44(Suppl. 12):217
10. Bidaux JE, Schaller R, Benoit W (1989) Study of the hpc-fcc phase transition in cobalt by acoustic measurements. Acta Metallica 37:803
11. Van Humbeeck J, Stoiber J, Delaey L, Gotthardt R (1995) The high damping capacity of shape memory alloys. Z Metallkunde 86:177
12. Pérez-Sáez RB, Recarte V, Nó ML, San Juan J (1988) Anelastic contributions and transformed volume fraction during thermoelastic martensitic transformations. Phys Rev B 57:5684
13. Mercier O, Melton KN, De Préville Y (1979) Low-frequency internal friction peaks associated with the martensitic phase transformation of NiTi. Acta Metallica 27:1467–1475
14. Van Humbeeck J (1996) Damping properties of shape memory alloys during phase transformation. Proceedings of ICIFUAS-96. J Phys IV France 6:371
15. Kustov S, Golyandin S, Sapozhnikov K, Van Humbeeck J, De Batist R (1998) Low-temperature anomalies in Young's modulus and internal friction of Cu–Al–Ni single crystals. Acta Mater 46:5117
16. Vandeurzen U (1982) Identification of damping in materials and structures-optimization of dinamyc behaviour- of mechanical structures. Ph.D. thesis. Leuven University, Leuven
17. Van Humbeeck J (1983) Studie en optimalisatie van de dempingseigenschappen van martensietische Koper-zink-aluminium legeringen. Ph.D. thesis. Leuven University, Leuven
18. Morin M, Guénin G (1983) Étude du frottement intérieur d`un alliage à transformation martensitique thermoélastique dans le Cu–Zn–Al. J Phys France 44:247
19. Koshimizu S, Mondino M, Benoit W (1979) Internal friction measurements during martensitic transformation in Cu–Zn–Al alloys at kHz frequencies. Proceedings of the ECIFUAS-3. Manchester, p 269
20. Granato A, Lücke K (1956) Theory of mechanical damping due to dislocations. II. Application of dislocation theory to internal friction phenomena at high frequencies. J Appl Phys 27:583
21. Zhu J-S, Wang Y-N, Shen H-M (1983) Ultrasonic study on martensitic transformation in Au–Cd alloy. J Phys France 44:235
22. Mercier O, Török E, Tirbonod B (1979) Internal friction peaks associated with the martensitic phase transformation of NiTi and NiTiCu alloys. Proceedings of Icomat. Massachusetts Institute of Technology, Cambridge, p 702
23. Tirbonod B, Koshimizu S (1981) Dislocation relaxation in the martensitic phase of the thermoelastic NiTi and NiTiCu alloys. J Phys France 42:1043
24. Sugimoto K, Mori T, Otsuka K, Shimizu K (1974) Simultaneous measurements of internal friction, Young's modulus and shape change associated with thermoelastic martensite transformation in Cu–Al–Ni single crystals. Scripta Metallurgica 8:1341
25. Suzuki K, Nakanishi N, Mitani H (1980) Effects of cooling rates on internal friction in Cu–Al–Ni ternary alloys. Japanese Institute of Metals, Vol. 44, p. 43
26. Dejonghe W, Delaey L, De Batist R, Van Humbeeck J (1977) Temperature and amplitude- dependence of internal friction in Cu–Zn–Al. Metal Sci 11:523
27. Burdett CF, Queen TJ (1979) The role of dislocations in damping. Metals Rev 4:44
28. Peguin P, Perez J, Gobin PF (1967) Amplitude-dependent part of the internal friction of aluminium. Metals Trans Am Inst Mining Metallurgical Petroleum Eng 239:438
29. Kajiwara S, Kikuchi T (1982) Dislocation structures produced by reverse martensitic transformation in a Cu–Zn alloy. Acta Metallica 30:589
30. Morin M, Guénin G, Gobin PF (1981) Internal friction measurements related to the two way memory effect in Cu–Zn–Al alloy exhibiting thermoelastic martensitic transformation. J Phys France 42:1013
31. Van Humbeeck J, Delaey L (1984) The influence of heat-treatment on the internal friction of Cu–Zn–Al martensite. Part II. The peaking effect. Z. Metallkunde 75:760
32. Morin M, Vincent A, Guénin G (1985) Internal friction time dependence of Cu–Zn–Al martensite. Proceedings of the International Conference on Internal Friction and Ultrasonic Attenuation in Solids (ICIFUAS-8). J Phys France 46:625
33. Van Humbeeck J, Hulsbosch J, Delaey L, De Batist R (1985) The influence ageing in the martensite phase on the internal friction in Cu–Zn–Al alloys. Proceedings International Conference on Internal Friction and Ultrasonic Attenuation in Solids (ICIFUAS-8). J Phys France 46:633

34. Ilczuk J, Delaey L, Van Humbeeck J (1987) The influence of martensite stabilisation on changes in dislocation density in Cu–Zn–Al alloys. Proceedings European Conference on Internal Friction and Ultrasonic Attenuation in Solids (ECIFUAS-5). J Phys France 48:553
35. Morin M, Haouriki M, Guénin G (1987) Study of the Cu-Zn-Al martensite ageing by internal friction measurements. Proceedings European Conference on Internal Friction and Ultrasonic Attenuation in Solids (ECIFUAS-5). J Phys France 48:567
36. Van Humbeeck J, Delaey L (1982) The evolution of the damping characteristics of Cu–Zn–Al martensitic alloys with time and temperature: the peaking effect. J Phys France 43:691
37. Simpson H M, Sosin A, Johnson D F (1972) Contribution of the defect dragging to dislocation damping. Phys Rev B 5:1382
38. Hasiguti R R, Iwasaki K (1968) Internal friction and related properties of the NiTi intermetallic compound. J Appl Phys 59:2182
39. Postnikov VS, Lebedinskiy VS, Yevsyokov VA, Sharshakov IM, Pesin MS (1970) Phase transformations in the intermetallic compound TiNi. Fiz Metallov 29:364
40. Liu Y, Van Humbeeck J (1997) On the damping behavior of NiTi shape memory alloy, presented in ESOMAT-97, July 1–5 1997, the Netherlands. J Physique IV France, p 519
41. Liu Y, Van Humbeeck J, Stalmans R, Delaey L (1997) Some aspects of the properties of NiTi shape memory alloys. J Alloys Compounds 247:115
42. Xie ZL, Liu Y, Van Humbeeck J (1988) Microstructure of NiTi shape memory alloy due to tension-compression cyclic deformation. Acta Materialia 46:1989–2000
43. Liu Y, Xie ZL, Van Humbeeck J, Delaey L (1998) Asymmetry of stress-strain curves under tension and compression for NiTi shape memory alloys. Acta Materialia 46:4325
44. BRITE-MANSIDE Project. Memory alloys for new seismic isolation and energy dissipation devices. Contract No. BRPR-CT95-0031. (1995–1999)
45. Aiken ID, Nims KD, Whittaker AS, James MK, M. EERI (1993) Testing of passive energy dissipation systems. Earthquake Spectra 9:335
46. Graesser EJ, Cozzarelli FA (1991) Shape memory alloys as new materials for aseismic isolation. J Eng Mech 117:2590–2608
47. Wittig PR, Cozzarelli FA (1993) Design and seismic testing of shape memory structural dampers. In: Proceedings of Damping 1993, San Francisco
48. Graesser EJ, Cozzarelli FA (1994) Effects of intrinsic damping on vibration transmissibility of Ni-Ti shape memory alloy springs. Metal Mater Trans A 26:2791
49. Hodgson DE, Krumme RC (1994) Damping in structural applications. In: Pelton AR, et al. (eds) Proceedings of the 1st International Conference on Shape Memory and Superelastic Technologies, California 7–10 March 1994. Shape Memory and Superelastic Technologies, Pacific Grove, pp 371–376
50. Wittig PR, Cozzarelli FA (1992) Shape memory structural dampers: materials properties, design and seismic testing. Technical report NCEER-92-0013. State University of New York, Buffalo
51. Whittaker SA, Krumme R, Hayses Jr R (1995) Structural control of building response using shape memory alloys. Technical report TR 95/22. US Army Construction Engineering Research Laboratories, Washington
52. Van Humbeeck J, Delaey L (1981) The influence of strain rate, amplitude and temperature on the hysteresis loop described during the pseudoelastic deformation of a Cu–Zn–Al crystal. J Phys France 42 (Suppl 10):1007–1011
53. Saburi T (1998) Ti-Ni shape memory alloys. In: Otsuka K, Wayman CM (eds) Shape memory materials. Cambridge University, Cambridge
54. Saburi T (1998) Structure and mechanical behaviour of Ti-Ni shape memory alloys. In: Shape memory materials (Proceedings Materials Research Society International Meeting on Advanced Materials, Tokyo, vol 9) Materials Research Society, Pittsburgh, pp 77–91
55. Van Humbeeck J (1991) Cycling effects, fatigue and degradation of shape memory alloys. J Phys IV France 1:189–197
56. Dauskardt RH, Duerig TW, Ritchie RO (1989) Effects of in situ phase transformation on fatigue-crack propagation in Ti–Ni shape memory alloys. In: Shape memory materials. (Proceedings Materials Research Society International Meeting on Advanced Materials, Tokyo, vol 9) Materials Research Society, Pittsburgh, pp 243–249
57. Brachet J-C, Olier P, Brun G, Wident P, Tournie I, Faucher C, Dubuisson P (1997) Superelasticity and impact properties of two way shape memory alloys: $Ti_{50}Ni_{50}$ and $Ti_{50}Ni_{48}Fe_2$. Journal de Physique, IV, Colloque C5, Suppl. J. de Phys. III, pp C5-561-566
58. Scherrer P, Bidaux J-E, Kim A, Manson JAE, Gotthardt R Passive vibration damping in an alpine by integration of shape memory alloys. Accepted for publication in J Phys IV, France 9, 1999, Pr 9-393–400

Physical and Biochemical Principles of the Application of TiNi-Based Alloys as Shape-Memory Implants

L. L. Meisner, V. P. Sivokha

1 Introduction

For medical treatment of various diseases and traumas, there is a wide distribution of the metallic, ceramic and polymer applications for the realization of different functions in living organisms. Such constructions are being called implants. Materials for medical implants (biomaterials) need to satisfy three important demands, i.e. (1) – the reliability of the mechanical functions, (2) chemical reliabilities – the resistance to deterioration of their properties in a biological medium, the resistance to expansion, dissolution, corrosion, and (3) biological reliabilities – biological compatibility, lack of toxicity and carcinogenicity, resistance to the formation of thrombus and antigens [1]. Biomaterials should be non-toxic during the implanted period in the body and, simultaneously, have rather high physical–mechanical characteristics. Because of these rigorous demands, only the following three metallic materials have been qualified to be available as implant materials, i.e. Fe–Cr–Ni, Co–Cr and Ti–Al–V [2]. However, shape memory alloys have been recently introduced to medicine, since they have unique functions such as shape memory effect, superelasticity and damping capacity.

Among the multitude of shape memory alloys, TiNi-based alloys are considered to be the best because of their excellence in mechanical stability, corrosion resistance, biofunctionality, and biocompatibility. Owing to the unique functions such as SME and superelasticity possessed by TiNi-based alloys, it is possible to realize governed complicated deformations in condition of their contacts with the living organism. Since the temperature interval of implant deformation should be narrow and be situated near 310°K (temperature of the living organism) in order to avoid thermal defeat of the tissue, it superimposes additionally very high requirements to the choice of the material for the implant among TiNi-based alloys.

TiNi shape memory alloys are investigated widely and in detail by various Russian materials science schools, in particular by scientists of the Tomsk Research center (Tomsk, Russia) since the 1970s [3–8]. On the basis of these results the Russian materials scientists have developed the basic physical-mechanical principles and biochemical standards for the choice of biomaterials for implants among the TiNi-based alloys. The purpose of the present work is to introduce some data on parameters of martensitic transformations and shape memory effect in TiNi-based alloys investigated by Russian materials scientists (Tomsk Research Center, in particular), which were used for creating these principles and standards.

2 Shape Memory Effect and Pseudoelasticity in TiNi-Based Alloys

The steadfast interest and popularity of TiNi-based alloys are due to unique pseudoelastic properties which is known to approximate those of biological tissues. The shape memory effect (SME) and the superelasticity properties of these alloys are of a great concern, and are described in the numerous reviews in detail [4–6, 9, 10]. We shall emphasize the most important aspects for particular use in medicine.

2.1 Role of the Chemical Composition

In alloys of the Ti-Ni system, unusual pseudoelastic properties (SME, superelasticity and others) occur in association with the various sequences of martensitic transformations (MT), they undergo when lowering the temperature, the parent phase (β) with a B2 structure (ordered on a type CsCl) can transform thermoelastically to a phase with a monoclinic B19' structure, or to a phase with the orthorhombic B19 structure, but more often to the trigonal phase (so called R-phase) and then to the B19' phase. Thus, the sequence of MT and the transformation temperatures depend on the chemical compound of the alloy [6, 10]. In turn, SME characteristics such as the temperature interval of SME (ΔO), the magnitude of the pseudoelastic reversible strain (ε), the recovery stress (σ_ε) developed when heating in the SME temperatures interval, and the critical stress for martensitic shear (σ_M) depend on the concrete scheme of the MT [4]. Differently, the quantitative and qualitative changes of TiNI-based alloys chemical compounds of the alloyage in particular result, in various sequences of MT and parameters of SME.

Examples of the alloyage influence on temperatures and sequences of phase transformations in Ti alloys with parent B2 structure are shown in Figure 1 for following cases: (a) the substitution of Ti by Al and the substitution of Ni by Cu [11]; (b) the substitution of Ni by Fe [12], (c) by Co and (d) -by Pt [13]; (d) the substitution of Ni by Pd and Fe, (e) by Pd and Co [13]; (f) the substitution of Ni by Pd [14] and (g) by Au [15]. In these diagrams are shown the start temperatures of B2→B19' (B19) transformations (I_s) and the start temperatures of R→B19' (B19) transformations (R_s). Since the B2→R transformation is similar to the second type phase transformation, then the two-phase range (B2+R) is practically absent. In other cases, for B2 (R)→B19 (B19') or B19→B19' phase transformations, the two phase (B2+B19 (B19')) range will change from 10°C up to 50°C in association with the chemical composition of the alloy. Schemes of the reverse martensitic transformation are similar, and temperature intervals of the two-phase ranges move correspondingly to hysteresis temperature [11–16].

Three-component alloys Ti–Ni–Cu have a practical interest. In these alloys, temperatures of MT do not change, being essentially near to room temperature (T_{room}; Fig. 2a) and the SME exhibit appreciably due to the complicated scheme of MT, even when the content of Cu is increased up to 30% [16].

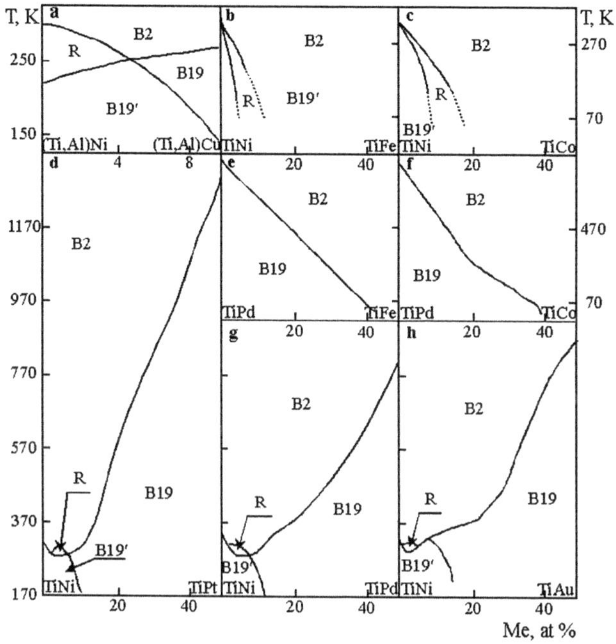

Fig. 1. Start temperatures of the B2→(R, B19′, B19), B19→B19′, R→ B19′ transformations.
a $Ti_{48.2}Al_{1.8}Ni_{(50-x)}Cu_x$ [11].
b $Ti_{50}Ni_{(50-x)}Fe_x$ [12].
c $Ti_{50}Ni_{(50-x)}Co_x$.
d $Ti_{50}Ni_{(50-x)}Pt_x$.
e $Ti_{50}Pd_{(50-x)}Co_x$ [13].
f $Ti_{50}Ni_{(50-x)}Pd_x$ [14].
g $Ti_{50}Ni_{(50-x)}Au_x$ [15]

In the three-component Ti–Ni–Zr alloys, the substitution of Ti by Zr does not lead to the complicated scheme of MT (B2↔B19'), but when the Zr-content is increased above 5–7 atomic %, the MT temperatures rise (Fig. 2b) [17, 18]. On the contrary, when the Ti-content is exactly 50 atomic %, the substitution of Ni by Zr does not change the B2↔B19' transformation scheme according to our results

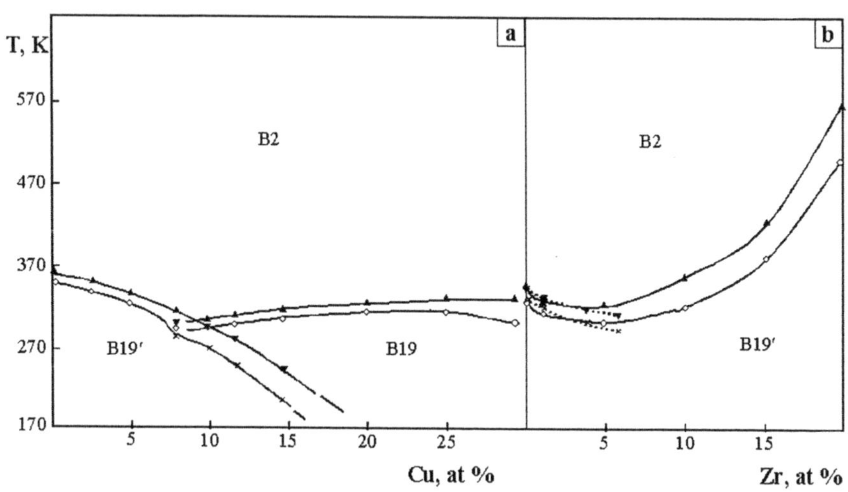

Fig. 2. Start (M_s) and finish (M_f) temperatures of the B2→B19(B19′), B19→B19′ transformations. a $Ti_{50}Ni_{(50-x)}Cu_x$ [16]. b *Solid line*: $Ni_{50}Ti_{(50-x)}Zr_x$ [17]; *dotted line*: $Ti_{50}Ni_{(50-x)}Zr_x$

(*dotted line*, Fig. 2b) [18a]. But the completeness of MT is reduced quickly when the Zr- content is increased in these alloys: in a $Ti_{50}Ni_{46}Zr_4$ alloy about 70 volume % of the parent B2 phase transforms into B19' martensite phase whereas in an alloy $Ti_{50}Ni_{44}Zr_6$, only about 10 volume % of B2 phase do it. At those volumetric parts of Ti_2Ni and $Ni_7(Ti, Zr)_2$, secondary phases are increased (for example, up to 60 volume % for the $Ti_{50}Ni_{44}Zr_6$ alloy).

The analysis of known triple and tetrad phase diagrams near the TiNi compositions shows that the maintenance of the preservation stoichiometry principle, when the Ti-analogue element is added instead of Ti, and the Ni-analogue element – instead of Ni, which is an important factor of the choice of the alloyage of TiNi-alloys. The non- maintenance of this principle leads to the restriction of the alloying element solubility into the B2 phase, to the formation of secondary phases and to the consequent degeneration of MT and SME, as in case of $Ti_{50}Ni_{(50-x)}Zr_x$ alloys (Fig. 2b).

The experience shows that the alloys can effectively operate the martensitic transformation temperatures and their schemes in TiNi-based alloys if Ti is substitutes by such elements, as Zr, Hf, Al, V and Ni is substituted by Fe, Co, Pd, Pt, Au, Cu, Cr, Mn [11, 18]. This means that there is a possibility to have control over the physical and mechanical properties of TiNi based alloys in the MT temperature intervals.

2.2 Role of the Phase Composition and the Thermomechanical Treatment

The Ti-Ni binary equilibrium phase diagram is characterized by a narrow region of the B2 phase at temperatures below 923 K [3]. In the vicinity of TiNi, there are certain temperatures ranges with their own decomposition schemes. For example, Ti-Ni alloys with nickel contents exceeding 50.5 atomic % decompose slowly on cooling from a high temperature or on aging at temperature below 970 K after quenching from a high temperature. The final product of decomposition is a mixture of $TiNi_3$ and TiNi. In any case, when the titanium contents exceed 50 atomic %, there are temperature regions with a mixture of Ti_2Ni (or Ti_4Ni_2O) and TiNi [19].

Thermoelastic martensitic transformations causing the phenomena of SME and superelasticity are only in TiNi-compounds. However, the above-mentioned decompositions lead to the stripping of the parent B2 phase with one of the components and then, to the modification of the character and schemes of structural transformations. For example, Figure 3 shows the partial phase diagrams of near-equiatomic TiNi compositions, which have been obtained by Lotkov [20] after quenching from 973 K of TiNi alloys (a) and by Khachin [21] after cooling slowly from same temperature for that alloy (b). As seen in Figure 3, in TiNi alloys with Ni content exceeding 50.5 atomic %, the MT temperatures are decreased which leads to the modification of the MT sequence.

The influence of the thermomechanical treatment on the MT character and sequence and on the SME parameters have been described in papers [4, 19] more deeply. It was established, that the heat aging of near-equilibrium TiNi alloys results in the intensive precipitation of particles of Ti_2Ni or/and $TiNi_3$ phases in

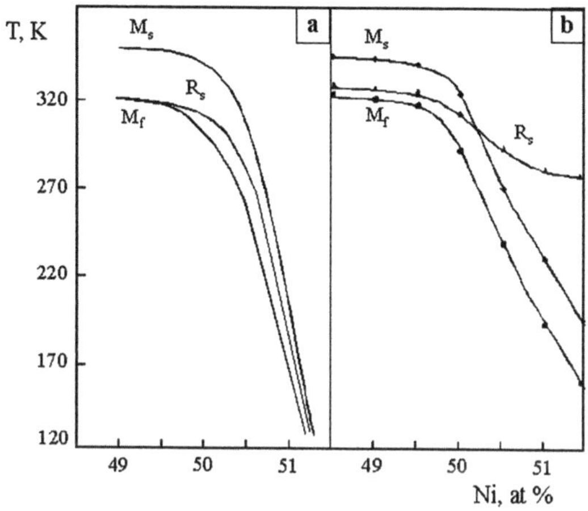

Fig. 3a, b. The partial phase diagrams of near-equiatomic TiNi compositions: (**a**) after quenching [20] and (**b**) after cooling [21] from 973 K

B2-matrix, leading to changes in the MT temperatures for instance. The especially the noticeable aging influence is observed in the TiNi alloys with Ni-content exceeding 50 atomic %. Figure 4 [20] is an example of the modification of MT temperatures B2→R (R_s) and B2→B19 (B19') (M_s) depending on aging parameters (temperatures and duration) in the Ti + 51 atomic % Ni alloy. As seen, the most sensitive transformation to heat aging is the R_s temperature of the B2→R transformation (*curve 1*, Fig. 4), and the less sensitive is the M_s temperature of B2→B19 (B19') (*curve 4*, Fig. 4). The complicated form of the M_s curve depending on the aging duration at T=523 K is due to formation of the Ti_3Ni_4 particles in the beginning of the aging process and to the exhaustion of B2 phase by Ni after that. That process is finished when the equiatomic TiNi content appear [19]. The presence of a secondary phase (Ti_2Ni, $TiNi_3$ or Ti_3Ni_4) particles in the B2 phase leads to change in the yield stress, strength and plastic properties, including SME and superelasticity.

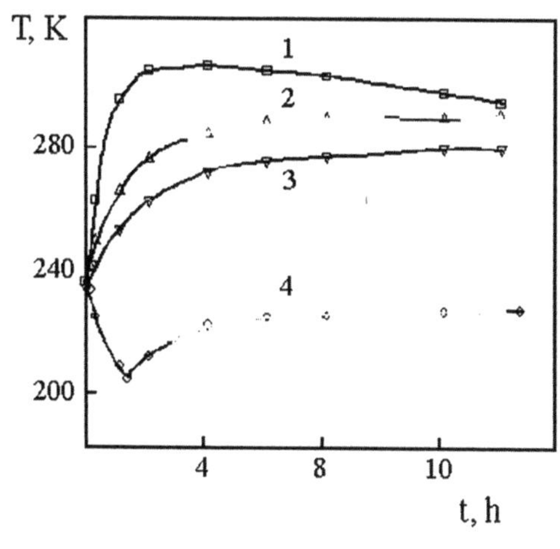

Fig. 4. Start temperatures of the B2→R (R_s) and R→B19′ (M_s) transformations depending on aging temperatures (T_{ag}) and aging time (t): T_{ag}=773 K – curve 1 (R_s), curve 3 (M_s); T_{ag}=523 K – curve 2 (R_s), curve 4 (M_s) [19]

Therefore, in TiNi-based alloys, the variation of SME parameters can be attained not only by various alloyage with different Ti and Ni contents, but also by thermomechanical treatment, by thermal cycling through the MT temperature interval (the so called phase hardening) and by strain hardening in the rate of the active plastic strain. Investigations results of the influence of thermomechanical treatment on the physical and mechanical properties TiNi-based alloys allow to formulate a series of important conclusions concerning the special pseudoelastic properties of that alloy. First of all, when heat aging TiNi-based alloys with the intensive participation particles Ti_2Ni and $TiNi_3$ in the B2 phase, the strain recovery ratio and maximum recovery stress hardly increase in alloys with Ti or Ni contents exceeding 50 atomic %. These parameters vary weakly for the alloy with an equiatomic B2 phase. In alloys with Ni contents exceeding 50 atomic %, there are wide intervals of heat aging temperatures where the maximum recovery stress is very high. In alloys with Ti contents exceeding 50 atomic % maximum recovery stress is very high only after heat aging in the temperature interval from 620 K to 820 K. Such distinction in the shape memory behavior of the above-mentioned alloys is associated with different precipitating kinetics of Ti_2Ni and $TiNi_3$ particles. In the near-equiatomic B2 phase of Ti-Ni alloys, the thermal treatment modification does not lead to the raise of maximum recovery stress. However, it is possible to raise the yield stress, if that alloy is subjected to phase hardening and then to strain hardening (up to 15%) in the rate of the active plastic strain. Such thermomechanical treatment results in superelasticity properties of these alloys.

2.3
Pseudoelastic Behavior of TiNi-Based Alloys

The possibility to realize different MT schemes, to vary the MT temperature intervals, to control mechanical properties, including SME and superelasticity, make TiNi-based alloys attractive for their application as biomaterials for medical implants. It is possible to solve a wide spectrum of medical problems using TiNi-shape memory implants. Just in these materials, researchers succeeded effectively in using all the unique functional properties: SME, all-round shape memory (two-way shape memory) and superelasticity. Some investigation data on these unique functional properties of Ti-Ni alloys particularly appropriate for SM implants, are submitted below.

2.3.1
Shape-Memory Effect

SME is exhibited more effectively, if there is a complicated series of MT in the alloy. For TiNi-based alloys, such series of MT are B2↔R↔B19’ or B2↔B19↔B19’. In these cases, the main SME parameters are optimum: the recoverable strain associated with the martensitic transformation (ε_M) is maximum, and the martensite shear stress (σ_M) is minimum [5]. Thus the recoverable strain ε_M may reach approximately 14% and the martensite shear stress σ_M may decrease 15–40 MPa in Ti alloys (Ni, Me; that is Me–Pd, Pt, Au, Cu at concentration

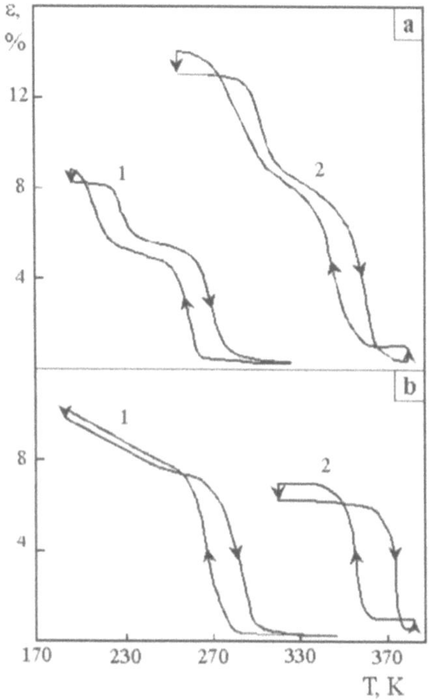

Fig. 5. Strain – temperature curves under constant load: σ=50 MPa: $Ti_{50}Ni_{38}Cu_{10}Fe_2$ (curve 1a), $Ti_{50}Ni_{39}Cu_{10}Fe_1$ (curve 1b) [5]; σ=80 MPa: $Ti_{50}Ni_{39}Au_{11}$ (curve 2a) [15]; σ=300 MPa: $Ti_{50}Ni_{30}Pd_{20}$ (curve 2b) [5]

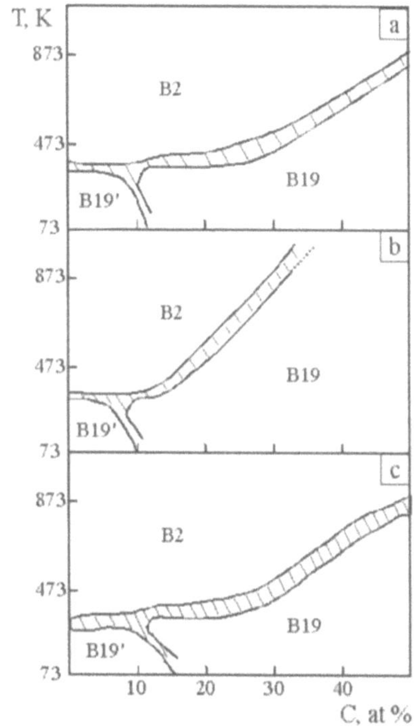

Fig. 6. Temperature ranges of the shape memory phenomena for various Ti-Ni alloys. **a** $Ti_{50}Ni_{(50-x)}Pd_x$ [11]. **b** $Ti_{50}Ni_{(50-x)}Pt_x$ [14]. **c** $Ti_{50}Ni_{(50-x)}Au_x$ [15]

Me from 7% to 12%), if the B2 phase transforms into B19' martensite by the scheme B2↔B19↔B19' [5, 13]. Figure 5 shows examples of the strain-temperature curves under constant load for different schemes of MT. It is seen in the figure that two steps appear on curves 1, 2a and 1b. The first step at R_s is due to the appearance of the R phase and the second at M_s is due to the appearance of the B19' martensite under the applied stress (the same for different alloys) on cooling. The reverse steps due to B19'→R and then R→B2 transformations. The only one step transformation appears at about 370 K on the curve 2b. That step at M_s is due to the appearance of the B19' martensite under the applied stress (50 MPa) on cooling. The reverse step is due to the reverse transformation from the B19' martensite directly to the B2 parent [5, 13]. The SME temperature intervals are represented as the shaded regions in Figure 6 for Ti (Ni, Me) alloys (Me–Pd, Pt, Au) [15].

The macroscopic shift modulus is known to have a small value in M_s(B19)–M_s(B19') temperature intervals for these alloys. That means that the B19-phase is unstable with respect to the yield stress during MT B2↔B19↔B19'. Unusual temperature dependencies of the recovery stress σ_r appear in that scheme of MT (Fig. 7, curves 1, 2). The generation of recovery stresses starts from a macroscopic deformation in the martensitic state B19' during subsequent heat-

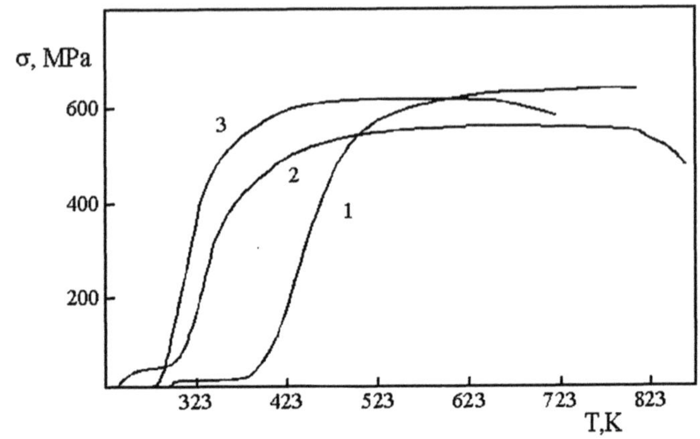

Fig. 7. The recovery stress – temperature curves for alloys: curve 1 – $Ti_{50}Ni_{37}Au_{13}$ [15], curve 2 – $Ti_{50}Ni_{40}Pd_{10}$ and curve 3 – $Ti_{50}Ni_{46}Pd_4$ [14]

ing. This shape memory property is the basis for some very successful shape memory applications. Nevertheless, the number of publications on this shape memory effect is rather limited [15, 22]. The one step recovery stress – temperature curve appear if the martensitic transformation schemes are different from the above-mentioned ones [15]. When these alloys are in the martensitic state, the magnification of the deformation degree leads to the displacement of SME temperature intervals towards the higher temperatures. Let's remark that the SME temperature interval associated with the B2→R transformation does not displace at the same thermomechanical treatment. It is necessary to take into account at the choice of the material for SM-implants.

2.3.2
All-Round Shape-Memory Effects

Specially treated TiNi-based alloys exhibit a two-way shape memory phenomenon or "all-round shape memory". That shape-memory property and its treatment conditions have been described in detail in many reviews [4, 6, 22, 23]. Figure 8 shows examples the reversible strain – temperature curves in alloys in which appear the all-round shape memory phenomena. First, the specimens were plastic deformed at temperatures above the B19→B2 (R→B2) transformation-finish temperature (the same in each case). Then, they were thermally treated by cycling throughout the following MT temperature intervals: B2↔R↔B19' (curve 1: $Ti_{49}Ni_{51}$), B2↔B19↔B19' (curve 2: $Ti_{50}Ni_{39}Au_{11}$) and B2↔R (curve 3: $Ti_{50}Ni_{45}Fe_5$). As seen in Figure 8, the two-steps reversible strain–temperature curves (1, 2) correspond to the complicated series of MT: B2↔R↔B19' or B2↔B19↔B19' with the resultant reversible strain reaching a significant value: 11–12%. The narrow hysteresis (3–5 K) and the small value of the reversible strain (~1%) on curve 3 is due to the closeness of the B2↔R transformation to the second-order transition [13].

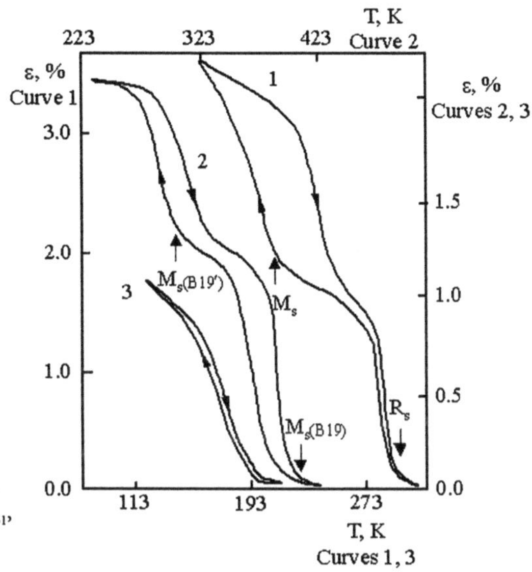

Fig. 8. The reversible strain – temperature curves for alloys: curve 1 – $Ti_{49}Ni_{51}$, curve 2 – $Ti_{50}Ni_{39}Au_{11}$ and curve 3 – $Ti_{50}Ni_{45}Fe_5$ [13]

There are several ways to cause the all-round shape memory effect in Ti–Ni based alloys [4]. For example:

1. When the B2 phase is transformed into the stress-induced martensite phase under the yield stress and then the alloy is cooled at temperature below the B2→B19 (B19') transformation-start temperature (M_s).
2. If we stimulate the reorientation of the martensitic planes along the direction of the applied yield force by plotting the yield stress more then the martensite shear stress.
3. If we plot the stress above the critical stress for slip, then one obtains the stress-induced martensitic transformation in the alloy.
4. If the parent B2 phase is deformed plastically and then the alloy is cooled in the free-stress or non-free stress condition.
5. By the multiple thermal cycling through the MT temperature intervals.
6. By the combination of these specified methods.

2.3.3
Superelasticity

When the SM alloy is tensile and tested at a temperature below A_f (A_f-finish temperature of B2↔B19' transformation), strain remains. They are recovered by heating to a temperature above A_f. This is the shape memory effect. When the SM alloy is tensile and tested at a temperature above A_f (A_f -finish temperature of B2↔B19' transformation), strains is recovered by unloading. This is superelasticity [4, 5, 9]. Under such conditions, the martensitic phase is transformed into the parent phase with the resultant reversible strain reaching 10%. The hysteresis on the stress-strain curve corresponds to that phenomenon.

Superelasticity is strongly obtained in the following groups of TiNi based alloys: in the near-equiatomic TiNi alloys with Ni-content exceeding 50 at. %; in Ti(Ni,Me) alloys with the content of Me – Fe, and Co lower than 3 at. %; in Ti(Ni,Me) alloys with the content of Me-Au, Pt, Pd, and Cu from 0 to 12 at. %. There are the MT B2↔R↔B19' and so called premartensitic phenomena in these alloys. The last one appears when the macroscopic shift modulus is decreased strongly throughout the wide temperature interval from $T>A_f$ to R_s. The superelasticity is poorly exhibited in alloys with MT B2↔B19 where the same softening of the macroscopic shift modulus does not appear [6].

3 Corrosion Properties and Electrochemical Behavior of TiNi-Based Alloys

The biochemical compatibility of the physiological solution and metal implants is substantially determined by the electrochemical interactions between them. Usually, it leads to the diffusion of metal ions into the body tissues and fluids. Nevertheless, it would be incorrect to evaluate the biological compatibility of the implant on the basis of chemical compound content alone without taking into account its own chemical properties. In particular, it concerns to corrosion characteristics of the Ti-Ni implants.

The medical and engineering applications of TiNi-based alloys stimulated a wide spectrum of investigations on their electrochemical and corrosion characteristics [1, 7, 24] with respect to aggressive solutions such as: $ZnCl_2$, HCl, K_2SO_4, H_2SO_4, water solutions, and sea and river water [25–28]. There are many papers on the study of the above-mentioned characteristics with respect to change of temperature and applied yield stress.

As shown in the paper [25], the most electrochemically passive are TiNi intermetallic compounds in comparison with the another ones (Ti_2Ni or $TiNi_3$) having the Ti-Ni equilibrium phase diagram. In the stressless state, the near-equiatomic TiNi alloys appear to have a high corrosion resistance, close to that of pure titanium. However, the deviation from stoichiometric composition (with Ni contents exceeding or depletion on ~ 2 atomic % and more) leads to a raise of the electrochemical activity and to a lowering of the corrosion resistance. The corrosion resistance increases in stressless TiNi alloyed with elements of the platinum group Ru, Rh, Os, Ir, Pd, Pt [28–30] or Mo [7, 24]. So, in the acidified 72%- solution of $ZnCl_2$, and at temperatures of ~370 K, the corrosion of Ti–Pd and TiNi alloys does not exceed 0.03 mm/years, whereas the corrosion rate of the pure titanium is ~50 mm/years under the same conditions [26].

TiNi alloyed with Mo exhibits a higher passivity similar to the poor Ti in water solutions of NaCl and HCl [7]. TiNi alloyed with Cu, Fe, Mn and Al result in the slight drop of the corrosion resistance [31, 32]. The electrochemical behavior of TiNi-based alloys is known to depend on the purity of the alloyable components and on the melting or casting processes. Granting this, the authors of [25–27] have investigated TiNi alloys with a Ni content from 0 weight % pure Ti up to 95 weight % – pure Ni prepared following two different methods: First, a sixfold high frequency induction melting of the chemically pure Ti and Ni

(99,99 weight %). Second, [8, 33] a self-propagating high-temperature synthesis (SHS). Raw materials were low purity Ti and Ni powders. The electrochemical behavior of the products was studied at room temperature in water solutions of H_2SO_4, HCl, HNO_3, HCOOH and KOH. The conditions of the active anode dissolution, of the active-passive behavior and of the stable passive state were determined for most of the investigated alloys. It was established that a higher corrosion resistant alloys appeared when prepared from pure components by the sixfold high frequency induction melting. On the contrary, the SHS products appeared to have the lower corrosion resistance due to their higher structural and chemical heterogeneity and impurity content. However, even the TiNi alloys prepared by the SHS melting exhibited high corrosion properties in comparison with stainless steel [33].

Authors of [25] have investigated the electrochemical and corrosion properties of pure titanium and the Ti–50 atomic % Ni alloy in 1% HCl water solution and the 1.5% NaCl water solution. In particular, it was established that the preliminary surface treatment of the TiNi alloy in the hot NaCl electrolyte (T≅100°C) leads to the magnification of the Ti oxide film thickness from 0.01 μm up to 1–3 μm. That film has effectively blocked the anode dissolution centers on the surface of the specimen. In part, it results in decreasing the dissolution activity of Ti and Ni elements from the alloys.

The effect of surface coating with the ionic beam influence, TiN and TiCN, has been investigated [2]. All these surface coating materials are effective in increasing the corrosion resistance and suppressing the dissolution of Ni from TiNi alloys.

When the surface coating is the Ti-oxide film, and that calcium and phosphorus are present in the biological solution, adsorption on the surface of the implant with formation of a phosphate film is similar to the apatite compound [26, 28]. This is important because it considerably raises the biochemical compatibility of TiNi-based alloys.

When choosing the species of the surface coating, it is necessary to take into account that TiNi implants may be in a loading state in the body and can remain in the loading-unloading cycling condition for a long time. Therefore, the coating does not separating during cycling, as a minimum, term of life.

Rather, the perspective for medical purposes is a surface treatment method by high-dozen ionic beam influence. This method is developed intensively at the Institute of Strength Physics and Materials Science of RAS (Tomsk city, Russia). The first results of this application on TiNi-based alloys [18b] allow a predicted success in the tasks of corrosion and cyclical stability of those materials.

References

1. Hagemeister N, Yahia L'H, Weynant E, Lours T (1995) Fatigue life of superelastic springs for an anterior cruciate ligament prosthesis. J Phys IV France 5:1223–1228
2. Miyazaki S (1998) Medical and dental applications of shape memory alloys. In: Otsuka K, Wayman CM (eds) Shape memory materials. Cambridge University, Cambridge, pp 267–282
3. Kornilov II, Belousov OK, Kachur EV (1977) Nickel titanium and others shape memory alloys. Nauka, Moscow
4. Likhachev VA, Kuzmin SL, Kamentseva ZP (1987) Shape memory effect. Leningrad State University, Leningrad

5. Khachin VN, Pushin VG, Kondrat'ev VV (1992) Nickel titanium: structure and properties. Nauka, Moscow
6. Pushin VG, Kondrat'ev VV, Khachin VN, et al. (1998) Pretransitional phenomena and martensitic transformations. Russian Academy of Science, Ekaterinburg
7. Gunter VE, Kotenko VV, Mirgazizov MZ (1986) Shape memory alloys in medicine. Tomsk State University, Tomsk
8. Gunter VE, Itin VI (1992) Shape memory effects and their medical applications. Nayka, Novosibirsk
9. Otsuka K, Shimizu K, Suzuki Y (1990) Shape memory alloys. Metallurgy, Moscow, pp 183–209
10. Otsuka K, Wayman CM (1998) Shape memory materials. Cambridge University, Cambridge
11. Khachin VN, Voronin VP, Sivokha VP, Pushin VG (1995) Martensitic transformation and shape memory effect in polycomponent TiNi-based alloys. J Physique France 8:765–769
12. Savvinov AS, Sivokha VP, Khachin VN (1983) Martensitic transformation in $Ti_{0.5}Ni_{(0.5-x)}Fe_x$. Izv VUZov SSSR (Fiz) 7:34–38
13. Khachin VN (1989) Martensitic transformation and shape memory effect in B2 intermetallic compounds of titanium. Rev Phys Appl 24:733–739
14. Khachin VN, Matveeva NM, Sivokha VP (1981) High-temperature SME in TiNi-TiPd alloys. Dokladi Akademii Nauk SSSR 257(1):167–169
15. Sivokha VP, Khachin VN (1986) Martensitic transformation and shape memory effect in TiNi–TiAu alloys. Fiz Met Metalloved 62:534–540
16. Tokarev VN, Savvinov AS, Khachin VN (1983) Shape memory effects causing by martensitic transformation in TiNi–TiCu alloys. Fiz Met Metalloved 56:341–344
17. Meisner LL, Sivokha VP (1996) Crystal lattice deformation under B2→B19' martensitic transformations in $Ni_{50}Ti_{(50-x)}Zr_x$. Fiz Met Metalloved 81:158–164
18. Meisner LL, Sivokha VP, Perevalova OB (1998) Formation features of fine structure of the $Ni_{50}Ti_{40}Zr_{10}$ alloy under different thermal treatment. Physica B 262:49–54
18a. Meisner LL, Sivokha VP (1999) Martensitic transformations in the TiNi-TiZr alloys. Fizika metallov i mtallovedenie. Russian, 88, 6, 59–62
18b. Meisner LL, Sivokha VP, Sharkeev Yu.P, Kulkov SN, Gritsenko BP (2000) Plastic deformation and fracture of the ion-implanted Ni50Ti40Zr10 alloy on meso- and macro-levels. Journal of Technical Physics. Russian, 70, 1, 32–36
19. Lotkov AI, Grishkov VN, Kuznetsov AV, Kulkov SN (1983) TiNi aging and its effect on the start temperature of the martensitic transformation. Phys Stat Sol 75:373–377
20. Lotkov AI, Grishkov VN (1985) Nickel titanium. Crystal structure and phase transformation. Izv VUZov SSSR (Fiz) 5:68–87
21. Khachin VN, Gyunter VE, Sivokha VP, Savvinov AS (1979) Lattice instability, martensitic transformations, plasticity and anelasticity of TiNi. In: Proceedings of the International Conference on Martensitic Transformations (ICOMAT-79), Cambridge, Massachusetts, USA. 24–29 June. Vol 5. pp 474–480
22. Van Humbeck J, Stalmans R (1998) Characteristics of shape memory alloys. In: Otsuka K, Wayman CM (eds) Shape memory materials. Cambridge University, Cambridge, pp 149–183
23. Saburi T (1998) Ti–Ni shape memory alloys. In: Otsuka K, Wayman CM (eds) Shape memory materials. Cambridge University, Cambridge, pp 49–96
24. Shabolovskaya SA (1995) Biological aspects of TiNi alloy surfaces. J Phys IV France 5:1199–1204
25. Tomashov ND, Ustinskaja TN, Chukalovskaja TV (1983) Electrochemical and corrosive behaviour of Ti_2Ni and TiNi intermetallic compounds in neutral and acid sulfate solutions. Zashchita metallov 19:584–586
26. Mamileecheena MV, Romanushkina AE (1978) Corrosion of titanium, Ti–Ni and Ti–Pd alloys into $ZnCl_2$ solution. Zashchita metallov 14:172–175
27. Stepanova TP, Krasnojarskii VV, Tomashov ND, Druzhinina IP (1978) Influence of Ni-content in Ti-based alloys on their anodal bechaviour in riverine water. Zashchita metallov 14:169–171
28. Kossiy GG, Trusov GN, Goncharenko BA, Micheev VS (1978) Corrosive and electrochemical characteristics of Ti–Ni intermetallic compounds in acid solutions. Zashchita metallov 14:662–666
29. Tomashov ND, Chukalovskaja TV, Chernova GP (1972). Influence of Ru, Rh, Os, Ir elements on corrosion behavior of TiNi-based alloys. Zashchita metallov 8:549–552
30. Tomashov ND, Kazarin VI, Micheev VS, Goncharenko BA (1976) Influence of the platinum group elements on corrosion behavior of Ti–Ni alloys. Zashchita metallov 12:268
31. Marshakov IK (1971) Rust protection of titanium-based alloys. In: Itogi nauki. Korroziya i zashchita ot korroziy. Moscow, VINITY Press, 1
32. Nevitt MV (1966) Electronic structure of transition metals and chemistry of their alloys. Metallurgy, Moscow, p 97
33. Gunter VE, Dambaev GT (1998) Medical shape memory materials and implants. Tomsk State University, Tomsk

Porous NiTi as a Material for Bone Engineering

Reed A. Ayers, Ted A. Bateman, Steven J. Simske

1 Introduction

The utility of nitinol as a superelastic, shape-memory alloy implant material has yet to be fully investigated. Nitinol, or porous, equiatomic NiTi shape memory alloy (approximately equal atomic masses of nickel and titanium), has recently been investigated as a material for craniofacial applications [1, 2]. In Russia, China and Germany, it has been in clinical use for approximately a decade in maxillofacial surgeries and other orthopedic procedures involving thousands of patients [3–5]. Porous nitinol can be produced by various manufacturing processes, including, but not limited to, sintering of molten NiTi and self-propagating-high-temperature-synthesis (SHS) [6, 7]. Such methods allow for a controlled range of NiTi porosity, and provide appropriately sized and interconnected (open) pores, creating an implant morphology similar to bone. A porous implant structure allows ingrowth of mineralized tissue, establishing a biological fixation of the implant. It has been shown that 50% porous NiTi provides greater initial bone ingrowth (as a percentage of the implant cross-section) than 30% porous hydroxyapatite, primarily due to the greater exposed surface area [1]. Moreover, NiTi in this porosity range provides a void space, after bone ingrowth, similar in percentage of cross-section to that of rabbit cranial bone further indicating NiTi's ability to at least architecturally mimic bone [1]. The shape memory property of NiTi also allows for the possibility of in situ implant shape in the case of injury to the implant or surrounding hard tissue.

The superelasticity and high strength material properties of nitinol also suggest its candidacy for orthopedic implantation. The superelastic properties allow the surgeon greater margin in sizing bony defects as the implant can be press-fitted into the bone without unduly damaging the surrounding bone or implant. In fact, such a press fitted superelastic, shape-memory alloy may naturally space surrounding bone through cyclic resorption. The high strength of NiTi (UTS of 895 MPa, annealed) allows for good initial fixation of the implant by withstanding the stresses induced by mastication or other imposed loads. With the incorporation of porosities into the NiTi, the potential for the matching of the mechanical properties of the implant to the surrounding bone becomes available, decreasing the prevalence and magnitude of subsequent stress shielding.

Metals and ceramics in current clinical use have a modulus of elasticity in the range of 100–400 GPa. This is in contrast to bone, which has an elastic modulus an order of magnitude less (20 GPa for cortical bone with approximately two-thirds mineral mass percentage of dry mass). The martensitic modulus of elasticity for solid NiTi is in the 28–41 GPa range (close to the modulus of bone). By making NiTi 50% porous, the apparent modulus of the implant is below the range of bone (14–20 GPa). If an exact match between a bone infiltrated implant and the surrounding bone is required to minimize stress-shielding, the low modulus of porous NiTi allows the possibility of significant ingrowth at this matching value. Itin et al. demonstrated further the ability of NiTi to mimic the mechanical properties showing 40–50% porous nitinol has a recoverable strain of 3.2% near physiologic temperatures, which is similar to the recoverable strain of bone at 2% (Itin et al. 1994). This important aspect of NiTi superelasticity suggests that if the surrounding bone is strained within its elastic region (less than 2%), the implant will deform with the bone and recover its original shape afterwards, preserving the implant/bone bond.

This review examines the most common types of porous biomaterials in clinical use for craniofacial applications, developing a hypothesis about what constitutes an effective porous orthopedic biomaterial. Next, it discusses the biocompatibility of NiTi. This, in turn, springboards a discussion about the advantages and disadvantages of NiTi as a porous biomaterial by comparing NiTi to commonly used orthopedic biomaterials. Future work necessary to characterize porous NiTi as a material for bone engineering is then presented.

2 Porous Biomaterials in Craniomaxillofacial Applications

The advantage of porous materials, in general, is their ability to provide biologic fixation of the surrounding bony tissue via the ingrowth of mineralized tissue into the pore spaces. This is accomplished by increasing the available surface area for apposition by having the interior of the implant accessible via pore spaces [8]. It has been established that mineralized tissue ingrowth requires pore sizes in the range of 100–400 µm [9, 10]. Such morphology allows for early rapid cartilaginous ingrowth and subsequent bone maturation over the lifetime of the implant. An open porosity (interconnected pores) allows for vascularization to support osseous tissue ingrowth and continued bone maturation [11]. This architecture is analogous to the perpendicular aspects of bone morphology, exhibited at the vascular level by Haversian and Volkmann's canals. Interconnected pores increase stability and cosmesis of the bone [12, 13] and increase resistance to fatigue loading [14]. The increased stability (defined for the puposes of this paper as micromotion under 150 µm [15, 16] reduces implant micromotion and the resultant resorption of adjacent bone [12] or inhibition of cartilaginous ingrowth [15].

Porous materials likely affect bone ingrowth into the implant pores by matching the mechanical properties of the interface to the surrounding bone, reducing stress-shielding through a graded transfer of the stresses which are imparted at the implant/bone interface [17–19]. As such, one can enhance the efficiency of the load transfer between the implant and surrounding bone by optimizing the porosity (in terms of pore size, gradient and percent) of the implant to the bone into which it is placed and the loading environment to which it is exposed. Recent experiments indicate that pore spaces also allow the delivery of appropriate healing and growth factors to the ingrowing tissue. Thus, porous materials allow one to address both biologic and mechanical aspects imposed upon orthopedic implants during the initial phases of mineralized tissue ingrowth and its continued maturation.

In general, the predominant implant materials clinically used in oralmaxillofacial and craniofacial applications are autogenous bone, bank bone (such as antigen extracted autolyzed bone) and porous block hydroxyapatite (Interpore 200 is a commercial example of such a material in clinical use). Autogenous bone is the most common porous material used in craniofacial reconstruction [20]. The use of this material has the significant advantage of reduced rejection by the patient. Donor sites for autogenous bone include the rib, crania and iliac crest [21]. Difficulties arise in the need for a secondary surgical site along with subsequent increases in operation time and the potential for donor site complications including, but not limited to infection, fracture and reduced patient ambulation [12, 2, 23]. Bank bone may be used to eliminate the need for a second surgical site, but there still remains the disadvantage of improper bonding between the host bone and the graft and the potential for infection [12]. Microhardness data indicates oven-ashed bone may provide an alternative [24]. Nevertheless, the resorption rates of autogenous and allogenic bone grafts are unpredictable leading to the possibility of implant instability and implant failure [12, 21, 20]. A graft should be resorbed in such a manner that it allows sufficient time and structure for vascularization of the porosities and subsequent bone ingrowth [20].

Slow resorption is a reason that ceramic biomaterials based on calcium phosphates (the mineral phase of bone) have gained favor. These materials include hydroxyapatite (HA) and tricalcium phosphate (β-TCP). They can be manufactured to provide for controlled resorption with appropriate porosity [12, 25, 26]. These ceramics have the disadvantage of being brittle and difficult to machine, but are strong enough to withstand the forces induced during mastication [13, 27, 28]. Dense hydroxyapatite in the form of porous block coralline HA is an effective material for use in craniofacial applications [13, 27–30]. It is also used as a porous coating for otherwise nonporous materials such as titanium, providing a large area for micromechanical fixation via osseointegration of the implant, increasing its stability during the early phases of bone ingrowth [31, 32].

In maxillofacial applications in humans, woven bone invades the porous HA in as early as 4 months up to 300 μm deep [17, 28]. This early woven bone is then remodeled into lamellar bone and, subsequently, Haversian type bone [13, 28, 29]. Bone ingrowth progresses until about 20 months reaching an asymptotic condition at all depths in the implant, with the relative amount of osseous tissue remaining constant [17, 28]. During this progression, the bone matures into Haversian-based bone, exhibiting its normal structural properties and metabolism [29]. The HA, meanwhile, may undergo modest resorption [28, 33].

The ideal implant for a variety of applications may have pore sizes that allow for rapid bone ingrowth and apposition with a porosity that matches the mechanical properties of the implant to the surrounding bone. This implant would also need to be bioinert, or preferably bioactive (osteoinductive and/or osteoconductive), and be resorbed over time at a rate that ensures stability and cosmesis of the surrounding bony structures. While porous NiTi is not resorbable, as the following discussion will highlight, it can be formed and treated to meet the other traits herein considered desirable in an orthopedic implant.

3 NiTi Biocompatibility

Numerous studies have examined the biocompatibility of NiTi in vitro and in vivo, with differing results. Rondelli, using human body simulating fluids, reported that NiTi has a localized corrosion resistance similar to Ti6Al4V, but when the passivation layer is abruptly damaged, NiTi's corrosion resistance is less than Ti6Al4V while is still being comparable to other austenitic steels (such as ASTM 316L) [34]. Putters et al., using the inhibition of mitosis in human fibroblasts cultured on nitinol, titanium and nickel substrates, stated that the results indicate that nitinol is comparable to titanium in its biocompatibility [35]. Sarkar et al. showed that NiTi had an earlier breakdown of its passive oxide layer than other implant materials such as titanium, stainless steel and cobalt-chrome alloys when subjected to potentiodynamic cyclic polarization tests in a sodium chloride solution [36]. It should be noted, these studies focused on the surfaces of solid NiTi, thus, it may be expected that porous NiTi may have diminished corrosion resistance by the fact of its greater surface area in contact with bodily fluids.

In vivo work is generally supportive of NiTi's biocompatibility. Simske and Sachdeva, and more recently Ayers et al. have demonstrated that bone ingrowth into porous nitinol in the crania of rabbits is evident as early as 6 weeks and that bone contact is made with the surrounding cranial hard tissue [1, 2]. A study using high purity nitinol alloy implanted in the femurs of beagles for 3, 6, 12 and 17 months showed no evidence of localized, or general corrosion on the surfaces of the implants and no metallic contamination of organs due to the implants [37]. Using quantitative histomorphometry, nitinol was shown to be progressively encapsulated by bony tissue in the tibiae of rats, albeit at a reduced rate when compared to pure titanium, anodic oxidized Ti and Ti6Al4V, over the course of a 168-day experimental period [38]. In a finding similar to Takeshita et al., Berger-Gorbet et al., using immunohistochemistry, showed NiTi screws implanted in rabbit tibia had slower osteogenesis with no close contact between implant and bone as compared to screws made of c. p. titanium, Vitallium, Duplex austenitic-ferritic stainless steel (SAF), and 316L Stainless Steel [39]. Clinical results of procedures using NiTi alloys in China and Russia state no significant detrimental effects of devices implanted in craniofacial bone [3, 4]. However, the specific studies upon which this conclusion is made are not readily obtainable, making replication difficult.

3.1 Mechanisms of NiTi Biocompatibility

The biocompatibility of NiTi derives from the formation of an oxide layer (TiO_2) on the surface of the implant. This is similar to the TiO_2 layer formed on pure titanium, which enhances its biocompatibility as an implant material [40]. The passivation layer can range in thickness from 2 nm–1 μm [40–42]. Resistance of this layer to damage correlates with the corrosion resistance, and hence biocompatibility, of the implant. Overall thickness of the passivation layer is less germane to biocompatibility than its uniformity [40]. Because the oxide layer is a brittle ceramic, the superelasticity of the NiTi substrate can induce stresses in the passivation layer as the implant deforms causing cracking and resulting in a pitting attack of the NiTi substrate [43]. Maintaining the integrity of the passivation layer is paramount with nitinol to prevent the potential release of metallic nickel into the body. It has been established in the literature that nickel *in vivo* is highly toxic, producing severe inflammatory responses, along with being a potential carcinogen.

In order to preserve the substrate from pitting corrosion numerous methods of manufacturing the oxide layer have been examined. The easiest method is simple aging of the material in air, allowing for a natural oxidation layer to form. An associated side effect, however, is that the oxide layer may contain metallic Ni and nickel-oxides at the NiTi surface [3, 40]. Steam or water autoclaving has been shown to reduce the presence of Ni, depleting it to a depth upwards of 10 nm into the NiTi substrate [3]. The resulting oxide layer contains primarily TiO_2 based oxides [3]. Heat treating the surface of NiTi in a nitrite/nitrate salt has been used to create a very thick oxide layer (~0.1 μm), as compared to other treatments [40]. However, this layer has been shown to contain a Ni rich region above the NiTi substrate, which could, if the oxide layer is damaged, result in dissolution of Ni from the implant [40]. Heat treating also carries the risk of altering the mechanical properties of the NiTi. Two methods that produce thin but very uniform oxidation layers are passivation of the NiTi surface with nitric acid solution and electropolishing [40]. Electropolishing significantly increases the corrosion resistance of NiTi [40].

Other methods for enhancing the corrosion resistance of NiTi involve the deposition of a non-metallic layer on the NiTi surface. This allows for the creation of thick (>1 mm) films on the NiTi substrate. One method that has shown promise is the plasma deposition of polymerized tetrafluoroethylene (PPTFE) [43, 44]. This method approximately doubled the passivation range of NiTi in physiological Hank's solution and decreased the pit diameter by an order of magnitude when used on osteosynthesis staples [43]. This passivation layer was also elastic enough to follow the large deformations induced by NiTi's shape memory effect without cracking [43].

Perhaps the most unique method of inhibiting the dissolution of Ni from the NiTi substrate involves creating a bioactive film. By creating a covalently bonded coupling layer between the Ti-oxide and immobilized human fibronectin, Endo was able to demonstrate increased corrosion resistance of the NiTi, along with the ability of the attached layer to withstand hydrolysis in solution at pH 4.0–7.0 [41, 42]. This offers a unique opportunity for bone engineering in which a material that may be considered to neither support or degrade bone ingrowth (an osteopermissive material) [2] can be made to be bioactive (similar to calcium phosphates such as HA). More importantly, this is a key extracellular matrix (ECM) compound upon which osteogenic cells attach and develop. Regardless, in the case of porous NiTi, whatever method is used to enhance the biocompatibility of NiTi it must be able to penetrate the interior pores of the material to ensure treatment of all of the implant's surfaces. The authors' have used steam autoclaving for 30 min. While the surface properties of the steam-autoclaved implants have not been analyzed, the implants prepared in this manner do allow for bone ingrowth and direct bone and implant contact (apposition).

4 Authors' Experience with NiTi

The authors' experiments have shown that porous nitinol is generally biocompatible when placed in the crania of rabbits, and deserves further study as a material for bone engineering. Studies conducted have examined the effects of NiTi porosity on rabbit cranial bone ingrowth at 6 weeks [2] and bone ingrowth over a 12-week period with indirect comparison to the well-characterized cranial implant material HA (in the form of Interpore 200) [1]. In both of these studies, porous NiTi implants were placed in the parietal bone of New Zealand White rabbits in defects machined to the specific geometry of the implant. In neither experiment were macrophage cells noted adjacent to, or within, the implants. Soft and connective tissues readily adhered to the implants post-surgically. Both studies used uncoated (other than the oxide layer induced during autoclaving) porous equiatomic nickel-titanium (nitinol) implants.

The study examining the effect of porosity on bone ingrowth after 6 weeks addresses two aspects of the use of nitinol in cranial bone defect repair. The first is the verification of substantial bone ingrowth into the implant after 6 weeks. The second is the determination of the effect of pore size on the ability of bone to grow into the implant during the early (6-week) post-operative period. Implant specimens with three different morphologies (differing in pore size and percent porosity) were implanted for 6 weeks.

A quick synopsis of the data (Table 1) shows mean pore size (MPS) of implant type 1 (353 ± 74 μm) differed considerably from that of implant type 2 (218 ± 28 μm) and implant type 3 (178 ± 31 μm). Quantitative histomorphometric measurements are presented in Table 2. There were no significant differences between implant types in the percentages of bone and void/soft tissue composition of the aggregate implants. The amount of bone ingrowth was also not significantly different between implant types. Implant 1 was significantly higher in pore volume and thus had a significantly higher volume of ingrown bone (2.6 ± 0.6 mm^3) than implant 3 (1.5 ± 0.7 mm^3); and a greater amount, but with-

Table 1. Porous nitinol implant morpholgy

Measurement	Implant 1 (n=7)	Implant 2 (n=6)	Implant 3 (n=7)
Thickness (μm)	644±21*	345±37	385±56
% Volume pore space (porosity)	42.9±4.0*	54.4±5.3	50.5±13.7
Mean pore size (μm)	353±74*	218±28	179±31
Available pore volume for ingrowth (mm^3)	6.9±0.6*	4.7±0.7	5.1±2.0

An asterisk denotes measurements statistically significantly ($P<0.05$, Tukey-Kramer HSD) different in implant 1 when compared to either implant 2 or implant 3

Table 2. Porous nitinol implant quantitative histomorphometry

Measurement	Implant 1 (n=7)	Implant 2 (n=6)	Implant 3 (n=7)
Percent implant (%)	57.1±4.0	45.6±5.3	49.5±13.7
Percent void (%)	26.9±3.8	33.6±5.1	35.1±10.9
Percent bone (%)	14.6±5.9	20.8±6.7	15.4±4.7
Percent ingrowth (%)	37.4±7.8	37.9±10.1	31.1±6.9
Bony apposition, exterior (%)	47.4±9.6#	41.6±9.2	32.0±9.1
Bony apposition, interior (%)	38.6±12.7	41.9±10.5	36.0±11.1
Total bone ingrowth (mm^3)	2.6±0.6#	1.8±0.5	1.5±0.7

Values are given as mean ± standard error of the mean for each of the three implant types. An asterisk indicates a significant difference ($P<0.05$) from implant 2. A pound sign indicates a significant difference ($P<0.05$) from implant 3.

out statistical significance, than implant 2 (1.8 ± 0.5 mm^3). The difference between implant types in total volume of bone ingrowth is ostensibly a function of the implant volume. Implant 1 had a greater volume available for bone ingrowth. The difference in implant 1's external bony apposition most likely reflects the greater surface area for bony contact of implant 1 as compared to the other implants.

In thin implants (i.e., implant thickness is on the same order of magnitude as pore size) pore size does not appear to affect the bone ingrowth during the cartilaginous (analogous to fracture repair) period of bone growth within the implant. This implies that over the commonly accepted range of implant porosities (100–400 µm), the bone ingrowth near the interface of nitinol implants at 6 weeks is similar. Surface contact (apposition) measurements were also used as gauge of the biocompatibility of the implants as this is an accepted general measure of biocompatibility (Ono et al. 1990; Simske and Sachdeva 1995). The measurements (Table 2) do not imply that nitinol is osteoconductive, but indicate that it does not inhibit bone ingrowth in the early healing phase of the defect.

In another study [1], geometrically equivalent (5 x 5 x 1 mm) uncoated porous nitinol and coralline hydroxyapatite (HA, Interpore 200) implants were placed 4 mm to either side of the midsection of the frontal bone and 4 mm anterior to the coronal suture of the cranial bone of New Zealand White rabbits. The rabbits were killed in postsurgical intervals of 2, 6 and 12 weeks, and the implants were evaluated for gross biocompatibility, bony contact and ingrowth.

Histologically, bony contact was present for both materials. Both materials made bone contact with the surrounding cranial hard tissue, and percent ingrowth increased with surgical recovery time. Measurements of microhardness

Table 3. Quantitative histomorphometry for porous nitinol and hydroxyapatite

Implantation time	Implant apposition (%)		Implant ingrowth (%)	
	HA	NiTi	HA	NiTi
2 weeks (n=2)	12.5±12.5	9.2±9.2	0.0±0.0	0.0±0.0
6 weeks (n=2)	39.0±4.8	34.9±0.5	6.7±6.7	12.2±0.5
12 weeks (n=3)	50.4±4.2	39.6±6.6	25.3±9.3	34.3±11.4

in conjunction with bone histological observations indicate that bone within and in contact with the implants is similar in site-specific structural properties to the surrounding cranial bone. Porous nitinol implants appear to permit significant cranial bone ingrowth after as little as 12 weeks, and thus nitinol appears to be suitable for craniofacial applications. Compared to HA, the nitinol implants demonstrated a trend for less total apposition and more total ingrowth after 6 weeks and 12 weeks of implantation (Table 3). These results may be due to the osteoconductive properties of HA [45, 46] or to the differences in the surface morphologies between the implants used in this study. The nitinol, with a greater surface porosity (50%) than the HA (30%), may have allowed readier access to the interior of the implant than the HA.

5 NiTi Versus Other Biomaterials

5.1 Mechanical Considerations

One of the primary concerns of bone engineering arises from the premise of "Wolff's Law": that bone not subjected to loading undergoes resorption. When an implant with an elastic modulus stiffer than bone is used, mechanical disuse causes the surrounding bone to resorb (stress-shielding), threatening the stability of the implant. Thus, matching the material properties of the implant to the bone for a given application may be paramount to the success of a porous metal implant. Material property matching is perhaps less important in craniofacial applications than in joint replacement (hip and knee arthroplasty), due to the different mechanisms governing bone growth [47]. Nonetheless, the mechanical aspect of craniofacial implantation must be considered [17].

It would be inappropriate to assign a single value to the elastic modulus of solid NiTi because the elastic modulus is nonlinear with respect to temperature. The martensitic elastic modulus follows the Clausius-Clapeyron equation:

$$\sigma_a/M_s=-{}_{\Delta}H/T\varepsilon_0 \qquad (1)$$

where σ_a is the applied stress, M_s is martensitic temperature, ε_0 is the transformation strain resolved along the line of the applied stress, ${}_{\Delta}H$ is the transformation latent heat and T is the temperature [48]. Thus, there is a family of stress-strain curves dependent upon temperature for a given specimen. When porous, determining the structural modulus of the implant is further complicated. For example, at a temperature of 293 K, the modulus of 40–50% porous nitinol is approximately 25 GPa (Itin et al. 1994). This compares to standard biomedical titanium alloys such as solid Ti6Al4V with a modulus of 110 GPa. Other metals such as ASTM 316L and CoCr alloys have elastic moduli of 200 GPa and 220 GPa, respectively if they are solid. Roughly, the metals used in clinical applications are an order of magnitude stiffer than bone, while 40–50% porous NiTi is similar to bone in stiffness.

5.2
Formation Considerations

Metals such as Ti6Al4V and CoCr are not normally manufactured in a porous form. They can be made "porous", however, by coating the outer surfaces with metal powders via plasma spraying either metal or ceramic powders onto the metal surface; or by double sintering metallic beads onto the heated metal substrate. Pore sizes can range from 150–300 μm using these techniques with percent porosity from 20–40%. While porous coatings may enhance the osseointegration of the implant, it has been shown that the bond between bone and coating is preserved better than the bond between the coating and the substrate, resulting in the possible failure at the implant coating/substrate interface [49, 50].

Ceramics occur naturally as porous materials (bone, coral, etc.) or can be manufactured to be porous via numerous methods including combustion synthesis, sintering, and plasma spraying. There are at least nine recognizable biodegradable bioceramics used in bone engineering: aluminum-calcium-phosphorous-oxides, glass fibers and their composites, corals, calcium sulfates, ferric-calcium-phosphorous oxides, hydroxyapatite, tricalcium phosphate, zinc-calcium-phosphorous oxides and zinc-calcium-phosphorous oxides [51]. In addition, Bajpai and Billotte list six bioinert ceramics including pyrolitic carbon coated devices, dense hydroxyapatites, dense nonporous aluminum oxides, porous aluminum oxides, zirconia and calcium aluminates. Surface reactive bioceramics include bioglasses and ceravital, dense and nonporous glasses and hydroxyapatite [51].

The elastic modulus of the bioceramics mentioned above range from 40–117 GPa for pure crystalline hydroxyapatite to as high as 400 GPa for corundum. These values can also be adjusted based upon the natural or manufactured porosity of the materials. For example, the elastic modulus of corals, which are predominately hydroxyapatite, changes by an order of magnitude over a porosity range of 0–50%; thus, a 100 GPa modulus can be reduced to 10 GPa in a highly porous form (30–50%). The apparent modulus of the porous forms of porous materials may be estimated via the equation

$$\boxed{E = E_s(V_s)^X}$$

where E is the apparent modulus, E_s is the elastic modulus of the solid; V_s is the volume fraction of the of the solid phase; X is a variable ranging from 1 to 2, being approximately 1 when V_s is approximately 1 and approximately 2 when V_s is approximately 0 (Lakes 1995). Given this, it is apparent that within an acceptable range of porosities, ceramic and glass materials can be manufactured to have apparent densities that of bone.

5.3
Machining

Machining considerations must also be taken into account when comparing these materials. This consideration arises from the need for the surgeon to be able to match the implant to the bony defect during the surgery to provide the best possible match between the implant and surrounding bone. Ceramics are very brittle, and are difficult to machine: warnings about the brittleness are prevalent in the literature. This is largely mitigated by the ability to form the ceramic into the appropriate shape beforehand, reducing the need for post-production machining. Porous metals formed by sintering or the plasma spraying of powders and diffusion bonding of metal fibers to a metal substrate can result in the damage to the underlying substrate and a coating that is also brittle and difficult to machine [52]. Self-propagating-high-temperature-synthesis (SHS) has, nevertheless, allowed the manufacture relatively complex shapes in nitinol (cones, polygons, etc.) reducing the need for post-production machining. The use of SHS in the formation of nitinol allows implants to be created very rapidly (on the order of seconds to minutes) in contrast to sintering or diffusion bonding processes, which can take hours to days to complete [7].

5.4 Biocompatibility

Ceramics and glasses such as HA, TCP and bioglasses are quite biocompatible. They promote the differentiation of the osteoblast phenotype from marrow stem cells, and are thus, osteoconductive in addition to being biocompatible. Another advantage of these ceramics over metals such as nitinol is their ability to degrade over time, allowing bone to fill in the implant space. While the biocompatibility of NiTi is still under study, it has been our experience that NiTi is bioinert in vivo. It acts as an osteopermissive (or bioinert, similar to pure Ti and its alloys) material simply providing a scaffold upon which the bone may grow, neither promoting bone formation nor preventing it. As has been discussed earlier, the passive oxide layer can render NiTi bioactive similar to HA, TCP and bioglass. There is enough clinical evidence where long-term implantation NiTi remains inert, while metals such as ASTM316L Stainless Steel, which have been optimized for corrosion resistance (hence biocompatibility), will corrode.

Porous NiTi formed and machined into an implant mimics the mechanical and material properties of bone. It is sufficiently ductile to be machined in an operating theater without the need for specialized equipment or processing. While it is not bioactive like many of the ceramics, there is the potential to make it so (via coatings, impregnating reagents, etc.). Perhaps the greatest draw back is that NiTi is not biodegradable. This can be an advantage, however, when repairing large defects caused by congenital bone diseases or non-union fractures wherein the normal mechanisms for bone growth are no longer present.

6 Present and Future Advantages of Porous NiTi

The advantages of NiTi over current implant materials are in its superelasticity at body temperature, ease of formation and versatility in creating graded open porosities. With a forming process such as SHS, one can readily create a wide variety of pore size and porosity combinations in almost any shape. While SHS can be used to create porous Ti, Ti alloys and other metals, NiTi again has the advantage of being a superelastic shape memory alloy. These properties allow the surgeon greater leeway in implant placement and better chance of saving the implant in the case of traumatic injury (i.e. fracture) in the area the implant is located (in situ implant shape recovery).

Porous NiTi's superelasticity is maintained even after bone ingrowth satisfying the need for biomechanical compatibility [6]. This advantage of NiTi over other implant materials opens several avenues of orthopedic treatment heretofore unavailable. The ability of 40–50% porous NiTi to undergo upwards of 3.2% recoverable strain means an implant is more likely to remain integrated with the bone when subjected to peak physiological stresses such as those noted during a stumble when climbing stairs (870% body mass), which may deform the bone beyond the elastic deformation limits of implant materials in current use (note that 3.2% is even greater than bone's own recoverable strain of approximately 2%). Superelasticity may also be used in limb elongation procedures. To accomplish this the implant is preloaded prior to implantation. Upon its osseointegration, thermoelectrical stimulus can be used to return the implant to its original shape. NiTi allows this to be done in small incremental steps with constant stress on the surrounding bone, reducing patient discomfort. A similar method is used in orthodontic archwires in humans [5] and in scoliosis correction in goats [53].

Other advantages of NiTi as a porous biomaterial arise from its ability to be produced via SHS. This method of formation relies on the exothermic reaction of nickel and titanium powders when heated to their combustion temperature of 1773 K (Yi and Moore 1990). When a gassifying reagent such as B_2O_3 is added, porosities are created. The pore size and porosity can be controlled based upon the amount of gassifying agent, pressure of the reaction chamber and/or gravitational forces. This process allows the creation of complex shapes (reducing the amount of secondary processing and machining) in very short time periods (order of minutes). Perhaps, in the future, the patient will undergo a CT scan at the specific site in need of repair, and a mold may be created using stereolithography or a similar technology. This mold would be filled with the appropriate mixture of nickel, titanium and gassifying agents and ignited, creating a custom implant for the specific patient application in a few days.

7 Future Work

Much work has yet to be done to fully characterize porous NiTi as a material for bone engineering. This work ranges from refining the formation and processing of NiTi to rendering NiTi bioactive. In the area of materials processing, it has been demonstrated that ceramics can be combined with NiTi to create a composite or aggregate material [54]. The incorporation of a superelastic shape-memory alloy enhances the tensile strength properties of the ceramic, while the ceramic provides the bioactivity for increased ingrowth of tissue [54]. It is very feasible that a NiTi core with a bioactive ceramic outer surface can be created using SHS. There would be no interface between the ceramic and NiTi, as the transition from one to the other would occur over a functional gradient. In so doing the material and mechanical properties of the surrounding bone are matched with the ceramic, providing a bioactive surface for osseointegration, reducing the time for mineralized tissue infiltration and consequently patient recovery time.

SHS production of NiTi allows one to quantify the nature of bone ingrowth into porous NiTi. In craniofacial applications, it has been proposed that in an approximately 65% porous block coralline HA implant with a mean pore size of 230 μm the mechanical transfer of loads occurs within the first millimeter of the implant surface [17]. If this is the case, are interior porosities needed? These questions may be answered by creating implants with functionally graded porosities, where the surface pore size is enough to allow for a rapid influx of tissue and scales down towards the center of the implant. Depending on the implant application, the interior could remain solid for implants subjected to high loading environments, or be porous, allowing for vascular tissue ingrowth and later bone maturation.

As has been discussed, NiTi offers the advantage of the implant being matched to the mechanical properties of the bone. On the other hand, NiTi is not considered to be as biologically advantageous as other implant materials; for example hydroxyapatite. However, cytokine infiltration of NiTi pore spaces and/or bio-coating the NiTi surface may bridge this gap between NiTi's osteopermissive nature and HA's osteoconductivity. Cytokine infiltration of implants is the addition of bone affecting proteins into the pore spaces of the implant. This offers the opportunity to improve the initial fixation at the bone/implant interface by enhancing the early development of osseous tissue. To highlight this case, biodegradable porous implants are beginning to be used as devices for the delivery of bone affecting proteins [55–57]. Porous NiTi infiltrated with bone affecting proteins could utilize a similar principal with a specific local response as the goal, given that reagents appropriate for the time course of bone growth in the implant are considered [58]. Of course NiTi is not biodegradable, thus its permanence at the repair site would need to be taken into consideration.

Infiltration into the implant pore spaces can use any bone-affecting reagent. The mechanisms for bone formation or inhibition of resorption would be possible target pathways. In other cases, controlled resorption in one area and formation at another may be desired. As such, release kinetics must be considered when choosing a target. An examination of bone morphogenic protein (BMP) release in microporous polylactic/polyglycolic acid (PLA-PGA) implants was examined in physiologic PBS for 72 days [59]. An initial BMP "burst" was released in the first four days. BMP continued to resorb from the PLA-PGA beyond two months at levels approximately an order of magnitude less than the initial burst. One may expect a similar temporal response in nitinol surface treated with the same BMP. With this in mind, anabolic bone proteins may be better candidate reagents to consider than anti-resorptives. An anti-resorptive would serve to prevent bone turnover at the interface between bone and implant or to prevent a stress shielding response. The above study [59] suggests that the kinetics of protein release would not be appropriate for preventing the longer-term resorption. However, long-term resorption should be mitigated by the very nature of the permanence of the NiTi implant and its structural/mechanical mimicry of mature bone.

BMP infiltration of implants is the most common protein currently being examined to promote growth of bone. Bone formation has been initiated using Plaster of Paris (PLP) infiltrated with bovine BMP improving the healing of human femoral non-union fractures in patients who had undergone unsuccessful surgeries to repair the defects [60]. Human demineralized bone allografts infiltrated with BMP-2 promote bone ingrowth into otherwise inactive implants [55]. BMP in a coral implant has been examined in the repair of a tibial defect in sheep [56]. Significantly increased bone ingrowth was noted in the first 6 weeks, as compared to coral controls. After 16 weeks of implantation, mechanical testing showed a trend towards decreased mechanical properties of the BMP impregnated implants as compared to controls. This was explained by the presence of high concentrations of anti-BMP antibodies suggesting an immunogenic reaction to the xenogenic BMP used [56]. This again suggests that the use of BMP infiltration of porous nitinol would be most valuable during the initial fracture healing stage post-implantation.

Reagent infiltration of NiTi has not yet been examined. This group has infiltrated porous $B_4C+Al_2O_3$ created with SHS with a bovine derived Bone Protein (Sulzer Orthopedics Biologics, WheatRidge, Colo.) in a rat skull on-lay model. Histologic analysis, bone ingrowth and surface contact measurements are currently being conducted. We are also currently in the process of implanting infiltrated porous NiTi using the same methods.

Biocoating of NiTi is an option for improving bony apposition. The surface characteristics of an implant play an important role in the rate and degree to which bone will bond with an implant [61]. Additionally, theoretical work has been done on how implant characteristics affect protein resorption. Human plasma Fibronectin (pFN) has been bonded to NiTi [41, 42]. This coating promoted fibroblast spreading in an in vivo system along with decreased implant corrosion [41, 42]. This modification offers a means to control or indeed reduce biological interactions with NiTi, with the possibility of making biocompatible materials bioactive, better mimicking the physiologic conditions. Biocoating in conjunction with reagent infiltration may be the best method of increasing both bone ingrowth and apposition during the initial phases of bone development in the porous implant.

Reagent infiltration, biocoating or the combination of the two may offer the opportunity to expedite the biological fixation of NiTi to bone. These methods may also cause bone infiltration into deeper pores and stimulate bone maturation and general health. Improving the biological behavior of porous metallic implants like NiTi can ultimately create a highly effective material for bone replacement.

8 Conclusions

There is, most likely, no one material or implant architecture that may be considered the ultimate bone replacement material. One must be cognizant of the application of the material, including its location in the body and subsequent loading environments. Porous NiTi does appear to be sufficiently versatile as a material to warrant its consideration in bone engineering. The potential for modification of NiTi's surface properties to create a bioactive implant is further encouragement.

Acknowledgement. The authors thank John J. Moore, Larry M. Wolford, Virginia L. Ferguson, Robert W. Norrdin, Mark Roedersheimer, Christa R. Nunes, Tom A. Schmeister, Mary L. Fleet, Jerry J. Broz for their valuable insights and support on the implant studies. We also would like to thank Dr. L'Hocine Yahia for the invitation to provide this review. This work was supported by the Colorado Advanced Technology Institute through a grant received from the Colorado Institute for Research in Biotechnology.

References

1. Simske SJ, Sachdeva R (1995) Cranial bone apposition and ingrowth in a porous nickel-titanium implant. J Biomed Mater Res 29:527–533
2. Ayers RA, Simske SJ, Bateman TA, Petkus A, Sachdeva RLC, Gyunter VE (1999) Effect of nitinol implant porosity on cranial bone ingrowth and apposition after 6 weeks. J Biomed Mater Res 45:42–47
3. Shabalovskaya SA (1996) On the nature of the biocompatibility and on the medical applications of NiTi shape memory alloys. Biomed Mater Eng 6:267–289
4. Dai K (1996) Studies and applications of NiTi shape memory alloys in the medical field in China. Biomed Mater Eng 6:233–240
5. Airoldi G, Riva G (1996) Innovative materials: the NiTi alloys in orthodontics. Biomed Mater Eng 6:299–305
6. Itin VI, Gyunter VE, Shabalovskaya SA, Sachdeva RLC (1994) Mechanical properties and shape memory of porous nitinol. Mater Characterization 32:179–187
7. Yi HC, Moore JJ (1990) The combustion synthesis of NiTi shape memory alloys. J Miner Metals Mater Soc 42:31–35
8. Greene D, Pruitt L, Maas CS (1997) Biomechanical effects of e-PTFE implant structure on soft tissue implantation stability: a study in the porcine model. Laryngoscope 107:957–962
9. Klawitter JJ, Hulbert SF (1971) Application of porous ceramics for the attachment of load bearing internal orthopedic applications. J Biomed Mater Res Symp 2:161–229
10. Hulbert SF, Young FA, Mathews RS, Klawitter JJ, Talbert CD Stelling FH (1970) Potential of ceramic materials as permanently implantable skeletal prostheses. J Biomed Mater Res 4:433–456
11. Van Eeden SP, Ripamonti U (1994) Bone differentiation in porous hydroxyapatite in baboons is regulated by the geometry of the substratum: implications for reconstructive surgery. Plast Reconstr Surg 93:959–966
12. Kent JN, Zide MF (1984) Wound healing: bone and biomaterials. Otolarynogol Clin North Am 17:273–319
13. Wolford LM, Wardrop RW, Hartog JM (1987) Coralline hydroxylapatite as a bone graft substitute in orthognathic surgery. J Oral Maxillofac Surg 45:1034–1042
14. Eppley BL, Sadove AM, (1990) Effects of material porosity on implant bonding strength in a craniofacial model. J Craniofac Surg 1:191–195
15. Bragdon CR, Burke D, Lowenstein JD, O'Connor DO, Ramamurti B, Jasty M, Harris WH (1996) Differences in stiffness of the interface between a cementless porous implant and cancellous bone in vivo in dogs due to varying amounts of implant motion. J Arthroplasty 11:945–951
16. Ramaurti BS, Orr TE, Bragdon CR, Lowenstein JD, Jasty M, Harris WH (1997) Factors influencing stability at the interface between a porous surface and cancellous bone: a finite element analysis of a canine in vivo micromotion experiment. J Biomed Mater Res 36:274–280
17. Ayers RA, Wolford LM, Bateman TA, Ferguson VL, Simske SJ (1999) Quantification of bone ingrowth into porous block hydroxyapatite in humans. J Biomed Mater Res 47:54–59
18. Pedersen DR, Brown TD, Brand RA (1991) Interstial bone stress distributions accompanying ingrowth of a screen-like prosthesis anchorage layer. J Biomech 24:1131–1142
19. Hollister SJ, Kikuchi N, Goldstein SA (1993) Do bone ingrowth processes produce a globally optimized structure? J Biomech 26:391–407
20. Phillips JH, Forrest CR, Gruss JS (1992) Current concepts in the use of bone grafts in facial fractures. Clin Plast Surg 19:41–58

21. Szachowicz EH (1995) Facial bone wound healing: an overview. Otolaryngol Clin North Am 28:865–880
22. Desilets CP, Marden LJ, Patterson AL, Hollinger JO (1992) Development of synthetic materials for craniofacial reconstruction. J Craniofac Surg 1:150–153
23. Motoki DS, Mulliken JB (1990) The healing of bone and cartilage. Clin Plast Surg, 17:527–544
24. Broz JJ, Simske SJ, Corley WD, Greenberg AR (1997) Effects of deproteinization and ashing on site-specific properties of cortical bone. J Mater Sci Mater Med 8:395–401
25. Eggli PS, Müller W, Schenk RK (1988) Porous hydroxyapatite and tricalcium phosphate cylinders with two different pore size ranges implanted in the cancellous bone of rabbits. A comparative histomorphometric and histologic study of bony ingrowth and implant substitution. Clin Orthop 232:127–138
26. Light M, Kanat IO (1991) The possible use of coralline hydroxyapatite as a bone implant. J Foot Surg 30:472–476
27. Holmes RE, Wardrop RW, Wolford LM (1988) Hydroxylapatite as bone graft substitute in orthognathic surgery: histologic and histometric findings. J Oral Maxillofac Surg 46:661–671
28. Nunes CR, Simske SJ, Sachdeva R, Wolford LM (1997) Long-term ingrowth and apposition of porous hydroxylapatite implants. J Biomed Mater Res 36:560–563
29. Ayers RA, Simske SJ, Nunes CR, Wolford LM (1998) Long-term bone ingrowth and residual microhardness of porous block hydroxyapatite in humans. J Oral Maxillofac Surg 56:1297–1301
30. Jahn AF (1992) Experimental applications of porous (coralline) hydroxylapatite in middle ear and mastoid reconstruction. Laryngoscope 102:289–299
31. Engh CA, Bugbee WD (1998) Extensively porous-coated femoral stems. In: Sedel L, Cabanela ME (eds) Hip surgery: materials and developments. Mosby, St. Louis, p 243
32. Ducheyne P (1998) Bioactive calcium phosphate ceramics and glasses. In: Sedel L, Cabanela ME (eds) Hip surgery: materials and developments. Mosby, St. Louis, p 75
33. Martin RB, Chapman MW, Sharkey NA, Zissimos AG, Bay B, Shors EC (1993) Bone ingrowth and mechanical properties of coralline hydroxyapatite 1 year after implantation. Biomaterials 14:341–348
34. Rondelli G (1996) Corrosion resistance tests on NiTi shape memory alloy. Biomaterials 17:2003–2008
35. Putters JLM, Kaulesar Sukul DMKS, deZeeuw GR, Bijma A, Besselink PA (1992) Comparative cell culture effects of shape memory metal (Nitinol), nickel and titanium: a biocompatibility estimation. Eur Surg Res 24:378–382
36. Sarkar NK, Redmond W, Schwaninger B, Goldberg AJ (1983) The chloride behaviour of four orthodontic wires. J Oral Rehabil 10:121–128
37. Castleman LS, Motzkin SM, Alicandri FP, Bonawit VL (1976) Biocompatibility of nitinol alloy as an implant material. J Biomed Mater Res 10:695–731
38. Takeshita F, Takata H, Ayukawa Y, Suetsugu T (1997) Histomorphometric analysis of the response of rat tibiae to shape memory alloy (nitinol). Biomaterials 18:21–25
39. Berger-Gorbet M, Broxup B, Rivard C, Yahia LH (1996) Biocompatibility testing of NiTi screws using immunohistochemestry on section containing metallic implants. J Biomed Mater Res 32:243–248
40. Trépanier C, Tabrizian M, Yahia LH, Bilodeau L, Piron DL (1998) Effect of modification of oxide layer n NiTi stent corrosion resistance. J Biomed Mater Res 43:433–440
41. Endo K (1995) Chemical modification of metallic implant surfaces with biofunctional proteins (part 1) molecular structure and biological activity of a modified NiTi alloy surface. J Dent Mater 14:185–198
42. Endo K (1995) Chemical modification of metallic implant surfaces with biofunctional proteins (part 2) corrosion resistance of a chemically modified NiTi alloy. J Dent Mater 14:199–210
43. Villermaux F, Tabrizian M, Yahia LH, Czeremuszkin G, Piron DL (1996) Corrosion resistance improvement of NiTi osteosynthesis staples by plasma polymerized tetraflouroethylene coating. Biomed Mater Eng 6:241–254
44. Yahia LH, Lombardi S, Piron D, Klemberg-Sapieha JE, Wertheimer MR (1997) NiTi shape memory alloys treated by plasma-polymerized tetraflouroethylene. Med Prog Technol 21:187–193
45. Neo M, Voigt CF, Herbst H, Gross UM (1998) Osteoblast activity at the interface between surface-active materials and bone in vivo: a study using in situ hybridization. J Biomed Mater Res 39:1–8
46. Ono K, Yamamuro T, Nakamura T, Kokubo T (1990) Mechanical properties of bone after implantation of apatite–wollastonite containing glass ceramic–fibrin mixture. J Biomed Mater Res 24:47–63
47. Rawlinson SCF, Mosley JR, Suswillo RFL, Pitsillides AA, Lanyon LE (1995) Calvarial and limb bone cells in organ and monolayer culture do not show the same early response to dynamic mechanical strain. J Bone Miner Res 10:1225–1232

48. Otsuka K, Wayman CM (1998) Mechanism of shape memory effect and superelasticity. In: Otsuka K, Wayman CM (eds) Shape memory materials. Cambridge University, Cambridge, p 27
49. Spector M (1987) Historical review of porous-coated implants. J Arthroplasty 2:163–177
50. Vercaigne S, Wolke JGC, Naert I, Jansen JA (1998) Histomorphometrical and mechanical evaluation of titanium plasma-spray-coated implants placed in the cortical bone of goats. J Biomed Mater Res 41:41–48
51. Bajpai PK, Billotte WG (1995) Ceramic biomaterials. In: Bronzino JD (ed) The biomedical engineering handbook. CRC, Boca Raton, p 552
52. Simske SJ, Ayers RA, Bateman TA (1997) Porous materials for bone engineering. In: Liu DM, Dixit V (eds) Porous materials for tissue engineering. Trans Tech Publications, Uetikon-Zuerich, p 151
53. Schmerling MA, Wilkov MA, Sanders AE (1976) Using the shape recovery of nitinol in the Harrington Rod treatment of scoliosis. J Biomed Mater Res 10:879–892
54. Itin VI, Shevchenko NA, Korosteleva, Tukhfatullin AA, Mirgazizov MZ, Gyunter VE (1997) "Bioceramic-titanium nickelide" functional composites for medicine. Tech Phys Lett 23:294–296
55. Schwartz Z, Somers A, Mellonig JT, Carnes DL, Wozney JM, Dean DD, Cochran DL, Boyan BD (1998) Addition of human recombinant bone morphogenetic protein-2 to inactive commercial human demineralized freeze-dried bone allograft makes an effective composite bone inductive implant material. J Periodontol 69:1337–1345
56. Gao TJ, Lindholm TS, Kommonen B, Ragni P, Paronzini A, Lindholm TC, Jalovaara P, Urist MR (1997) The use of a coral composite implant containing bone morphogenetic protein to repair a segmental tibial defect in sheep. Int Orthop 21:194–200
57. Guicheux J, Gauthier O, Aguado E, Heymann D, Pilet P, Couillaud S, Faivre A, Daculsi G (1998) Growth hormone-loaded marcroporous calcium phosphate ceramic: in vitro biopharmaceutical characterization and preliminary in vivo study. J Biomed Mater Res 24:47–63
58. Hollinger J (1993) Strategies for regenerating bone of the craniofacial complex. Bone 14:575–580
59. Agrawal CM, Best J, Heckman JD, Boyan BD (1995) Protein release kinetics of a biodegradable implant for fracture non-unions. Biomaterials 18:1255–1280
60. Meng-Hai B, Xing-Yan L, Bao-Feng G, Chao Y, Dong-An C (1996) An implant of a composite of bovine bone morphogenetic protein and plaster of Paris for treatment of femoral shaft non-unions. Int Surg 81:390–392
61. Kieswetter K, Schwartz Z, Dean DD, Boyan BD (1996) The role of implant surface characteristics in the healing of bone. Crit Rev Oral Biol Med 7:329–345

Ti–Ni–Mo Shape-Memory Alloys for Medical Applications

Tae-Hyun Nam

1 Introduction

Near equiatomic Ti–Ni shape memory alloys have been known to transform from the B2 (cubic) parent phase to the B19' (monoclinic) martensitic phase. Elongation transformation and hysteresis transformation associated with the B2–B19' transformations were reported to be large. The former is about 7% and the latter is about 50 K [1, 2]. In contrast to the B2–B19' transformation, elongation and hysteresis transformations associated with the B2–R (rhombohedral) transformation, were reported to be small. The former is about 0.8% and the latter is about 2 K [3]. The B2–R transformation is known to be induced as follows:

1. thermal cycling the near equiatomic Ti–Ni alloys [4],
2. annealing the Ni-rich Ti–Ni alloys [5],
3. thermo-mechanically treating the near equiatomic Ti–Ni alloys [6],
4. adding Al, Fe, Mo to the near equiatomic Ti–Ni alloys [7–9].

The B2–R transformation is of great promise for medical applications of Ti–Ni shape-memory alloys because of its small transformation hysteresis and constancy in transformation temperature. The small transformation hysteresis means a rapid response to an environmental change, which is an important aspect for medical applications. In many shape memory alloy implants, shape recovery is required to occur near the temperature of human body. Therefore the reverse transformation finish temperature (Af) of shape memory implants should be near the body temperature. The above mentioned transformation behaviors, i.e., the B2–B19' and the B2–R can meet this requirement. In order to make medical implements using shape memory alloys, however, many metallurgical processes, i.e., casting, hot working, cold working and heat treatment, are involved. During the processes, especially cold working and heat treatment, transformation temperatures of the B2–B19' transformation change largely [6]. In contrast to the B2–B19' transformation, during the processes, changes in transformation temperatures of the B2–R transformation are relatively small [6]. This means that it is most convenient to use the B2–R transformation for making medical implements of shape memory alloys.

As mentioned previously, there are many methods for inducing the B2–R transformation in Ti–Ni based alloys. Among them, Ti–Ni–Mo alloys seem to be very attractive for medical applications because of their superior corrosion resistance. According to Gunter [9], the corrosion resistance of Ti–Ni–Mo alloys is much better than pure Ti, which is widely used as an implant material. In this chapter, we are concerned with phase transformation behaviors, deformation characteristics and shape memory characteristics of Ti–Ni–Mo alloys.

2
Phase Transformation Behaviors of Ti–Ni–Mo Alloys

Figure 1 shows differential scanning calorimetry (DSC) curves of Ti–Ni–Mo shape-memory alloys. Specimens were annealed at 1123 K for 3.6 ks, and then quenched in iced water. Figures 1a and 1b are DSC curves of 51Ti–48.5Ni–0.5Mo and 51Ti–48.3Ni–0.7Mo alloys, respectively. Two peaks separated clearly on cooling curves in both alloys are seen, while they are not separated clearly on heating curves. In order to explain the two peaks on cooling curves, X-ray diffractions were carried out, and the results obtained are shown in Figure 2. Diffraction peaks corresponding to the R phase are seen in the pattern obtained at 298 K.

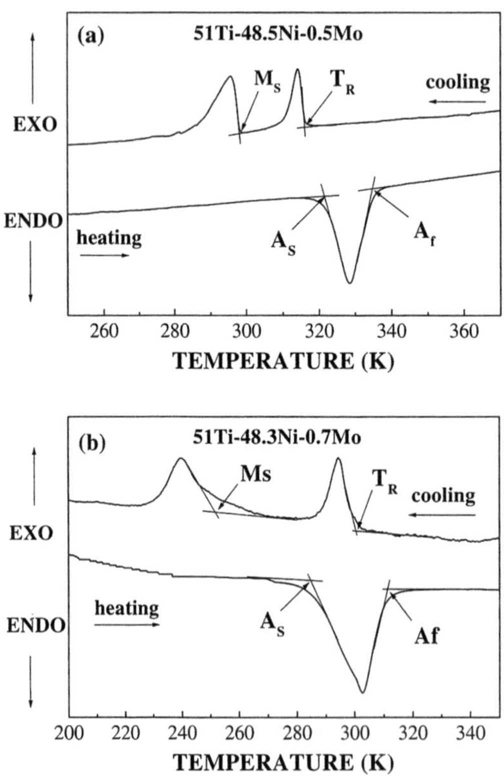

Fig. 1a,b. Differential scanning-calorimetry curves of Ti–Ni–Mo alloys

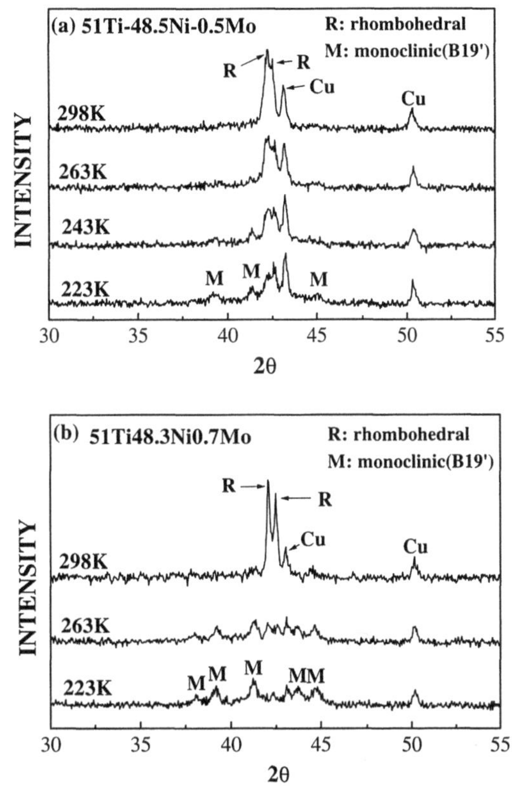

Fig. 2a,b. X-ray diffraction patterns of Ti–Ni–Mo alloys

With decreasing temperature, the intensity of the R-phase diffraction peaks decreases, while the B19' martensite increases. The two peaks appeared in cooling curves of Figure 1 and, therefore, are ascribed to the B2–R and the R–B19' transformations, respectively. Figure 3 shows transmission electron micrographs of the 51Ti–48.3Ni–0.7Mo alloy. Figure 3a is a bright field image where three kinds of R phase variants are seen, and Figure 3b is an electron diffraction pattern which is a typical one for the R phase with the zone axis of [10] B2. From Figures 1–3, it is concluded that Ti–Ni–Mo alloys transform in two stages on cooling, i.e., the B2–R and then R–B19'.

Figure 4 shows changes in transformation temperatures of Ti–Ni–Mo alloys by thermo-mechanical treatment conditions. For comparison, purposes, for the 51Ti–49Ni binary alloy are shown also in the figure. Thermo-mechanical treatments were carried out by 25% cold drawing, and then heat treating at 623–1023 K. As seen in the figure, Af changes largely in the Ti-Ni binary and 0.5 atomic % Mo alloys with heat treatment temperature. In the 51Ti–48.3Ni–0.7Mo alloy, however, Af remains constant at about 310 K, which is the temperature of human body.

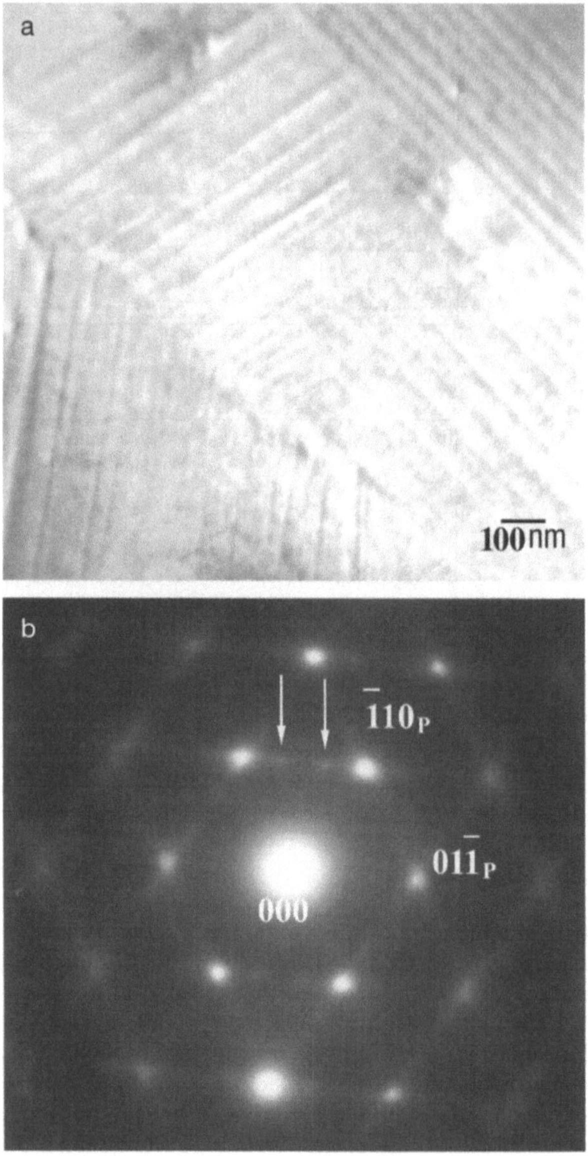

Fig. 3a, b. A bright-field image of the R phase (**a**), and corresponding electron-diffraction pattern (**b**)

For medical applications of Ti–Ni based shape memory alloys, for example staples, clamps, stents et al. [11], they should be deformed into wire. Since strain hardening exponent of Ti–Ni based shape memory alloys is very high, intermediate annealing during deformation process is necessary. Transformation temperatures of binary Ti–Ni alloys change largely during deformation and intermediate annealing process. In order to apply binary Ti–Ni shape-memory alloys to medical field, very precise control of deformation and intermediate

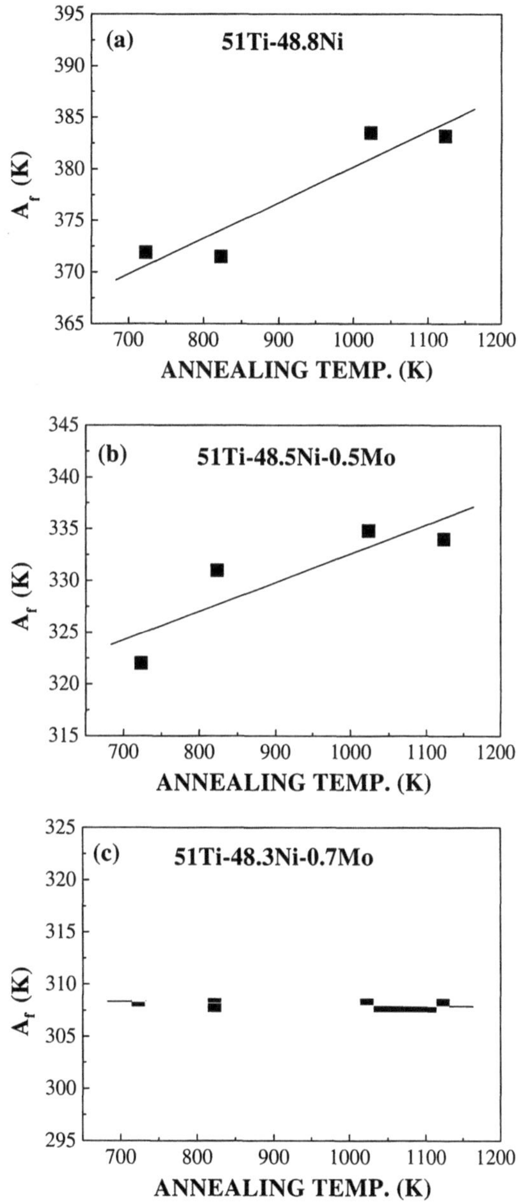

Fig. 4a–c. Relationships between annealing temperatures and Af in Ti–Ni and Ti–Ni–Mo alloys

annealing conditions should be done, since the Af of the alloys is very sensitive to the amount of cold working and annealing temperature. As seen in Figure 4, however, Af of the 51Ti–48.3Ni–0.7Mo alloy does not depend on the conditions of deformation and annealing temperature. Therefore, it is concluded that the 51Ti–48.3Ni–0.7Mo alloy is very suitable to medical applications.

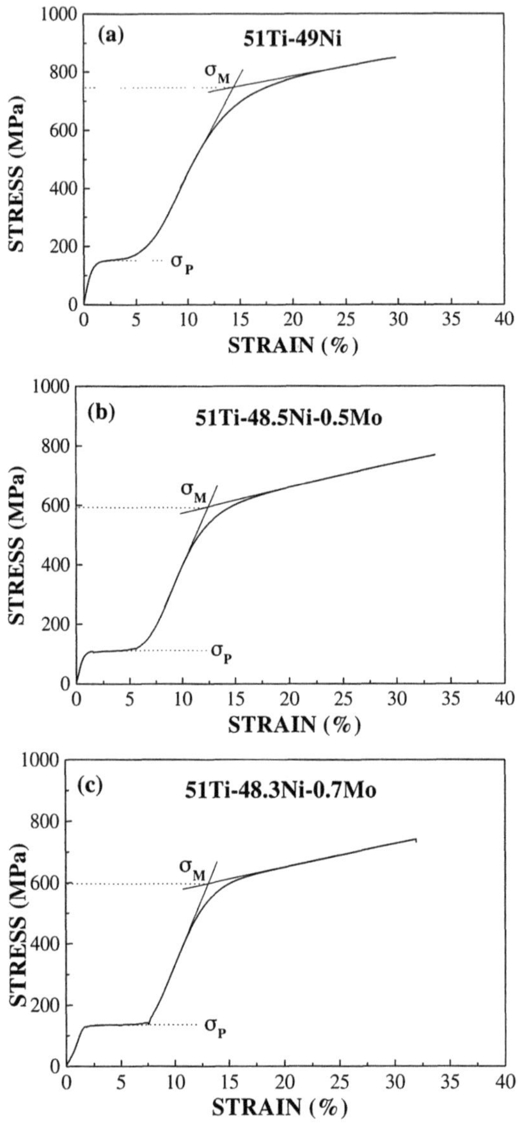

Fig. 5a–c. Stress versus strain curves of Ti–Ni and Ti–Ni–Mo alloys

3
Deformation Characteristics of Ti–Ni–Mo Alloys

Typical stress vs. strain curves obtained at 298 K of Ti–Ni binary and Ti–Ni–Mo ternary alloys are shown in Figure 5. All specimens were cold drawn by 25%, and then annealed at 823 K for 0.6 ks. At 298 K, the Ti–Ni binary alloy is transformed into the B19' martensite completely before loading and so the plateau in the stress vs. strain curve corresponds to a rearrangement of the B19' martensite variants.

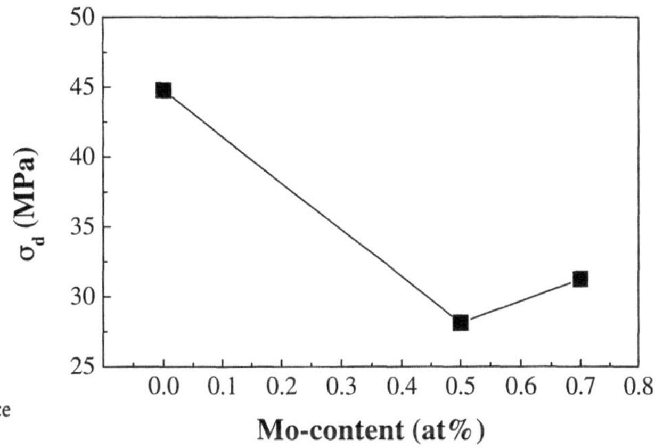

Fig. 6. Mo-content dependence of wire-drawing stress

In the 0.5 atomic % Mo alloy, only the R phase exists before loading and so the plateau in the curve corresponds to the stress induced R–B19' transformation. In the 0.7 atomic % Mo alloy, the B2 parent phase and the R phase co-exist before loading and so the plateau in the curve corresponds to the stress-induced B2–R and R–B19' transformations. As seen in the figure, the stress (σ_p) corresponding to the plateau in stress vs. strain curve of the Ti–Ni–Mo alloy is lower than that of the Ti–Ni binary alloy. It is a well-known fact that the stress required the rearrangement of martensite variants is much lower than the stress required to induce a martensitic transformation. As mentioned before, σ_p in the stress vs. strain curve of the Ti–Ni–Mo alloy is the stress required to induce the B19' martensitic transformation, while σ_p in the stress vs. strain curve of the Ti–Ni alloy is for the rearrangements of martensite variants. This means that Ti–Ni–Mo alloys are deformed more easily than the Ti–Ni binary alloys.

On increasing applied stress over the stress corresponding to the plateau, the thermally and/or stress induced B19' martensite is deformed elastically, and then the macroscopic plastic deformation starts to occur at the stress of σ_M. While wire drawing's being done, large amounts of plastic deformation by slip occur. With the above mentioned three alloys being drawn at room temperature, i.e., cold drawing, slip deformation should occur by plastic deformation of the B19' martensite. When examining stress vs. strain curves in Figure 5, it is found that the σ_M of the Ti–Ni–Mo alloys is lower than the σ_M of the Ti–Ni binary alloy. From these results, it is expected that the stress (σ_d) required for wire drawing of the Ti–Ni–Mo alloys is lower than that of the Ti–Ni binary alloys. Figure 6 shows the σ_d of the Ti–Ni and Ti–Ni–Mo alloys. It is clear that the σ_d of the Ti–Ni alloy is larger than that of the Ti–Ni–Mo alloy. Therefore, it is concluded that the workability of Ti–Ni alloys is improved largely by Mo addition. By applying tensile load to specimens, they elongate about 30% as can be seen in Figure 5, irrespective of alloy composition.

As mentioned previously, since the strain hardening exponent of Ti–Ni based alloys is very large, an intermediate annealing process to release internal stress

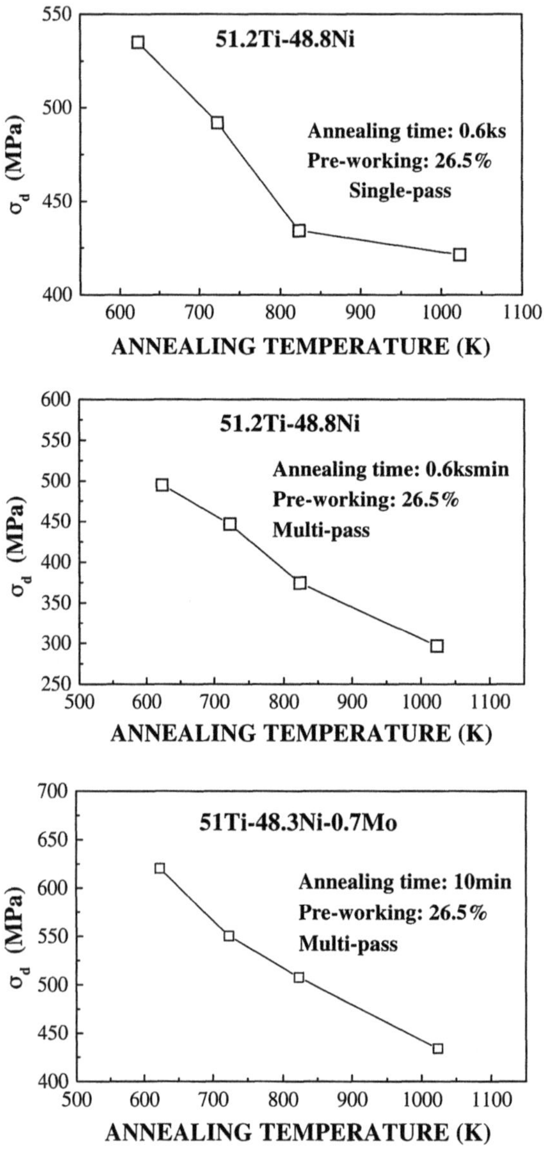

Fig. 7a–c. Annealing-temperature dependence of wire-drawing stress in Ti–Ni and Ti–Ni–Mo alloys

introduced by cold drawing is necessary during wire drawing. The amount of internal stress released during the intermediate annealing process depends on annealing temperature and time. In Figures 7 and 8, σ_d is plotted against annealing temperature and time, respectively. As seen in Figure 7, σ_d decreases with increasing annealing temperature, and σ_d of a multi-pass drawing is smaller than that of a single-pass drawing at the same annealing temperature. Also, as seen in

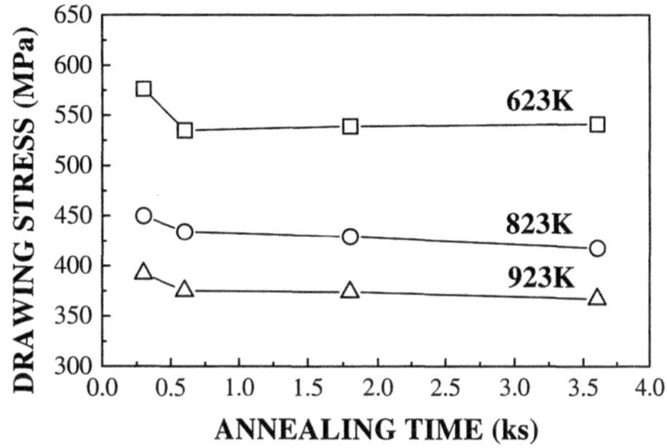

Fig. 8. Annealing-time dependence of wire-drawing stress

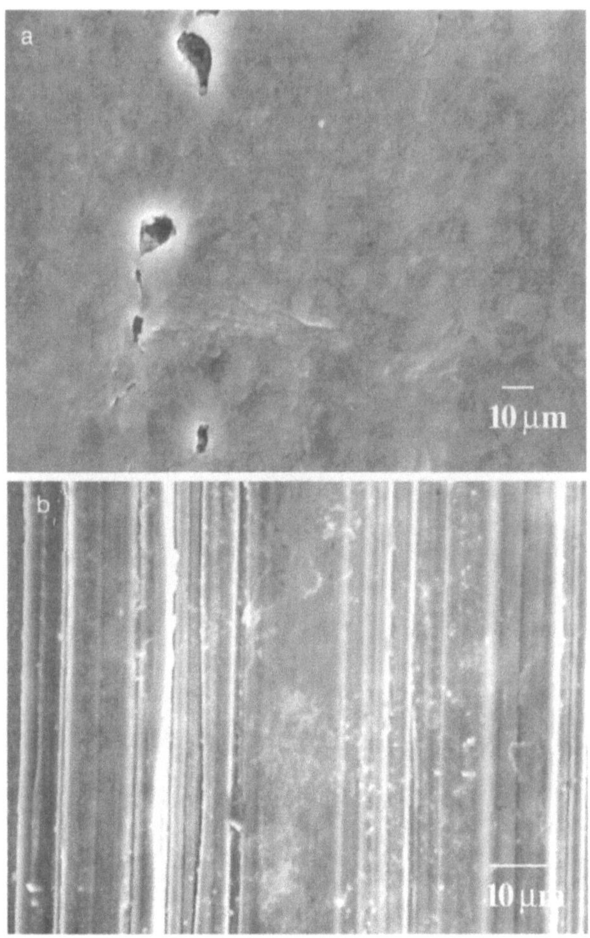

Fig. 9a–d. Scanning electron micrographs of a Ti–Ni–Mo alloy showing an oxide film formed by annealing in air

Figure 8, σ_d decreases with increasing annealing time up to 0.6 ks, and then appears to keep constant with further increasing annealing time. From Figures 7 and 8, we can conclude tentatively that a high annealing temperature above 823 K with about 0.6 ks is desirable for a good drawability.

The intermediate annealing process is generally made in air, and so oxidation on the surface of Ti–Ni alloys is inevitable. It was reported that an oxide film on Ti–Ni alloys is detrimental to their shape memory effect and pseudoelasticity [10]. Oxidation behaviors depend on annealing temperature and time [12]. Therefore, it is necessary to investigate oxidation behaviors of Ti–Ni based alloys to determine an optimum wire drawing condition. The surface morphologies of the 51Ti–48.3Ni–0.7Mo alloy obtained by scanning electron microscopy are shown in Figure 9. Figure 9a is a micrograph of the alloy annealed at 823 K for 0.6 ks. It seems that the surface of the alloy is covered a dense oxide film. After being annealed, the alloy was drawn by 20% at room temperature. Figure 9b is a micro-

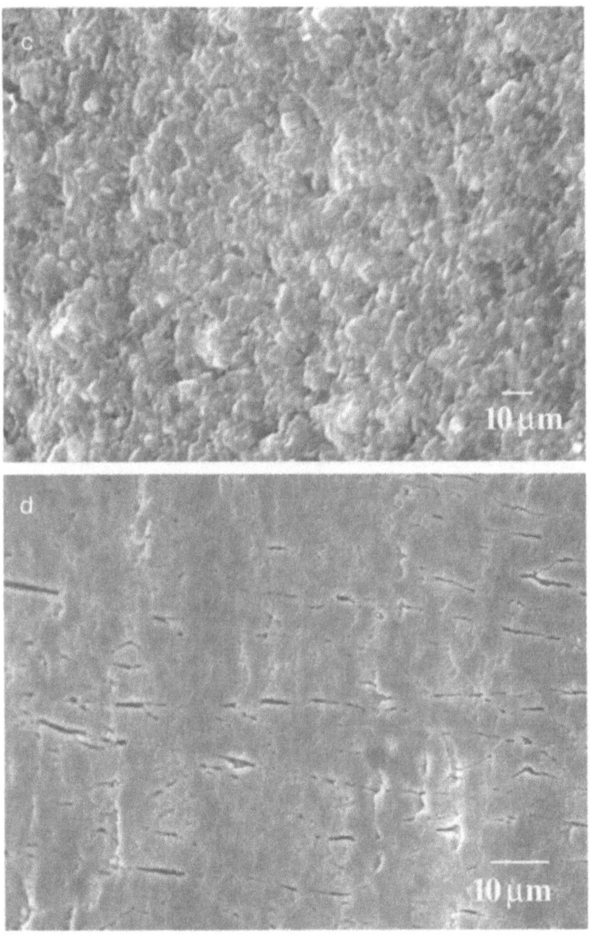

Figure 9c, d

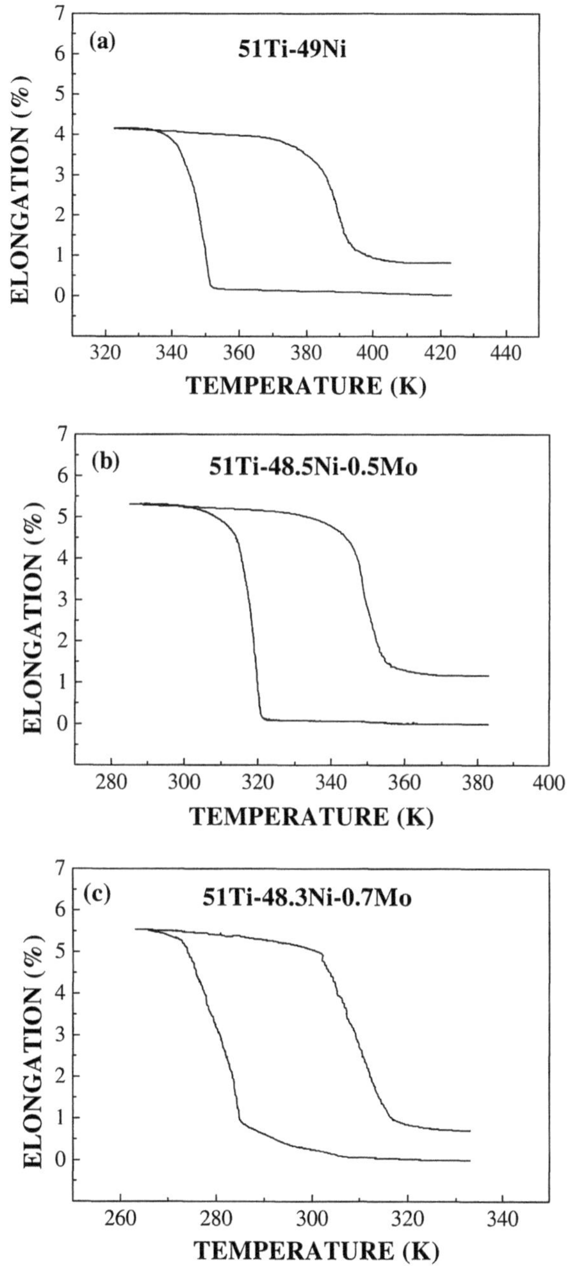

Fig. 10a–f. Elongation vs. temperature curves of Ti–Ni and Ti–Ni–Mo alloys

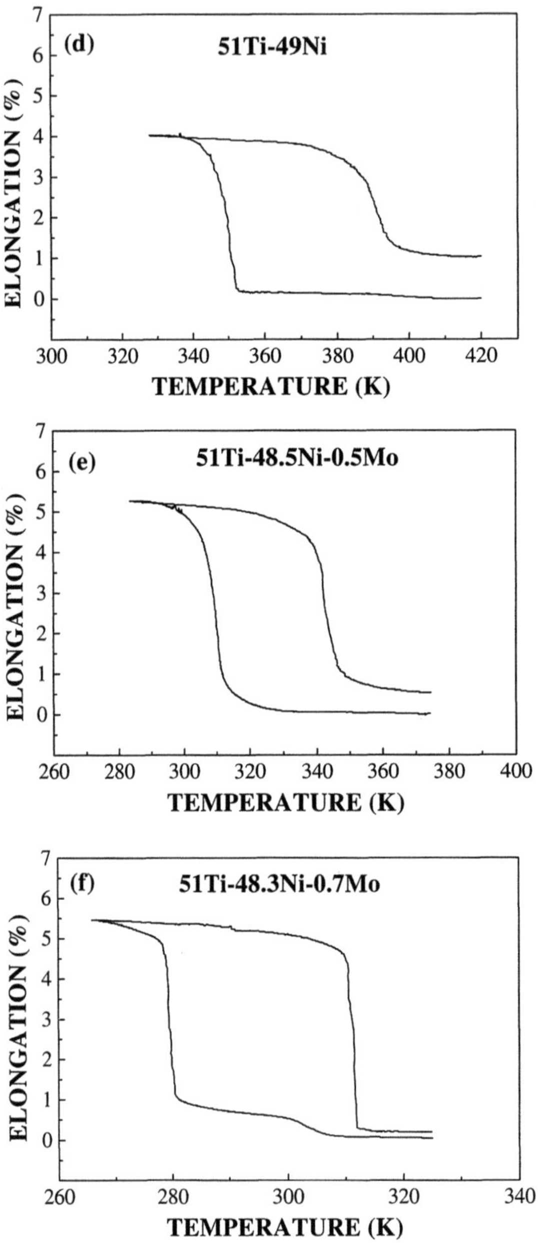

(d)
51Ti-49Ni
ELONGATION (%)
0 1 2 3 4 5 6 7
300 320 340 360 380 400 420
TEMPERATURE (K)
(e)
51Ti-48.5Ni-0.5Mo
ELONGATION (%)
0 1 2 3 4 5 6 7
260 280 300 320 340 360 380 400
TEMPERATURE (K)
(f)
51Ti-48.3Ni-0.7Mo
ELONGATION (%)
0 1 2 3 4 5 6 7
260 280 300 320 340
TEMPERATURE (K)

Figure 10d–f

graph obtained from the alloy drawn by 20%. It should be noted here that most of the oxide film formed during the annealing process is removed by cold drawing. The lines aligned along the vertical direction (drawing direction) in the figure are scratches formed during cold drawing. Figure 9c is a micrograph of the alloy annealed at 1023 K for 0.6 ks. A porous oxide film seems to cover the entire surface of the alloy. As seen in Figure 9d, the porous oxide film is removed by cold drawing. However, a dense oxide film remains even after cold drawing, although some microcracks exist. That is, an oxide film made by annealing at 823 K is removed, while the one made by annealing at 1023 K is not removed by cold drawing. The unremoved oxide film is detrimental to the shape memory effect and pseudoelasticity of Ti–Ni–Mo shape-memory alloys.

In order to determine the optimum wire drawing condition of Ti–Ni–Mo shape-memory alloys for medical application, transformation temperature, drawability and shape memory effect should be considered simultaneously. As mentioned before, the Af of the 51Ti–48.3Ni–0.7Mo alloy is near the human body temperature and it does not depend on the intermediate annealing condition. Although drawability increases with increasing intermediate annealing temperature, an oxide film, which is detrimental to shape memory effect and pseudoelasticity remains even at annealing temperatures above 823 K. Therefore, it is concluded that the optimum intermediate annealing temperature is 823 K and annealing time is 10 min. On increasing annealing time exceeding 0.6 ks at 823 K, porous oxide film similar to Figure 9c is formed [13].

4 Shape-Memory Characteristics of Ti–Ni–Mo Alloys

Figure 10 shows elongation vs. temperature curves of Ti–Ni and Ti–Ni–Mo alloys obtained from thermal cycling tests under constant load. All curves were obtained under the applied stress of 80 MPa. Curves a, b and c are results from specimens annealed at 1123 K for 3.6 ks. In curves a and b, the large elongation appearing on cooling is due to the B2–B19' transformation and it is recovered partly on heating. The residual elongation, which is not recovered on heating, is due to the plastic deformation introduced during cooling. In curves c, elongation appears in two stages upon cooling. The small elongation is ascribed to the B2–R transformation and the large elongation, to the R–B19' transformation. Irrespective of Mo content, considerably large residual elongation appears. Curves d, e and f are results from specimens annealed at 823 K for 0.6 ks. A one-stage elongation appears in curves d and e, while a two-stage elongation appears in curve f. The 51Ti–49Ni and 51Ti–48.5Ni–0.5Mo alloys annealed at 823 K for 0.6 ks show two-stage transformations, i.e., B2–R–B19' without applied stress. This is because stability of the R phase decreases with increasing applied stress. In fact, the elongation associated with the B2–R transformation appears in the 51Ti–49Ni and 51Ti–48.5Ni–0.5Mo alloys under applied stress less than 60 MPa. On the other hand, the two-stage transformation is shown in 51Ti–48.3Ni–0.7Mo alloys under applied stress of 80 MPa. This means that Mo increases the stability of the R phase in Ti–Ni binary alloys.

5 Summary

Ti–Ni–Mo alloys are very attractive for medical applications because of its relatively small thermal hysteresis, small dependence to transformation temperatures on wire drawing condition, superior corrosion resistance, good workability and shape memory characteristics. For medical applications of Ti–Ni–Mo shape memory alloys, for example staples, clamps, stents etc., they should be deformed into wires. In order to determine the optimum wire drawing condition of Ti–Ni–Mo shape memory alloys, drawability and shape memory effect should be considered simultaneously. The optimum annealing temperature is found to be 823 K, above which an oxide film detrimental to shape memory effect and pseudoelasticity remains after cold drawing, even though the drawability of Ti–Ni–Mo alloys increases with increasing annealing temperature.

Acknowledgement. Shin Han Metal Co. kindly supplied some of the specimens used in this study, and it is greatly acknowledged.

References

1. Todoroki T, Tamura T (1986) Deformation behavior of thermally cycled Ti–Ni alloy coil under an applied stress. J Jpn Inst Metals 50:538–545
2. Eucken S, Duerig TW (1989) The effects of pseudoelastic pre-straining on the tensile behavior and two-way shape memory effect in aged NiTi. Acta Metallica 37:2245–2252
3. Stachowiak GB, McCormic PG (1988) Shape memory behaviour associated with the R and martensitic transformations in a NiTi alloy. Acta Metallica 36:291–297
4. Airoldi G, Rivolta B, Turco C (1986) Heats of transformations as a function of thermal cycling in NiTi alloys. In: Proceedings of ICOMAT-86. pp 691-696
5. Nishida M, Wayman CM (1988) Electron microscopy studies of the "premartensitic" transformations in an aged Ti-51 atomic % Ni shape memory alloy. Metallography 21:255–273
6. Todoroki T, Tamura H (1987) Effect of heat treatment after cold working on the phase transformation in TiNi alloy. Materials Transactions JIM 28:83–94
7. Hwang CM, Meichle M, Salamon MB, Wayman CM (1983) Transformation behaviour of a Ti50Ni47Fe3 alloy. Phil Mag A 47:31–62
8. Edmond KR, Hwang CM (1986) Phase transformations in ternary TiNiX alloys. Scripta Metallurgica 20:733–737
9. Gunter VE (1998) Medical materials and implants with shape memory effect. Tomsk University, Tomsk
10. Lin HC, Wu SK, Yen YC (1995) A study on the wire drawing of TiNi shape memory alloys. In: Proceedings of PRICM-2, Kyongju, pp 1167–1672
11. Lipscomb IP, Nokes LMD (1996) The application of shape memory alloy in medicine. Paston, London
12. Chu CL, Wu SK, Yen YC (1996) Oxidation behavior of equiatomic TiNi alloy in high temperature air environment. Mater Sci Eng A 126:193–200
13. Nam TH, Jung DW, Lee HW (2000) Phase transformation and drawability of Ti–Ni–Mo alloys. Metals & Materials (submitted)

Orthopaedic Applications

Ti–Ni–Mo Shape-Memory Alloys for Medical Applications

Kerong Dai

1 Introduction

The unique shape memory effect (SME) of shape memory alloy (SMA) was first described by Buehler in 1965 and was then employed in industry. The application of SMA in the medical field started in orthodontics with the extracorporeal application of orthodontic wires, at this time the superelasticity of the material, not the shape memory effect was the property that was looked for. Intracorporeal application of SMA in humans began in orthopaedics, where the progress was the most rapid and the technology best developed.

On the basis of a series of fundamental studies [1–3], Dai and his colleagues first developed shape memory compression staples and applied them inside the human body for the treatment of transarticular fracture in 1981 [4]. Soon, the scope of application of NiTi shape memory alloys in orthopaedics increased, and the use of SMA even extended to other clinical aspects, including plastic surgery, oral surgery, gynecology and obstetrics, thoracic surgery, biliary surgery and uro-surgery with satisfactory results [5]. Up to now, there are more than a thousand patients who received SMA implants that remained in their body for even more than 10 years. Good therapeutic effects were registered without adverse reactions. In this chapter, only the application of NiTi shape memory alloys (NTSMA) in orthopaedics will be considered.

2 The Basic Principles and Requirements

2.1 Biocompatibility and Mechanical Properties

The shape-memory effect exists in at least a dozen of metals, among which NTSMA is so far the only one that can be used in the human body. The main reason is that to date, only NTSMA has the biocompatibility required for long-term presence in the human body [5]. Studies show that the NTSMA implant is neither cytotoxic, allergic or genotoxic, nor apparently rejective. It is not carcinogenic and possesses an excellent resistance to corrosion. The histological observations and trace element determinations of the tissues surrounding the NTSMA implants do not show any obvious adverse reactions either [1–3, 6–8].

Table 1. Mechanical properties of NT-2 shape-memory alloy

Property	Value
Density	6.45 g/cm^2
Elastic modulus	56.84 x 10^3 N/mm^2
	79.36 x 10^3 N/mm^2
	83.30 x 10^3 N/mm^2
Poisson's ratio	0.33
Shear strength	882 N/mm^2
Impact strength	19.6 N/cm^2
	U-notched specimen 5
Tensile strength	800–900 MPa
Corrosion speed	0.00059 mm/year in Hank's solution
Magnetic conductivity	<1.000225 Gs/Oe
Fatigue strength, compression-compression (0–480.2 N/cm^2), high frequency fatigue cycles	>2.5 x 10^7
Fatigue strength, constant (489.2 N/cm^2), rotation bending fatigue cycles	>2.5 x 10^7

Compared to other metals such as stainless steel and Co–Cr–Mo alloy, which are often used to make implants in orthopaedics, NTSMA has the further advantages of high strength, low rigidity, low specific gravity and good fatigue resistance (Table 1).

2.2 Transformation and Recovery Temperatures

For SMA implants used in orthopaedics, the transformation temperature is set at 4–7°C, which can be attained by immersing the implant in sterile ice-cold saline. At this temperature, the shape of the device can be changed (not more than 6–8%) without any harm to the metal structure, and it will not cause frostbite in the surrounding tissue when implanted. The recovery temperature is generally 37°C: when it reaches this temperature, the device will totally restore to its original shape.

2.3 The Recovery Force

In the course of warming up the SMA implant, the recovery force increases gradually (Figure 1). Making good use of the recovery force is the key to bringing the features of SMA into full play.

The shape of the implant can be transformed only when the temperature is lowered to 4–7°C. And the extent of transformation in any part of the implant shall not exceed 6–8%, or the shape memory effect of the material will be partially or completely damaged. Hence, the shape transformation of the implant shall not be limited to one spot only. If the above criteria are met, theoretically the transformation and recovery of the implant will be repeated for a million times. But in practice, each implant is only used for once, it is discarded after removal. The reason is that it is very hard to reduce the temperature to the point required by transforma-

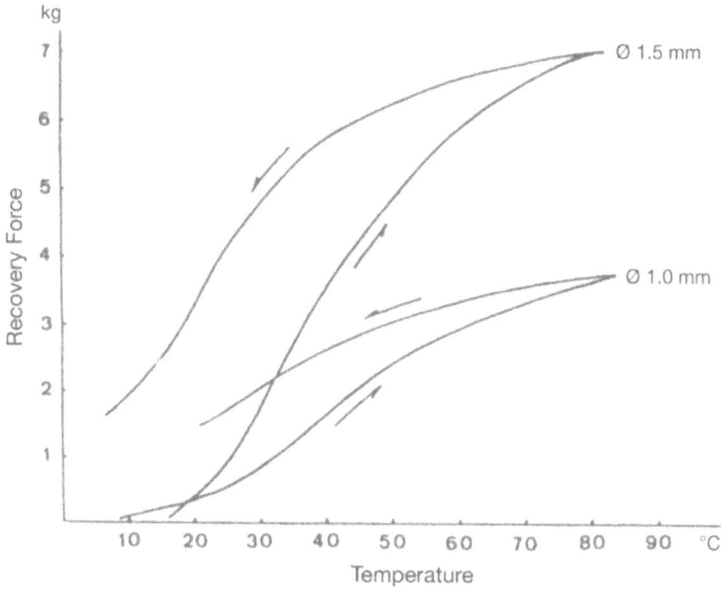

Fig. 1. Relationship between recovery force and temperature (compression staple of 1.5 mm and 1.0 mm diameter wire)

tion when removing the implant, while prizing the implant out, the transformation usually focuses at a certain point leading to some damage to the material.

Recovery force is generated when the SMA implant is warmed up. If there has been no resistance, the recovery force will vanish when the implant has restored its original shape. But when there has been resistance in the course of shape recovery, the implant cannot be restored to its original shape and the recovery force will sustain. Therefore, the above principle must be adhered to when choosing shape memory implants. For instance, the internal diameter of the cross section of an embracing internal fixator should be smaller than the external diameter of the long bone to be fixed; for the intramedullar fixator for long bone, the original extent of opening should be greater than the internal diameter of the medullar cavity, thereby exerting the action of fixation through embracing and expanding.

As shown in Figure 1, when compression staples made of 1.0- and 1.5-mm diameter wires are heated to 37°C, the recovery force is 1.7 kg and 3.0 kg respectively. But when heated to a higher temperature and then reduced to 37°C, the recovery force is 2.5 kg and 5.6 kg respectively. When performing internal fixation of fracture, this hysteresis phenomenon should be utilized to the full in order to bring about the action of fixation by the implant to the maximum.

Since the biocompatibility of NTSMA is good, its product can stay in the body for a rather long time. If it is desirable to remove it, the implant is exposed and immersed in sterile ice-cold saline or wrapped up in ice-cold saline gauze for more than 3 minutes, and then prized out with an elevator. As the elastic modulus of the implant is closely related to temperature, it can be prized out even if the temperature has not yet fallen to 4–7°C.

3
Shape-Memory Implants in the Treatment of Transarticular Fracture

3.1
Compression Staples

In the treatment of transarticular fracture, the basic principles include anatomical reduction, reliable fixation and early ambulation. But the fracture fragments tend to be separated and displaced as a result of pulling by a tendon or a muscle. In addition, the fragments are rather small and the coverage of periarticular soft tissue is thin, so it is rather difficult to apply an ordinary bone plate to the fracture. The shape memory compression staple developed by Dai [4] has its unique advantages in the treatment of transarticular fracture. The staple is U-shaped, having two straight legs (segments to be inserted) connected by a transversal wavelike segment (compressive segment) with included angles of 70°. This design helps to raise the compressive force at the distal ends of the legs to prevent the fracture from eccentric compression and the staple from slipping (Figure 2a). The staple's force of compression is mainly produced in the compressive segment. In practice, the compressive segment is expanded at a low temperature (4°C or below) to reduce the wavelike curvature and increase the length, and at the same time, expand the included angles to 90° in order to elongate the span between the legs (Figures 2b, 3). To do so, the fracture is reduced, expanded span of the staple(s) is measured, and holes are drilled with this distance apart across the distal and proximal segments of the fracture, and then staples are inserted across the fracture line into the prepared holes. The next step is to apply a hot compress of hot saline gauze around the operation site to raise the temperature. The staple(s) will restore to its original shape and generate a recovery force to tightly hold the fracture fragments in place, exerting sustained compression on the fracture and resisting muscular pull or the tension formed in the fracture when the position of the joint is changed. Postoperatively, early exercise is encouraged. Thanks to the small size, simple procedure of application, slight surgical

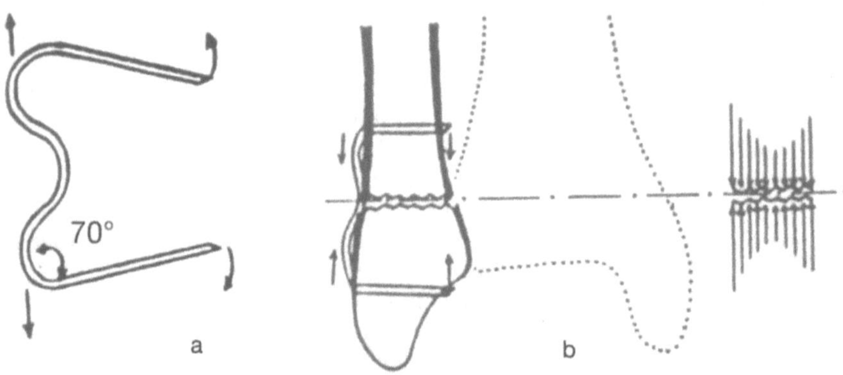

Fig. 2a, b. Schematic diagram of the shape memory compression staple. **b** The staple is distracted at low temperature to fix the fracture; the staple produces recovery force at restoration temperature, exerting compression on both sides of the fracture

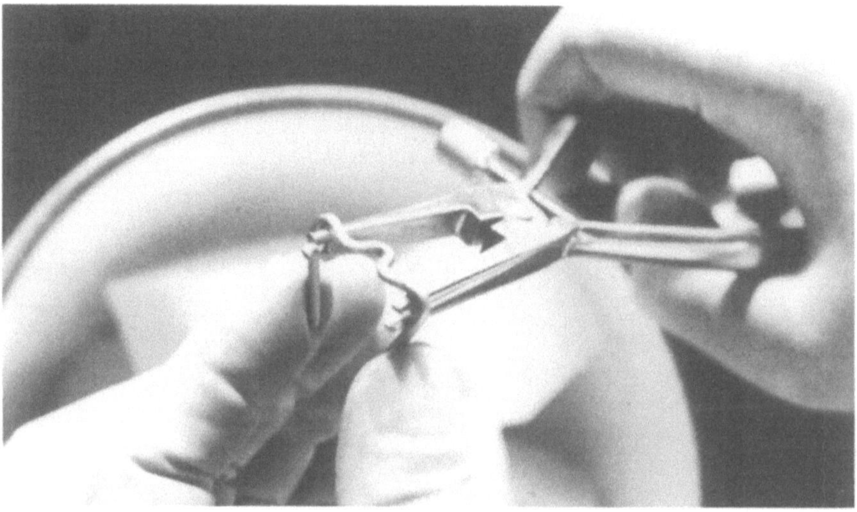

Fig. 3. Compression staple is expanded at low temperature by means of a special device

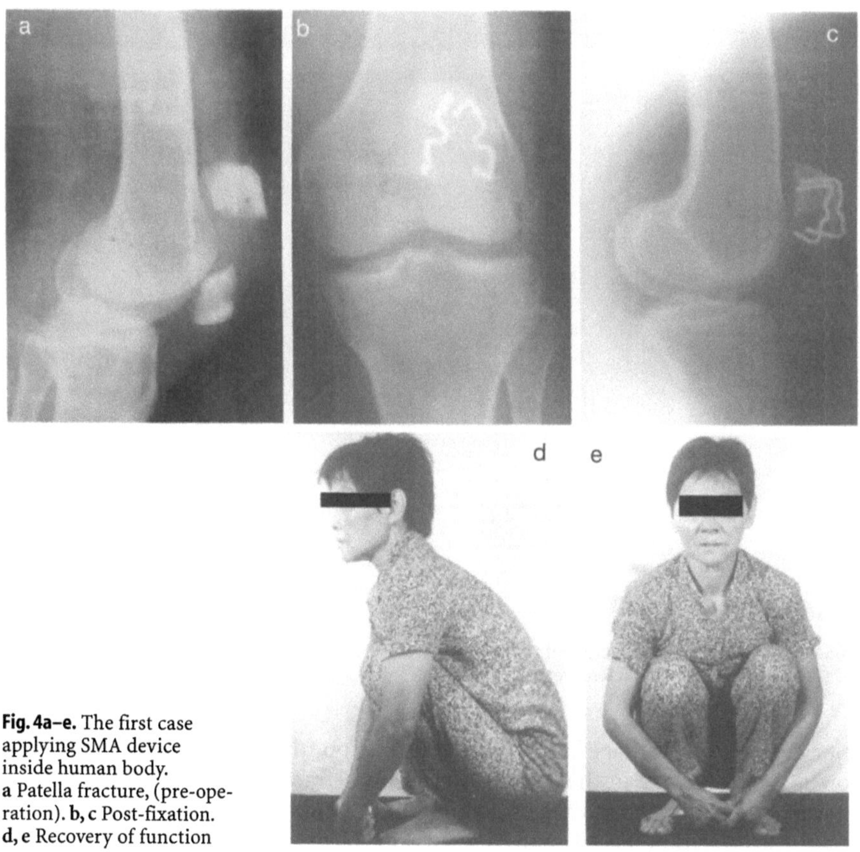

Fig. 4a–e. The first case applying SMA device inside human body. **a** Patella fracture, (pre-operation). **b, c** Post-fixation. **d, e** Recovery of function

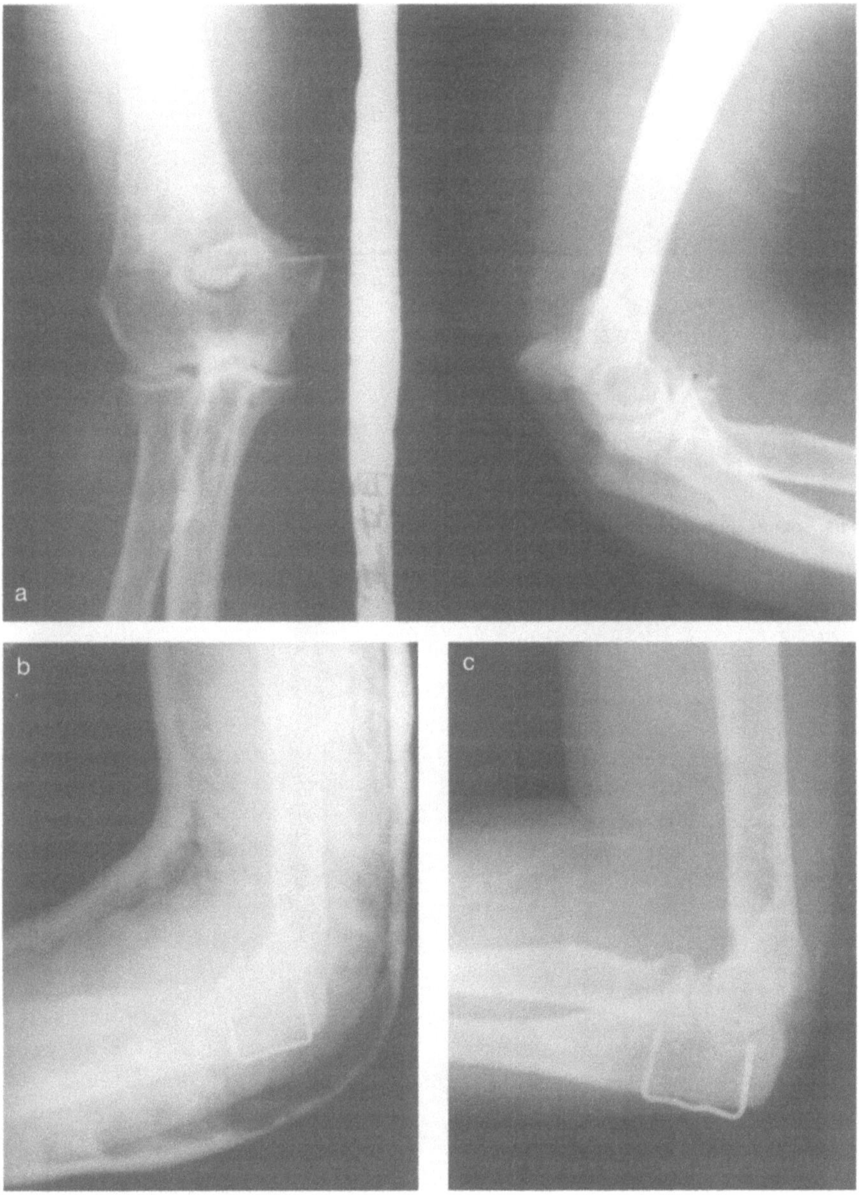

Fig. 5a–d. Radiographs of a fracture of the olecranon. **a** Pre-operative. **b** Post fixation. **c** 2 months post-operation, fracture healing. **d** Recovery of function

injury and the advantage of compressive fixation, the compression staples are now in extensive use in the treatment of patella fracture (Figure 4), fracture of internal and external condyles of femur, fracture of the malleolus, fracture of the neck or greater tubercle of humerus, fracture of medial and lateral condyles of

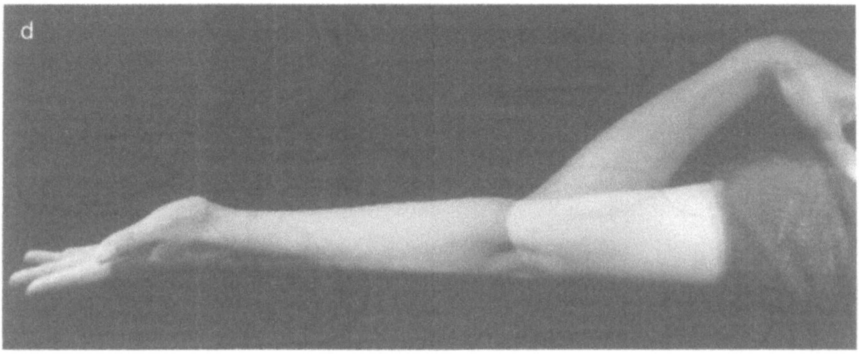

Figure 5d

humerus, olecranal fracture (Figure 5), distal fracture of the radius, transversal and short oblique fractures of the short bones in the hand and the foot. They are also applied to the supracondylar osteotomy of humerus and femur, upper tibial and the first metatarsal osteotomy correction, and the triple arthrodesis of foot. However, the staples are not suitable for the fixation of the fracture of long bone shaft, where the surrounding muscles are rather strong.

3.2
Patellar Fixator

Zhang designed a shape memory internal fixation reductor for patellar fracture (NT-patellar fixator), which has been used with success (Figure 6) [9, 10]. The patellar fixator consists of a few claws and a connecting waist, while the curvature formed by the claw and the connecting waist is lesser than the anatomical curvature of the patellar surface. In clinical use, the claws are gradually stretched at a low temperature in order to reduce the curvature formed by the claws and the waist to make the distance between the claws slightly greater than the longitudinal diameter of patella; after reduction of fracture, and the free ends of the claws are stuck into the space around the patella. Warming up with hot saline then restores the shape of the fixator, generating convergent compression from around the patella to counter the tension of quadriceps and to stabilize the fracture while allowing articular movement. The fixator exerts a compressive force on the fragments to hold them together and thus promotes fracture healing. The device is suitable for several types of patella fractures, its effect being outstanding specifically in the fixation of comminuted fracture of patella (Figures 7, 8).

3.3
The Shape-Memory Screw

The shape memory screw was designed by Zhao and used in the treatment of fresh fractures of femur head [11]. This device looks like a cancellous screw.

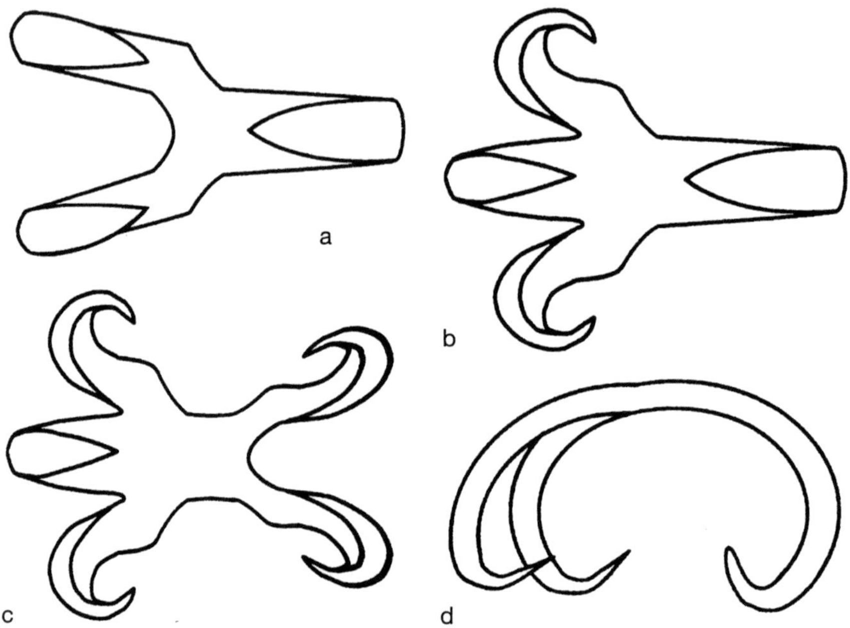

Fig. 6a–d. NT-patellar fixator. **a–c** Different types of fixator. **d** Lateral view.

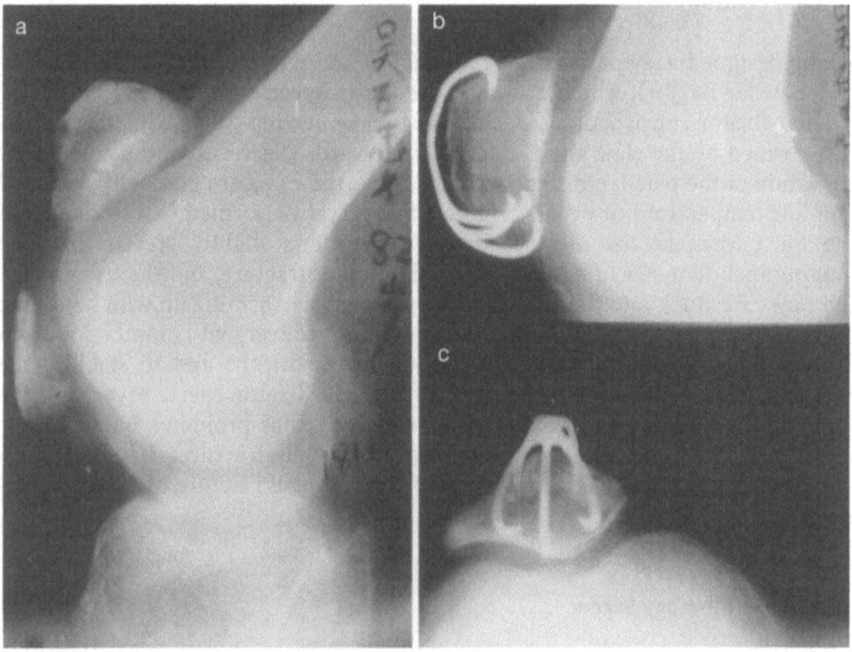

Fig. 7a–c. Radiographs of a transverse fracture of the patella. **a** Pre-operation. **b, c** Post fixation (provided by Chuncai Zhang)

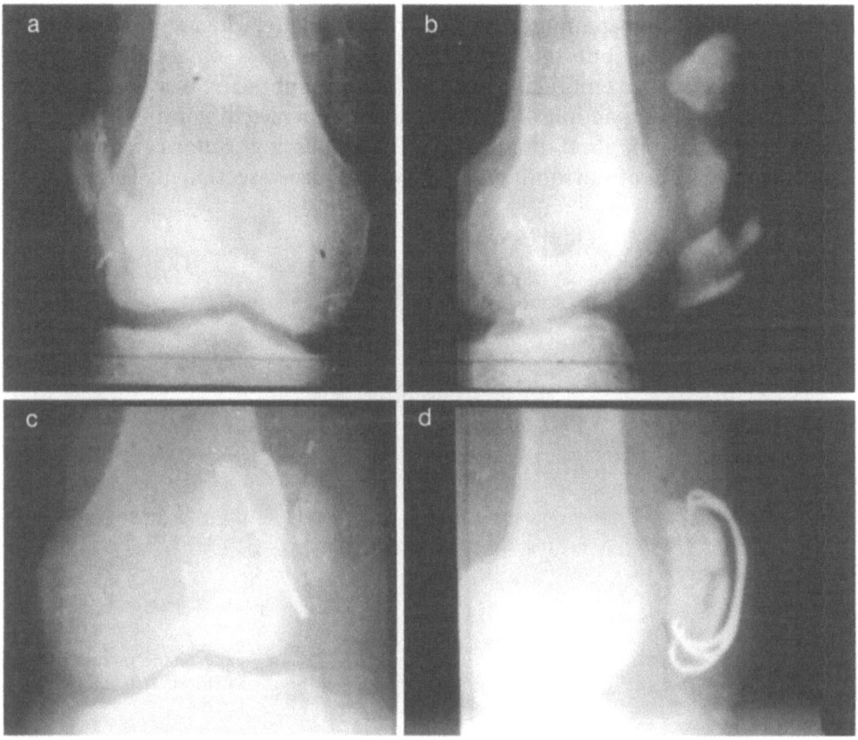

Fig. 8a–d. Radiographs of a comminuted fracture of the patella. **a, b** Pre-operation. **c, d** Post fixation (provided by Chuncai Zhang)

The screw is 5 mm in diameter, with its front part threaded. Half of the front part of the screw is longitudinally split into two halves, which are opened to form a 40°-angle and fix the fracture. Two screws are used. Their halves are pressed together at a low temperature, and then screwed into the femoral neck with the directions of their openings perpendicular to each other. At body temperature, the halves open and fix the fracture. The device can effectively prevent the fracture from displacement and control the rotation of the femur head.

4 Shape-Memory Implants in the Treatment of Long-Bone Shaft Fractures

4.1 Shape-Memory Sawtooth-Arm Embracing Internal Fixator

In addition to reliable fixation, the treatment of long-bone shaft fracture requires effective resistance to the stress of bending, shearing and torsion while retaining certain stress from axial compression in order to stimulate fracture healing and bone remodeling. The common compressive bone plate, when applied properly, offers stable fixation, but the stress shielding effect is obvious. The shape memory sawtooth-arm embracing internal fixator (embracing fixator) developed by Dai has a grasping action of fixation, and has achieved good results in the treatment of long bone shaft fracture [12]. The embracing fixator is made up of a body and a few pairs of sawtooth-arms (Figure 9). Cross-sectionally (Figure 10a),

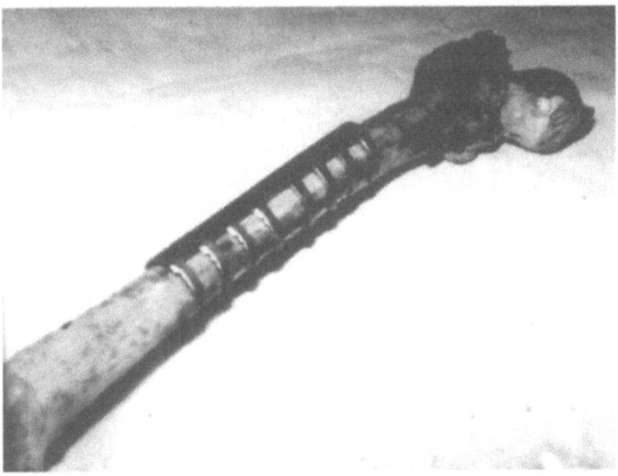

Fig. 9. Shape-memory sawtooth-arm embracing fixator

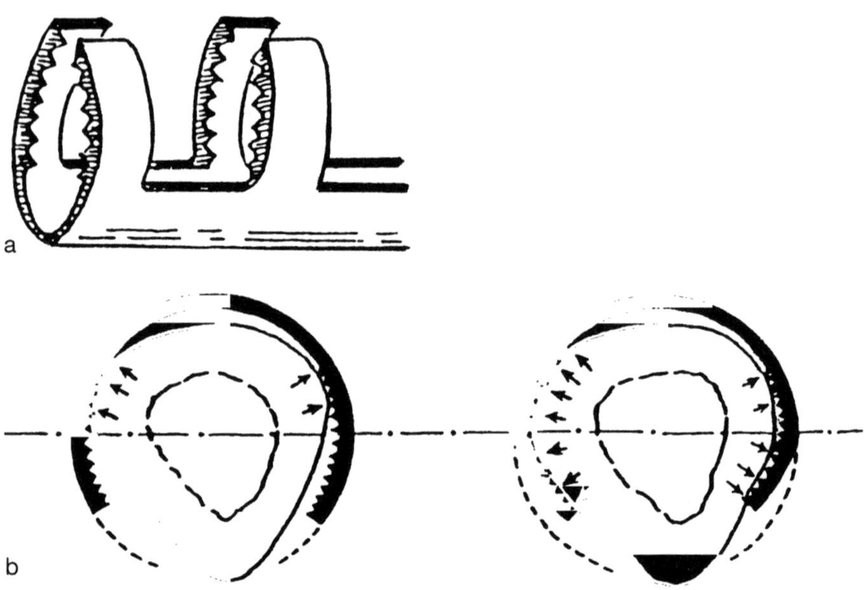

Fig. 10a, b. Cross section of embracing fixator. The portions of the arms beyond the semicircle bend more inwards, so that fixation points are distributed on both sides of the bone (*below, right*)

a

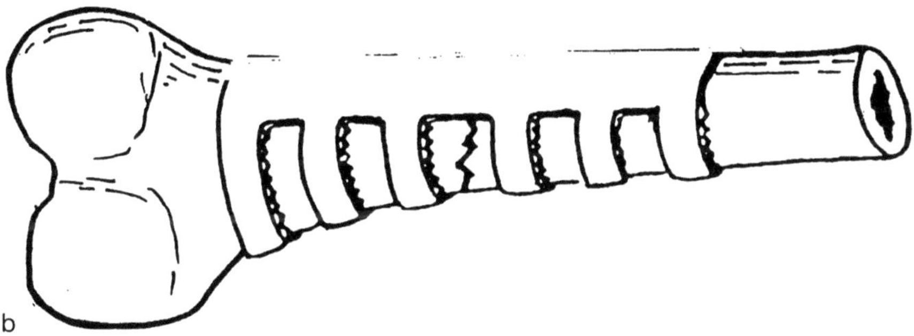
b

Fig. 11a, b. Embracing fixator. **a** Cylindrical type. **b** Conical type

the arms and the body are curved to form an arc of two thirds of a circle. The portions of the arms beyond the semicircle bend more inwards, so that fixation points can be established on the opposite side of the body of the fixator (Figure 10b). The embracing fixator can be cylindrical (Figure 11a) or conical (Figure 11b), indicated for use in the mid-segment and the upper or lower third of the long-bone shaft, respectively. There are different sizes, lengths and widths, with four, six or eight pairs of embracing arms. In practice, the arms are first expanded at a low temperature. After reduction of the fracture, the body of the fixator is placed on the tensile side of the bone, and the temperature is raised with hot saline. With the shape memory effect, the sawtooth-arms close up, grasping the distal and proximal segments tightly. During mechanical testing, the bending strength of the fixation system is similar to that of a bone plate, but lower than of an intramedullar nail. The torsional strength is also similar to that the bone plate, but obviously higher than the intramedullar nail. Stress shielding is markedly decreased compared to bone plating [12–16]. Furthermore, it does not cause any additional damage to the intramedullary blood supply and inner periosteum, and promotes fracture healing. This technique has also been successfully used in the fracture of the extremities, including the femoral shaft (Figure 12), humeral shaft, ulna, radial and clavicular fractures.

The embracing fixator has been shown to be a simple and effective technique in the treatment of periprosthetic femur fracture. Dai and his collaborators applied the sawtooth-arm embracing fixator to fix periprosthetic femoral fracture in 12 cases during or after hip joint replacement (Figure 13). The results are satisfactory and the fixation of the fracture is stable without any external fixation.

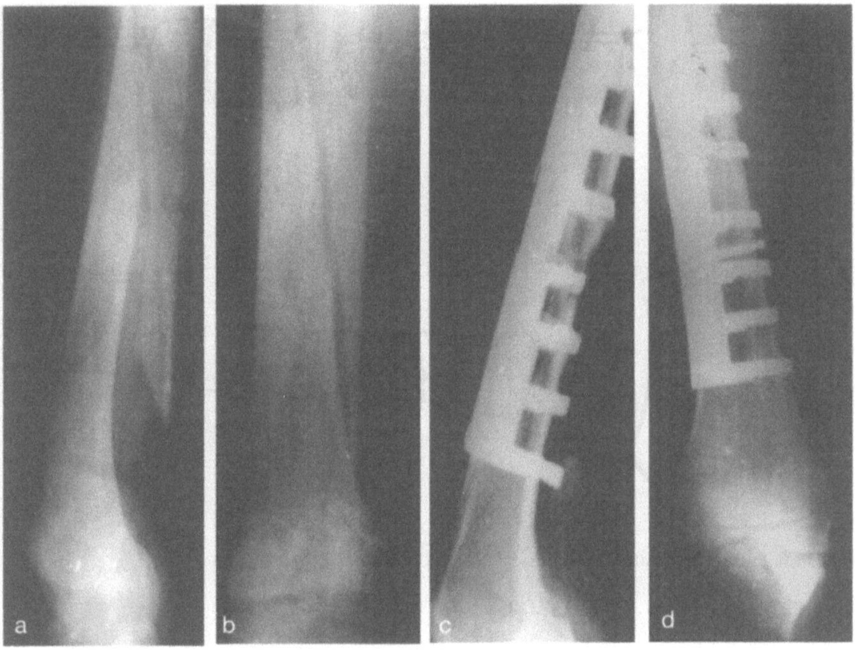

Fig. 12a–e. Radiographs of a spiral femoral fracture. **a, b** Pre-operation. **c, d** Post fixation. **e** Recovery of function

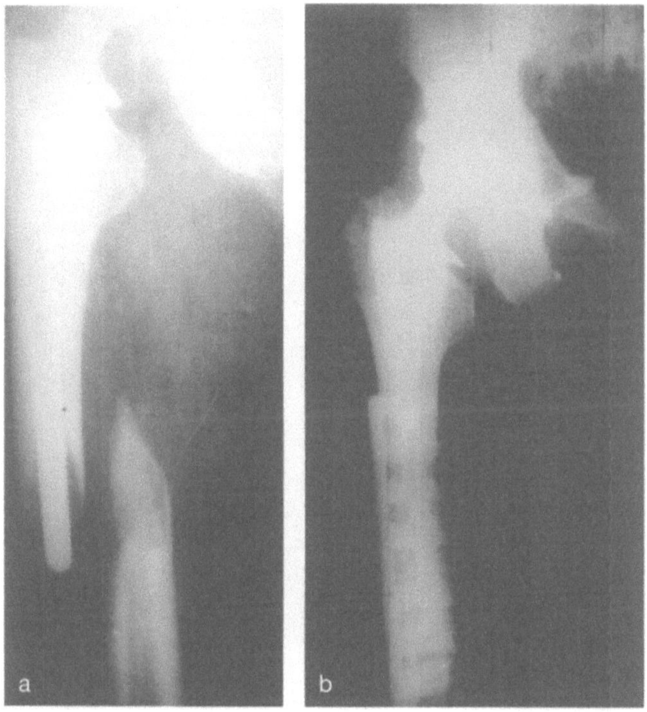

Fig. 13a,b. Periprosthetic fracture of the femur. **a** Pre-operation. **b** Post fixation (provided by Zhirong Li)

4.2
Fork-Like Shape-Memory Intramedullar Nail and Bow-Shaped Compressive Osteo-Connector

Zhang designed and used the fork-like shape memory intramedullar nail to treat humeral fracture [17]. On raising the temperature after intramedullar fixation with the nail placed across the fracture, the bifurcated part opens to prevent rotation, strengthening the stability of the fracture. Zhang further designed a shape memory bow-shaped compressive osteo-connector, which includes two longitudinal compression branches, several semicircular fixation branches and a connection part [18]. It is used in the fixation of the fracture of upper limb bone shaft. The osteo-connector produces a sustained longitudinal compression along the bone shaft, which improves fracture healing (Figure 14).

5
Hand Surgery

The fracture of short tubular bone of the hand is predisposed to displacement due to the pull by tendon, and in most of the cases, surgical internal fixation is

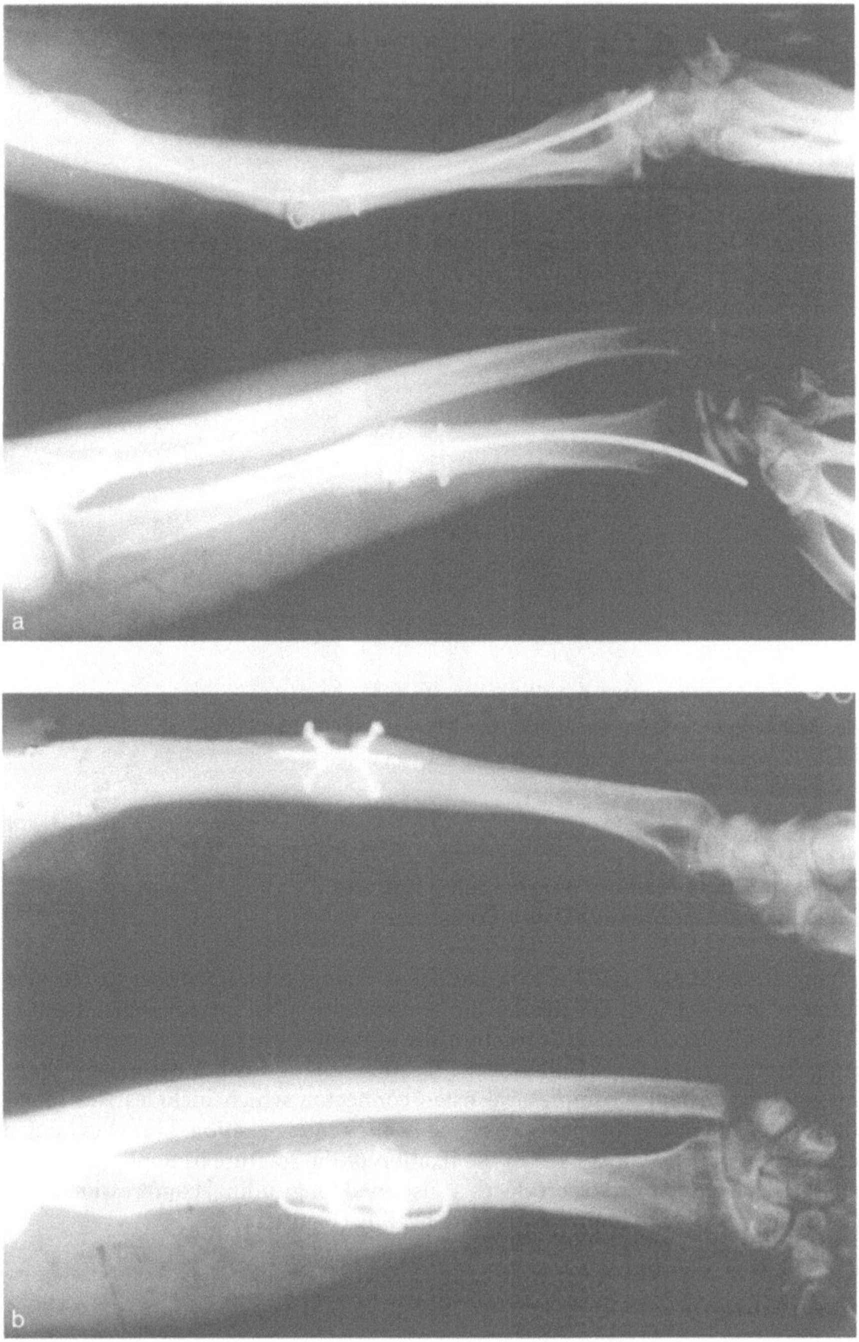

Fig. 14a, b. **a** Non-union of fracture of the radius after an inappropriate treatment. **b** Healing was obtained 4 months after fixation with a bow-shaped compressive osteo-connector (provided by Chuncai Zhang)

necessary. For transverse or short oblique fracture of metacarpal or phalangeal shaft, fixation with a small compression staple will render good results.

5.1 Shape-Memory Compression Plate

Yang designed a shape memory compression plate (with screws) and achieved satisfactory results in the fixation of metacarpal and phalangeal fractures [19]. It is a small bone plate, the two ends of which are bent into hooks perpendicular to the plate. There are two holes in the plate for screws placement. Between the holes the plate is horizontally wavy in shape (Figure 15). Before application, the wavy segment of the plate is stretched out at a low temperature. A hole is drilled on each side of the fracture to insert the hooks. Holes are drilled in the bone through both plate holes and the screws are inserted. With the shape memory effect, the plate returns to its original shape when warmed up, exerting a sustained axial compression at the fracture site.

5.2 Clamping Plate

In addition, Yang designed a clamping plate for internal fixation, which is a small embracing fixator, composed of a trunk and several arcuate arms covering two thirds of a circle [20]. With its shape memory property, it has been used to fix metacarpal and phalangeal shaft fractures of 64 cases with success.

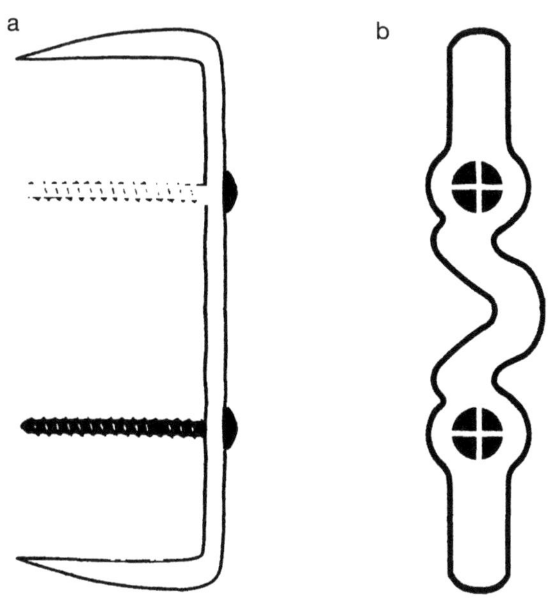

Fig. 15a,b. Shape memory compressive plate. **a** Lateral view. **b** Top view

6 Spinal Surgery

6.1 Ω-Shaped Intravertebral Artificial Joint

In cervical spondylosis, especially with secondary myelopathy type, anterior decompression and invertebral fusion are often necessary in patients with persistent and serious symptoms. The latter may be relieved after surgery, but degeneration will be accelerated or induced in the adjacent disc spaces. Zhao [21] developed a shape memory Ω-shaped intervertebral artificial joint for the cervical spine, and this device serves to replace the traditional fusion with bone grafting and to reconstruct the height and moving function of cervical spine after decompression. In practice, anterior decompression is performed first, and the Ω-shaped bow arms are flattened at low temperature (Figure 16a) and placed in the center of the intervertebral space (Figure 16b). Under the influence of the body temperature, the arms will recover their original shape and expand, maintaining the height of the intervertebral space (Figure 16c), this arthroplasty facilitates the patient's recovery of function, minimizes the degeneration in the adjacent discs and avoids the pain and complications associated with autogenous bone graft collection.

6.2 Shape-Memory Expansion Clamp

Another complication of anterior decompression and fusion with bone grafting is the displacement of the bone graft resulting in a pseudarthrosis, which cannot be eliminated completely through improvements in the shape of bone graft and

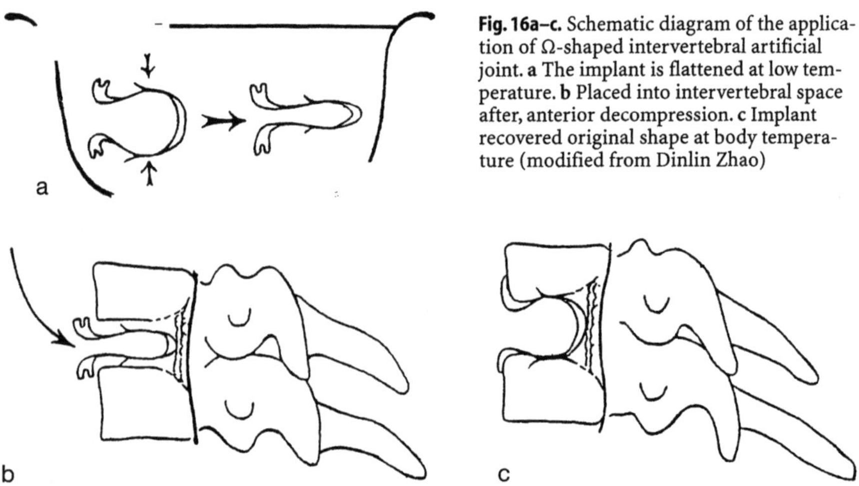

Fig. 16a–c. Schematic diagram of the application of Ω-shaped intervertebral artificial joint. **a** The implant is flattened at low temperature. **b** Placed into intervertebral space after, anterior decompression. **c** Implant recovered original shape at body temperature (modified from Dinlin Zhao)

techniques of fusion. The shape memory expansion clamp designed by Mei in 1990 can effectively prevent the bone graft from displacement during the treatment of a cervical spine injury. It provides good fixation of the annular bone graft following anterior decompression. The expansion clamp consists of two obliquely open arms and a horizontal arm. The two oblique arms are open at a low temperature and inserted in the center of the annular bone graft obtained with a trephine. The arms are rotated 90° and placed in the intervertebral space together with the bone graft. At body temperature, the arms automatically close and produce an expansive force, which effectively fixes the bone graft. With the shape memory clamp, the injury and complications caused by anterior plate fixation and bone graft collection can be avoided, resulting in satisfactory fusion of all 30 cases further reported by Mei [22].

6.3 Shape-Memory Device Used in Scoliosis

Scoliosis can be corrected in many ways, a common shortcoming of the procedures being only one single correction available during the operation. Sustained correction is unavailable and late loss of deformity correction is present. Lu developed the shape memory rod for the correction of scoliosis, the appearance of which resembles the Luque rod [23]. This correction rod is equipped with antirotation stoppers to prevent the rod from rotating. In application, the rod is bent in advance at a low temperature to a certain degree according to the lateral curvature of spine, and then is fixed on the vertebral laminae, on both sides of the spinal column, with wires. By raising the temperature with hot saline, the rod restores its straightness and produces a recovery force to correct the lateral curvature. As there may be some remaining curvature in the spinal column and the SMA rod has not yet restored its exact original straightness, the correction force on the spinal column is maintained, a factor contributing to the sustained postoperative correction of lateral curvature and prevention of the loss of deformity correction. A total of 32 cases of idiopathic scoliosis have been treated. The preoperative Cobb angle averaged 53.5° (35–105°), and the postoperative average was 25.8° (9–67°), the ratio of angular correction being 53.8% (23.4–75%). In average, the follow-up lasted 17.2 months (1–36 months), and no obvious loss of angular correction was found.

7 Arthroplasty

7.1 Shape-Memory Double-Cup Prosthesis of Hip

The concept of surface replacement of hip joint was initiated by Charnley [24], and Paltrini reported six cases in 1971. This procedure removes a small portion of bones, the anatomical relation and stress distribution of the hip joint remain near-normal, the implanted material are few, leaving sufficient room for secondary remedial surgery. This is quite suitable for young patients with articular cartilage and subchondral disorders. However, clinical practice shows that there are quite a few complications, including loosening or displacement of the prosthesis, ischemic necrosis of femoral head, and late fracture of femoral neck. The main cause is the shape of the prosthesis: with the traditional prosthesis, the femoral head cup is like a hemisphere, so that it can be placed on the femoral head, which is predisposed to loosening, rotation and luxation (Figure 17a), with the ideal prosthesis, where the opening is larger than its base, is possible to limit

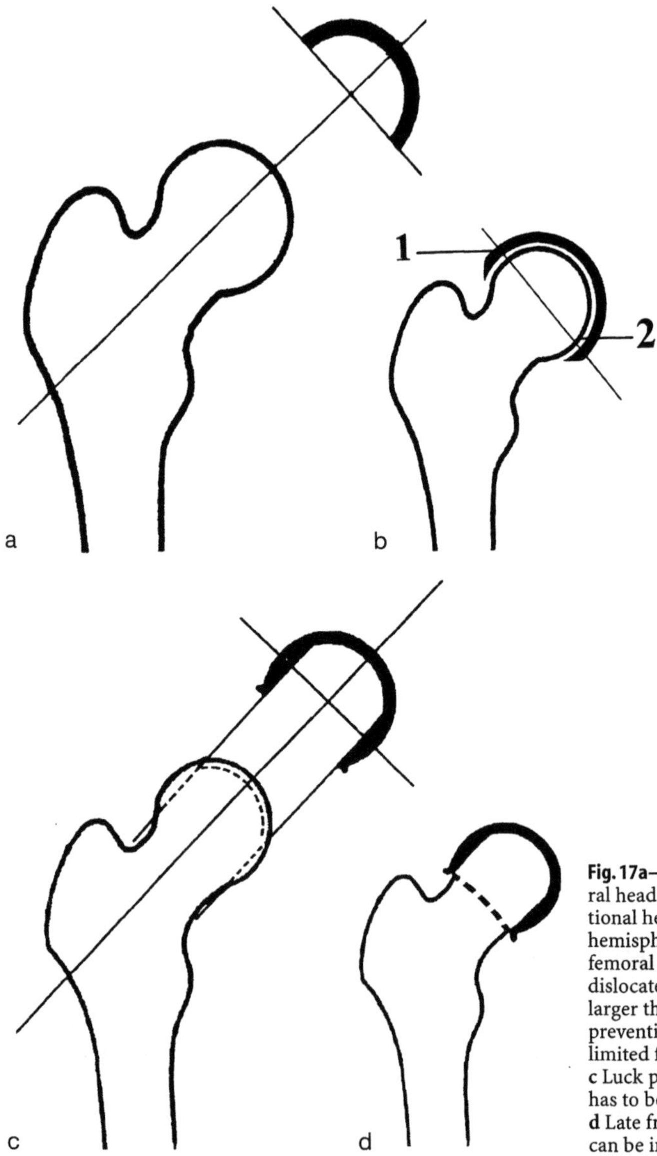

Fig. 17a–d. Shape memory femoral head cup prosthesis. **a** Traditional head prosthesis like a hemisphere: easy to place on the femoral head and also easy to be dislocated. **b** Ideal prosthesis: larger than a hemisphere for preventing dislocation and limited femoral head excision. **c** Luck prosthesis: femoral head has to be shaped like a cylinder. **d** Late fracture of femoral neck can be induced

femoral head excision and also to prevent dislocation (Figure 17b). In the pursuit of a solution, Luck used a femoral head cup of greater depth, and his design was cylindrical on the inside of the cup to reduce the incidence of internal or external rotation of the cup (Figure 17c). In order to do this, the femoral head had to be shaped like a cylinder, which resulted in annular damage of the bone cortex at the junction of head and neck, leading to some stress concentration and then late fracture of femoral neck (Figure 17d). Dai developed the SMA femoral head cup, which broke through the limits in the design that the opening of the cup should be equal to or larger than the diameter of the hemisphere [25]. The opening of the cup held out six anchor flukes (Figure 18). With the shape memory effect, the flukes could be expanded when implanted and closed after implantation. The procedure begins with the flukes of the cup expansion at a low temperature (Figure 19) and placing it on the femoral head with the cartilaginous surface excised. The head slightly trimmed if necessary. Raising the temperature with hot saline will close the flukes, which firmly fix the cup prosthesis on the femoral

Fig. 18. Shape memory double cup prosthesis of hip

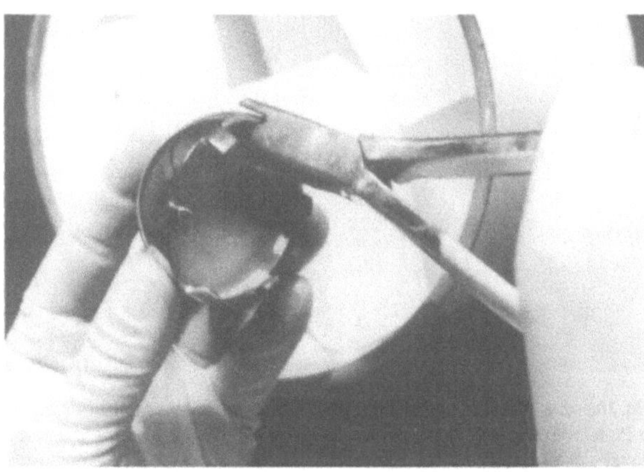

Fig. 19. Flukes of the prosthesis are expanded by special instrument

head, effectively preventing the head cup from luxation. The acetabular cup is made of ultra-high molecular weight polyethylene and fixed with bone cement. In the 23 cases followed up for over 10 years, two were operated again because of loosening of the acetabular cup (Figure 20a). No head cup loosening or fracture

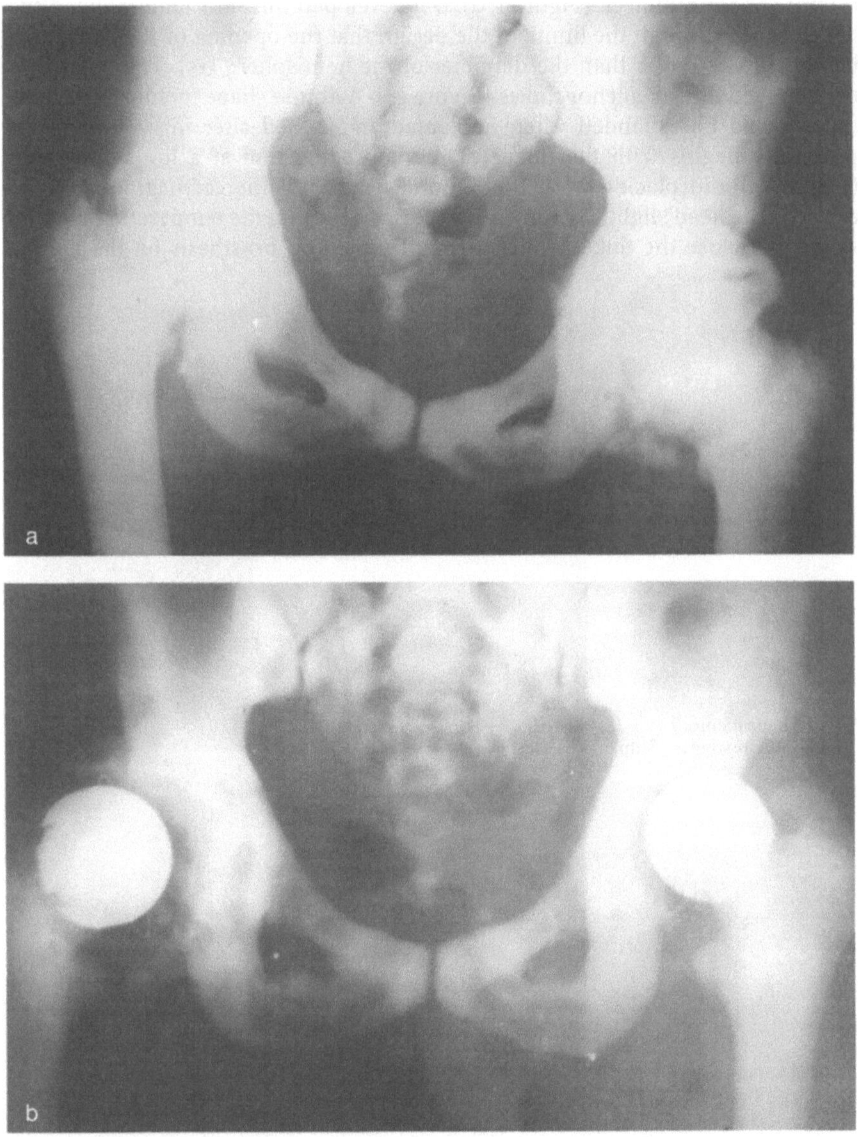

Fig. 20a–d. **a** Male, 20 years old: marble disease, stiffness of both hip joints, medullar cavity was closed. **b** Post-operation of bilateral surface replacement by shape memory double-cup prosthesis. **c, d** Recovery of function

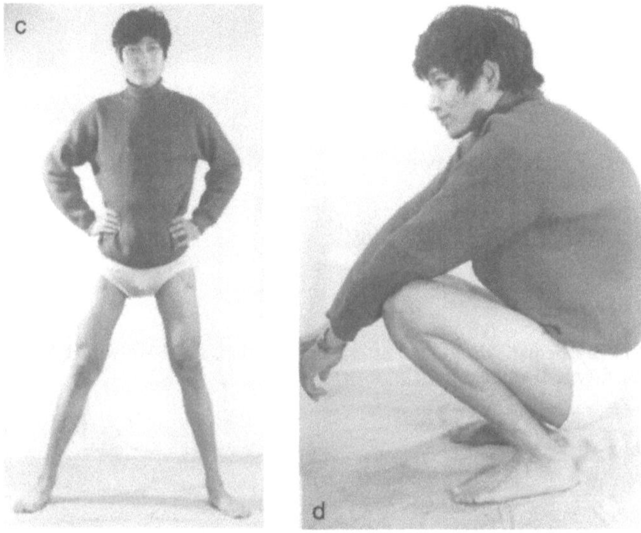

Figures 20c, d

of the femoral neck has been found. This shape memory double-cup prosthesis is suitable for young patients, especially for those with deformity or occlusion in the femoral marrow cavity (Figures. 20b, 20c), in which the traditional total hip replacement is difficult or impossible.

7.2 Other Applications

Compression staples and shape memory mini-compression bone plates have been used to fix fracture of the jaw and the mandible [26]. Shape memory rivets have been used in the fixation of custom-made cranial bone prosthesis to repair skull defects.

8 Future Studies

Following the systematic studies of the biological properties of NiTi SMA by Xue and his coworkers in Shanghai in 1978, the good biocompatibility of NTSMA has been repeatedly proved in experimental studies on orthopaedic implants. Since 1981, NTSMA implants have been used in thousand cases in Mainland China, and no biological adverse reaction was found, which has laid a good foundation for its further application in orthopaedics. It is expected that in the beginning of the twenty-first century, the application of SMA in orthopaedics will register substantial progress.

In addition to the two-way shape memory property in arterioembolization devices [27], other NTSMA implants used in clinical orthopaedics are also able to remember their shapes in high and low temperature respectively. Therefore they

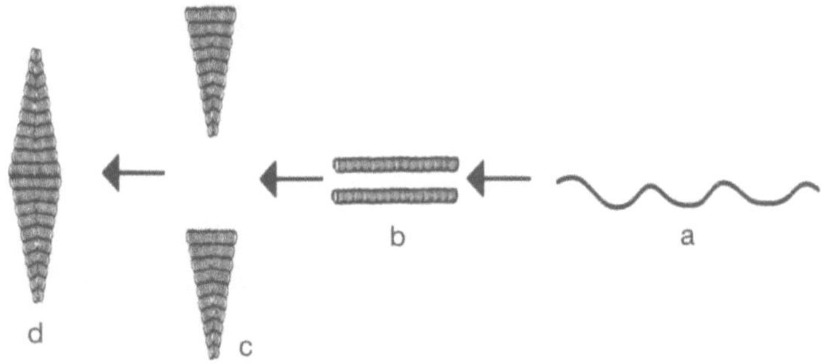

Fig. 21a–d. Schematic diagram of arterioembolization. **a** NTSMA wire. **b** Spring-tube. **c** Type-I embolizator. **d** Type-II embolizator

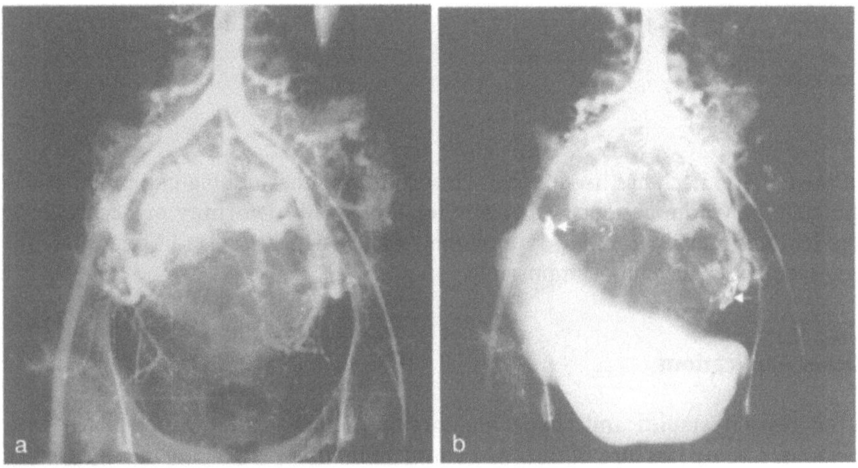

Fig. 22a, b. Arterioembolization. **a** Female, 17 years old: aneurysinal bone cyst of sacrum. X-ray film of angiography before embolization. **b** Post embolization of bilateral internal iliac arteries. *Arrows* indicate the shape memory embolizator (provided by Chuncai Zhang)

can change shape by modifying the temperature. It promises a broadened future of application of orthopaedic implants, but imposing higher demands on the design of implants and the technology of manufacture (Figures 21, 22).
Superelasticity is another special physical property of NTSMA; it has been employed in orthodontics and the improvements in home appliances such as the cellular phone antenna. The development and the applications of this alloy, such as superelastic guide wires, cables for long and thin devices like scissors and clamps, and the joint application of elastic behavior and shape memory effect, deserves equal emphasis [28].

Apart from the NiTi SMA, is there any other shape memory material with good biocompatibility and promising future in the medical field? At this time, the basic studies in the exploration of shape memory ceramics and other shape memory materials such as TiNiCu, TiNiCo are under way. It is a new trend worth laying stress on [6, 26]. The porous SMA will have the advantages of possible ingrowth of bone, the rigidity and density approximating the ones of the skeleton. The studies on surface modification will, in time, possibly lead to the development of better corrosion-resisting, wear-resisting and/or nickel ion escape-proof SMA products, which is of great significance for permanent implants [29–32].

The application of shape memory implants is closely related to the control of the ambient temperature by means of immersion and compress with cold or hot saline. Other simpler and more precise methods of intraoperative and postoperative (extracorporeal) temperature control have yet to be devised. Some scientists have attempted the use of radio frequency. For long implants, such as scoliosis correction systems, techniques to make segmental transformation available may also be contemplated. All these will further improve the operability and precision of the application of SMA products in clinical orthopaedics.

References

1. Xue M, Chen XX, Li YM, et al. (1981) Basic study of NiTi shapes memory alloy-stimulating corrosion test. Stomatology 1:40–43
2. Xue M, Pan JS, Chen XX, et al. (1982) Basic study of NiTi shape memory alloy-stimulating corrosion test and histological observation. Med. T. china 62: 728–731
3. Xue M, Li YM, Gu GZ, et al. (1983) Studies on NiTi shape memory alloy with histological test. Chin J Biomed Eng 2:28–33
4. Dai KR, Zhang XF, Yu CT (1983) Orthopaedic applications of shape memory compression staple. Chin J Surg 21:343–345
5. Dai KR, Chu YY (1996) Studies and applications of NiTi shape memory alloys in the medical field in China. J Biomed Mater Res 6:233–240
6. Wever DJ, Veldhuizen AG, Sanders MM (1997) Cytotoxic, allergic and genotoxic activity of a nickel-titanium alloy. Biomaterials 18:1115–1120
7. Ryhanen J, Niemi E, Serlo W, et al. (1997) Biocompatibility of nickel-titanium shape memory metal and its corrosion behavior in human cell cultures. J Biomed Mater Res 35:451–457
8. Ryhanen J, Kallioined M, Tuukkanen J, et al. (1998) In vivo biocompatibility evaluation of nickel-titanium shape memory metal alloy: muscle and perineural tissue responses and encapsulate membrane thickness. J Biomed Mater Res 41:481–488
9. Zhang CC, Liu ZS, Gao JZ (1989) Treatment of patellar fracture with internal fixator of shape memory alloy. Chin J Surg 27:692–695
10. Zhang CC, Wang JL, Xiao J, et al. (1996) The research of NiTi patellar concentrator and biomemory mechanical characteristics for treatment of every type of patellar fracture. J Bone Joint Inj 2:78–81
11. Zhao WK, Ge ZS, Zhou SL, et al. (1995) The use of shape memory alloy screw for the treatment of fresh fracture of the femoral neck. Chin J Orthop 15:142–144
12. Dai KR, Ni C, Wu XT, et al. (1994) An experimental study and preliminary clinical report of shape memory sawtooth-arm embracing internal fixator. Chin J Surg 32:629–632
13. Wu XT, Dai KR, Ni C, et al. (1995a) The experimental observation of fracture healing fixed by shape memory sawtooth-arm embracing internal fixator. J Bone Joint Inj 10:164–166
14. Wu XT, Dai KR, Qiu SJ, et al. (1995b) A comparative study of effects on bone healing and remodelling between embracing fixator and bone plate. Chin J Surg 33:481–484
15. Wu XT, Dai KR, Ni C, et al. (1995c) A comparative observation of the effect on bone collagen fibers between embracing internal fixator and stainless steel plate. Chin J Traumatol 11:89–91
16. Zu XS, Dai KR, Wu XT, et al. (1995) Investigation on stability of shape memory sawtooth-arm embracing internal fixator. J Appl Biomech 10:40–46
17. Zhang CC, Wang JL, Zhu LH, et al. (1990) Design and application of Nitinol arc-fork intramedullar nail for the humans. In: Proceedings of the International Conference on Medical Applications of Shape-Memory Alloys. Shanghai, China. pp 144–150

18. Zhang CC, Liu CJ, Chen TY, et al. (1993) Design of the scorpion-like dynamic Nitinol-osteo connector and its use in treatment of fracture and nonunion of the upper extremity tubular bones. Chin J Surg 31:269–271
19. Yang PT, Tao JC, Zhang YF (1987) The application of NiTi shape memory plate and screw in hand surgery. J Hand Surg 3:2–4
20. Yang PT, Tao JC, Ge MZ (1991) Clinical application of clamping internal fixation plate. Chin J Traumatol 7:69–71
21. Zhao DL, Zhang WM (1984) Application of the NT-2 shape memory alloy artificial joint of the cervical intervertebral articulation for cervical spondylosis on anterior decompressive operation. Chin J Surg 22:410–412
22. Mei FR, Ren XJ, Wang WD (1997) The biomechanical effect and clinical application of a Ni–Ti shape memory expansion clamp. Spine 22:2083–2088
23. Lu SB, Guo JF, Wang JF (1986) Treatment of scoliosis with shape memory alloy rod. Chin J Surg 24:129–132
24. Charnley J (1961) Arthroplasty of the hip: a new operation. Lancet 1:343–344
25. Dai KR, Zhang XF, Yu CT, et al. (1983) Application of NiTi shape memory alloy in double-cup prosthesis of hip. Chin J Surg 21:540–542
26. Drugacz J, Lekston Z, Morawiec H, et al. (1995) Use of TiNiCo shape-memory clamps in the surgical treatment of mandibular fractures. J Oral Maxillofac Surg 53:665–672
27. Zhang CC, Jiang ZX, Gao JZ, et al. (1992) Arterioembolization with shape memory angioembolus: experimental study and clinical application. Chin J Surg 30:111-114
28. Cuschieri A (1991) Variable curvature shape-memory spatula for laparoscopic surgery. Surg Endosc 5:179–181
29. Nakamura T, Shimizu Y, Ito Y, et al. (1992) A new thermal shape memory Ni–Ti alloy stent covered with silicone. ASAIO J 38:347–350
30. Yahia LH, Lombardi S, Piron D, et al. (1996) NiTi shape memory alloys treated by plasma-polymerized tetrafluoroethylene. A physicochemical and electrochemical characterization. Med Prog Technol 21:187–193
31. Villermaux F, Tabrizian M, Yahia LH, et al. (1996) Corrosion resistance improvement of NiTi osteosynthesis staples by plasma polymerized tetrafluoroethylene coating. Biomed Mater Eng 6:241–254
32. Trépanier C, Tabrizian M, Yahia LH, et al. (1998) Effect of modification of oxide layer on NiTi stent corrosion resistance. J Biomed Mater Res 43:433–440

The Surgical Correction of Scoliosis with Shape-Memory Metal

Dirk Jan Wever, Albert G. Veldhuizen

1 Introduction

Because of its unique properties, shape memory metal (NiTi) can be put to excellent use in a device for the surgical correction of scoliosis [4, 37, 38, 52–54, 59]. Before such a device may be applied clinically (after being developed), it should meet two important criteria: biocompatibility and biofunctionality [11, 66]. Biocompatibility relates to the ability of the device to remain biologically nontoxic during the implantation period and biofunctionality refers to the ability of the device to perform the purpose for which it is designed. In this chapter both aspects of a new scoliosis correction system, based on shape memory metal are discussed [53, 59].

1.1 Scoliosis

Scoliosis is a three-dimensional deformity of the trunk, characterized by lateral deviation and axial rotation of the spine. The deformity is usually accompanied by thoracic deformation. The cause of most cases of scoliosis is unknown ("idiopathic"). The remaining 20–30% of the cases is caused either by a neuromuscular or congenital disorder.

Idiopathic scoliosis usually begins in middle childhood and tends to increase progressively during skeletal growth. The main concern in the patient with scoliosis relates to curve progression, resulting in cosmetic deformity and finally in a reduction of pulmonary function. The current knowledge of the natural history of idiopathic scoliosis indicates that there is a large number of patients with minor degrees of curvature. Some of these patients can be expected to have curvatures that will increase levels requiring brace treatment [7, 9, 36, 50]. In some cases, the scoliotic curve increases to such an extent that surgical treatment becomes necessary.

1.2 The Current Surgical Treatment of Scoliosis

Operations for idiopathic scoliosis are designed to correct the deformity and to prevent its progression by achieving a solid fusion. In 1962, Harrington introduced instrumentation for the internal correction and stabilization of scoliosis

[21]. The Harrington instrumentation consists of a distraction rod at the concave side and a compression rod at the convex side of the curve. The device only has a temporary function: it should maintain the correction achieved during surgery until the fusion is completed. A good fusion is obtained by means of decortication and excision, and grafting of facet joints. To increase the fusion mass, autologous or homologous bone grafting is also performed. Spinal fusion is established within 3–6 months. An insufficient fusion may lead to a high incidence of metal fatigue and implant failure.

The application of distraction forces in Harrington's system has the important drawback that it decreases normal thoracic kyphosis, sometimes resulting in a flat back [56]. Moreover, a postoperative brace is necessary to prevent pseudoarthrosis and to prevent the system from breaking out. Since the introduction of more segmental systems, such as the Cotrel-Dubousset, TSRH and Isola instrumentation, this postoperative external immobilization is no longer necessary [3, 6, 13, 14, 57, 58]. The more stable fixation of these systems is achieved by using hooks, sublaminar wires and cross-links between both rods at various levels of the spine (Figure 1). The application of translation forces and selective distraction and compression forces on the scoliotic spine and the rigid fixation makes these systems superior to the Harrington system.

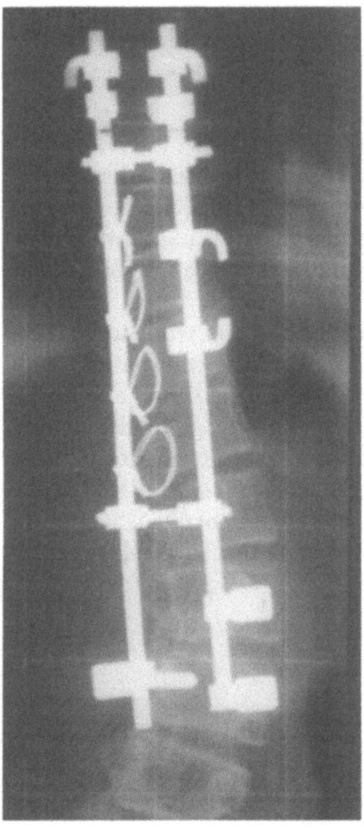

Fig. 1. Standing anteroposterior radiograph of a scoliotic curve corrected with a segmental dorsal system consisting of two rods fixated to the spine with hooks, sublaminar wires and pedicle screws

However, if one considers the 3-D deformity of scoliosis, no optimal results are achieved using these systems. Significant corrections are obtained in the frontal and sagittal plain, but axial vertebral rotation and rib cage deformity are hardly corrected. Although a number of new systems claim to correct axial rotation to some extent through the application of the so called "derotation manoeuvre", the correction of the rotation found in postoperative 3-D radiological measurements is hardly significant [19, 35, 58]. For example, Labelle et al demonstrated that the Cotrel-Dubousset procedure is effective in producing three-dimensional improvement by an "en bloc" displacement of the instrumented spine rather than by vertebral axial derotation [32].

One of the causes of a failure to obtain optimal corrections is the presence of structural deformities of the intervertebral discs, vertebral bodies and the rib cage [12, 16, 62]. The anterior approach to the thoracolumbar spine via a thoracotomy or a thoracoabdominal retroperitoneal approach allows resection of the deformed disci and vertebral bodies in order to produce a considerable mobility of the scoliotic segment. This allows a larger degree of correction over a shorter segment than by posterior surgery. The anterior procedure with anterior instrumentation was first introduced by Dwyer in 1969 [18]. Modification of this system has led to the development of other anterior instrumentation systems (Zielke, Webb-Morley and Kaneda) [30, 67]. By carrying out a resection of rib heads and its ligament-capsular structures on the convex side of the curve as well as dissectomies, Kaneda reported significant corrections of the apical vertebral rotation [30].

Apart from the fact not always achieving optimal corrections, the so-called "crankshaft phenomenon" may cause further curve progression when the posterior approach is used [17, 51]. This phenomenon is the continued progression of the deformity due to uninterrupted growth at the vertebral end plates of the anterior spinal column. Therefore, in patients with a significant growth expectation (Risser score of zero), it is recommended to carry out an anterior fusion as well, to arrest the growth of the vertebral growth plates and to remove the driving force behind further progression.

Another cause of a substantial correction loss at follow up is related to the viscous properties of the spine. This negative effect occurs when the forces applied by the instrumentation are gradually absorbed by the relaxation of the bones and soft tissues. Nachemson and Elfstrom reported that axial forces decrease by 20% within 30 min after distraction with a Harrington rod. After about 10 days the recorded force stabilized at about one-third of the maximum force during surgery. They also reported that additional correction might be achieved by a renewed distraction, which required a second operation [43]. The procedure with the shape memory metal scoliosis correction device is based on this principle: using the viscous behavior of the spine by correcting the scoliotic spine with a continuous force, also postoperatively, it is expected that an additional correction will be obtained [53, 59].

2 Biomechanical Aspects of the Correction of Scoliosis with Shape-Memory Metal

The optimal correction of scoliosis can be obtained by placing correction forces along the vertebral bodies in the curve, counteracting the progression forces. Knowledge of the forces that play a part in the progression of scoliosis is therefore crucial for the development of new techniques in brace treatment and surgical correction of scoliosis.

2.1 The Force System in the Scoliotic Spine

The spine may be regarded as an inherently unstable system which requires support of the musculo-ligamentous structures to maintain its posture and to provide motion [31, 46, 49, 65]. The majority of these musculo-ligamentous structures is located at the posterior side of the spine and these contribute to the stability of the spine by resisting primarily tension forces. On the contrary, the anterior spinal column, which consists of the vertebral bodies and intervertebral discs has a special role in transmitting compressive forces. Nachemson et al measured the in vivo lumbar intradiscal pressure and found large compressive forces in different body positions. They also demonstrated pre-stress in the intervertebral disk which was mainly preserved by the musculo-ligamentous structures of the posterior spinal column [41, 42, 65]. Except for the compressive and tension columns, the rib cage is also thought to influence the stability of the spine [45].

In the normal spine the stability of the physiological sagittal curve, i.e., the thoracic kyphosis and lumbar lordosis, is primarily maintained by a balance between the compressive forces of the anterior column, and the tension forces in the posterior column. Resulting transversal components of the compressive column in the sagittal plane due to kyphosis or lordosis will primarily be resisted by the powerful musculo-ligamentous structures of the posterior spinal column [44].

In a scoliotic curve the direction of the compressive forces in the anterior spinal column has changed. In the frontal plane, the compressive forces at the apical level are inclined at an angle, which results in a shear force towards lateral (Figure 2a). To achieve an equilibrium in the scoliotic spine the lateral force components in the anterior column must be counteracted by the musculo-ligamentous structures of the posterior column as well as bone elements such as the facet joints and ribs (Figure 2b). However, should these structures fail to stabilize the scoliotic spine, for example during periods of growth, curve progression will occur.

The biomechanical difference between the anterior pressure column and the posterior tension column with scoliotic deformity was already made by Meijer in 1866 [39]. Other authors also stress the importance of "posterior tethering" with regard to the geometrical and morphological configuration of the scoliotic deformity [23]. Various physical models and cadaver models are described in which the posterior column functioned as a tension column with a strong tendency to shorten [2, 16, 28, 33, 44, 48]. For example, it was postulated that in scoliosis the

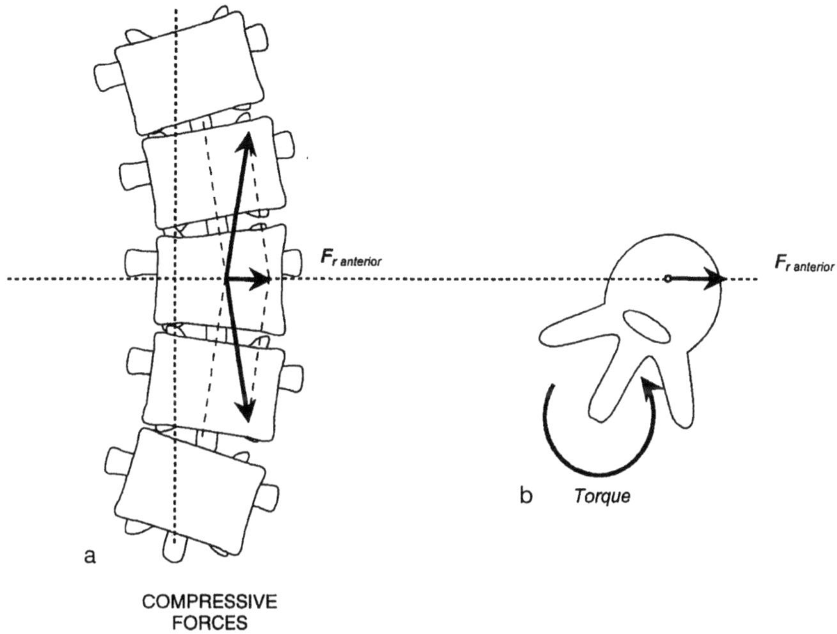

Fig. 2a,b. The force pattern in the scoliotic spine. The compressive forces of the anterior column result in a force which drives the apical vertebral body out of the midline (**a**) whereas the torque, provided by the posterior column, attempt to keep the posterior complex in the normal position (**b**). $F_{r\ anterior}$=resulting anterior force

growth rate of the anterior column is not in equilibrium with the lengthening of the musculo-ligamentous structures of the posterior column. This relative lengthening of the anterior components compared to the posterior elements is thought to result in a rotation in which the anterior column deviates more than the posterior column [40].

Concluding, one might state that the system of forces which results from scoliosis and which is responsible for the progression consists of an anterior shear force that tends to drive out the vertebral body to the convex side of the curve. A torque produced by the musculo-ligamentous structures of the posterior spinal column attempts to balance this anterior force, but is insufficient in some cases, as in the case of a large curve and during periods of rapid growth. Therefore, in the case of a posterior approach of the spine a maximal lateral force and a torque is necessary at the apical vertebra in order to correct the scoliotic curve.

2.2 Force-Controlled Correction of Scoliosis with Shape-Memory Metal

The application of shape memory metal in a device for the correction of scoliosis is not new. In 1976, 13 years after Buehler and Wang described the characteristics of the shape memory alloy for the first time and 14 years after the introduction of the Harrington instrumentation Schmerling already experimented with a memory metal rod to replace a standard Harrington rod in a human cadaver [8, 21, 54]. Preliminary investigations were also conducted with an anterior system: a shape memory metal wire was used with Dwyer instrumentation [4]. In China, Lu reported on surgical procedures in patients with an idiopathic scoliosis using shape memory metal rods instead of Luque rods. They reported good correction without post-operative loss of correction and no complications [37]. In 1993, Sanders et al. reported on experiments with animals in which a previously induced scoliotic curve in goats was corrected with a shape memory metal rod. In this study a good curve correction was achieved, but the fixation of the rod to the spine with transspinous process wires caused increased nickel and titanium levels in the surrounding tissues, possibly through fretting between these wires and the rod [52]. A comparable study was conducted on monkeys with two shape memory metal rods, which were fixated with sub-laminar wires. It was concluded that the memory metal device could induce a gradual correction and that the correction rate could be controlled through the rod's temperature. This study also showed higher level of nickel possibly caused by fretting of the sub-laminar wires [38] (Sect. 3.2).

Sanders et al. have developed the device discussed and tested in this book chapter [53, 59]. The device is unique in that it is engineered from the start with the aim to correct three-dimensional scoliosis with a shape memory metal rod. The dorsal system consists of a memory metal rod with a transition temperature of 25°C and a square cross-section of about 6 mm. Correction loads are generated if the memory metal rod is deformed and subsequently is heated above its transition temperature. This will cause the rod to regain its original shape. This original shape can be programmed pre-operatively by using a "bender" which bend the rod beyond its reversible limit of about 6–8%. The rod may then be bent to the desired sagittal curve and scoliotic rest curve. The device further consists of an adequate anchoring system, which ensures the transfer of the correction loads. These loads consist of lateral and anterior-posterior forces perpendicular to the spine, which induce a bending moment to correct the lateral and sagittal curves and a torque to correct the axial rotation. A good torque transfer is ensured by the square cross section of the shape memory metal rod. The size of the rod's cross-section has been chosen with respect to the selected magnitude of the load applied by the device: based on data from the literature a maximum bending moment of 7.5–10 Nm was chosen for the correction of the lateral deviation and a torque of 2–5 Nm for the correction the axial rotation. It is expected that with these moments the spine remains force loaded post-operatively and that this will take advantage of the viscous behavior of the spine in order to obtain extra correction.

From extensive biomechanical tests of the shape memory metal rod in a specially designed apparatus it appears that the desired correction moments are achieved when the deformed rod is heated to a temperature between 40°C and 50°C. In this situation the rule applies that the higher the overshoot in temperature above the body temperature, the larger the correction forces: for example, 10°C increase in temperature realizes an extra bending moment of about 4 Nm. During the procedure the rod should therefore be heated to the desired temperature. Resistive heating was chosen, using a low voltage, high frequency current. This current is supplied by a specially developed heating apparatus. The temperature of the rod is registered by two thermocouples and is fed back into the heating device. This enables the surgeon to only set the desired end temperature of the rod. The apparatus automatically heats then the rod to that temperature.

Just like the current operative scoliosis correction methods, the new system seeks to achieve a post-operative fusion. This fusion should prevent failure of the system in the long term. It is expected that additional post-operative correction may be obtained before this vertebral fusion takes place.

3 Biocompatibility Aspects of the Shape-Memory Metal Scoliosis-Correction Device

Although the biomedical application of shape memory metal in a scoliosis correction device is intriguing, the implantation of nickel containing materials in human body requires caution [11, 66]. Nickel is nutritionally essential, but it is well known that nickel is capable of eliciting toxic and allergic responses. Therefore, in the biocompatibility estimation of the shape memory metal scoliosis correction device, attention has been focussed to the potential nickel release, especially when such nickel-titanium device is implanted into young patients which results in a potentially longer host exposure [63, 64].

3.1 The in Vitro Biocompatibility of Shape-Memory Metal

The good anticorrosive properties and associated good tissue compatibility of titanium and its clinically used alloy Ti6Al4V can be ascribed to a chemically stable oxidized passivation layer [10, 34, 66]. Because titanium has a high affinity for oxygen, these beneficial surface oxide films form spontaneously in an oxygenated environment. Consequently, oxide films may also be expected on the metal surface of the nickel-titanium, shape memory alloy. Depending on its composition, structure, stability and thickness, this layer may act as a barrier against nickel diffusion. In addition, it is known that calcium phosphate surface films are naturally formed on titanium alloys in a biological environment. It is thought that the specific reaction of TiO_2 with calcium and phosphorus may serve as a further barrier against ion diffusion [20, 22].

As with any corrosion resistant alloy, despite the presence of a passive oxide film, a small release of ions will still be present. The rate of this release, the solubility and protein binding capacity of the ions and their cytotoxic, allergic and genotoxic activity will finally determine the biological safety of an implant material [66].

In an extensive test programme of the shape memory alloy used in the scoliosis correction device the following aspects were investigated:

a) The elemental composition and thickness of the surface film formed on the shape memory alloy, prior to and after immersion in a physiological solution, determined by X ray photoelectron spectroscopy (XPS)

b) The stability of the surface film formed on the shape memory alloy, determined by electrochemical potentiodynamic measurements
c) The passive diffusion of nickel from the shape memory alloy in a physiological solution, measured by atomic absorption spectrophotometry
d) The toxic effects of the shape memory alloy determined by four short-time biocompatibility tests: a cytotoxicity test (1), a sensitization test (2) and two genotoxicity tests (3, 4). These tests were carried out in compliance with the rules of the International Organization of Standardization, ISO 10993/EN 30993 (the biological evaluation of medical devices) [24–27]

The shape memory alloy was obtained from AMT, Swiss Metal (Herk-de-Stad, Belgium) and consisted of 50 atomic % nickel and 50 atomic % titanium with minimal amounts of trace elements in the final alloy (total of carbon, oxygen and other trace elements <0.5 weight %). The studies were compared with two clinical reference controls, i.e. the implantable grade of stainless steel, AISI 316 LVM, with a nickel content of 13–15% and the well known and biocompatible alloy, Ti6Al4V.

a) A S-probe spectrometer (Surface Science Instruments, Mountain View, CA, USA) equipped with an aluminium anode (10 kV, 22 mA) and a quartz monochromator was used for the XPS analyses. The analyses of the shape memory metal samples before immersion in physiological solution confirm the presence of a passive surface film. This passive film consists of a mainly TiO_2 based oxide, comparable with Ti6Al4 V. Minimal amounts of nickel were found in the outermost surface layers. After immersion of the samples in a physiological solution a calcium-phosphate layer was formed on the surface oxide. An increase in Ca and P atomic percentages was found with longer immersion in the physiological solution. During the formation of this calcium-phosphate layer lower percentages of titanium were found in the first surface layers (Figure 3).

b) Potentiodynamic tests were performed on the samples in a Hanks' physiological solution on a Princeton Applied Research potentiostat according to ASTM G5-94 procedures. The shape memory alloy exhibited passive electrochemical behavior comparable to other clinically used alloys. A more pronounced diffe-

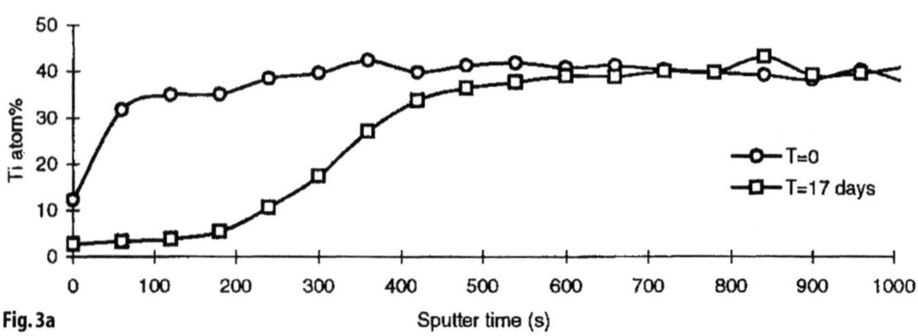

Fig. 3a

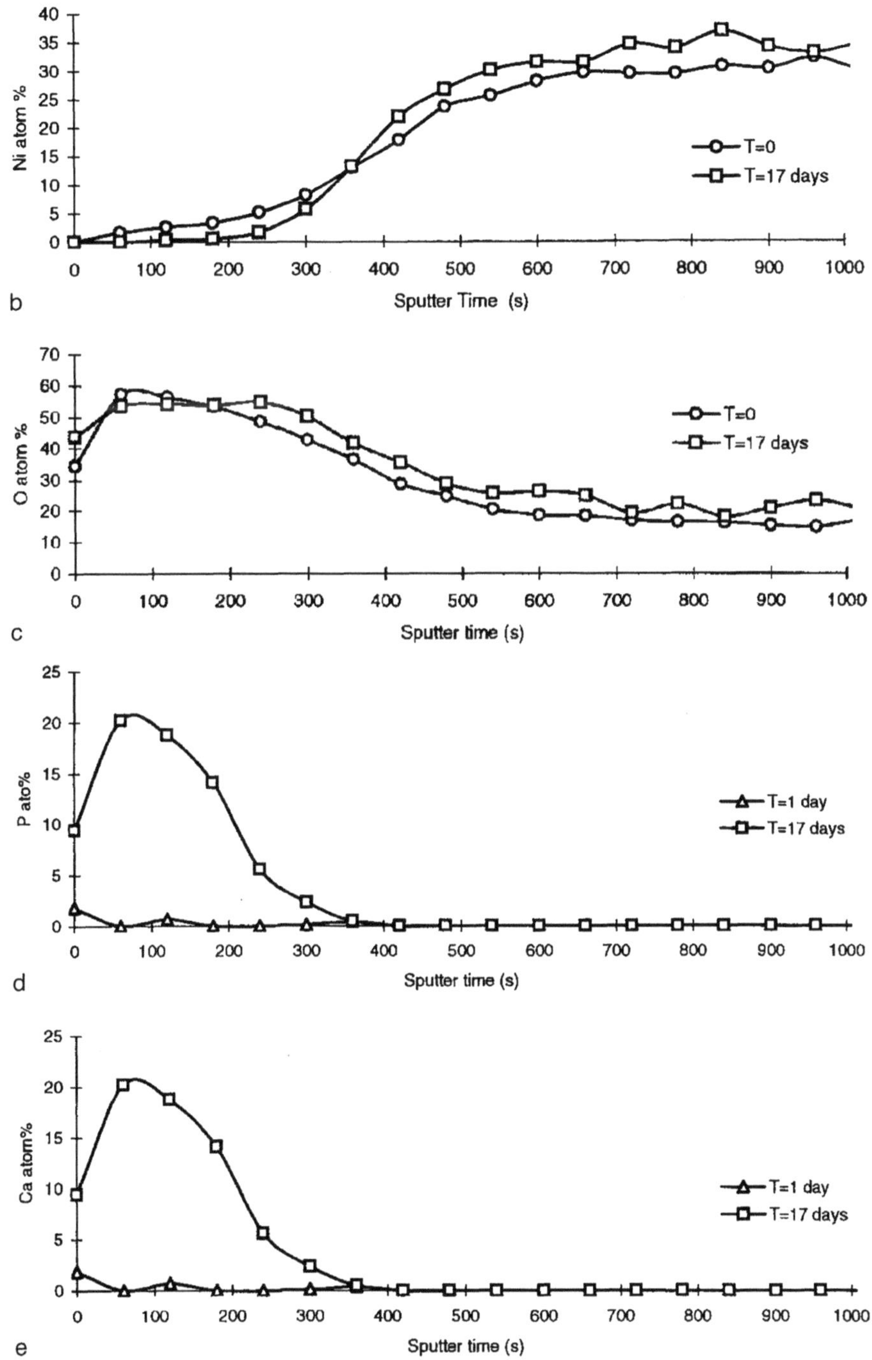

Fig. 3a–e. XPS depth profiles of the shape memory metal (NiTi) samples prior to, and after immersion in a Hanks' solution; showing the depth distribution of titanium (**a**), nickel (**b**), oxygen (**c**), phosphorus (**d**) and calcium (**e**) versus sputtering rate

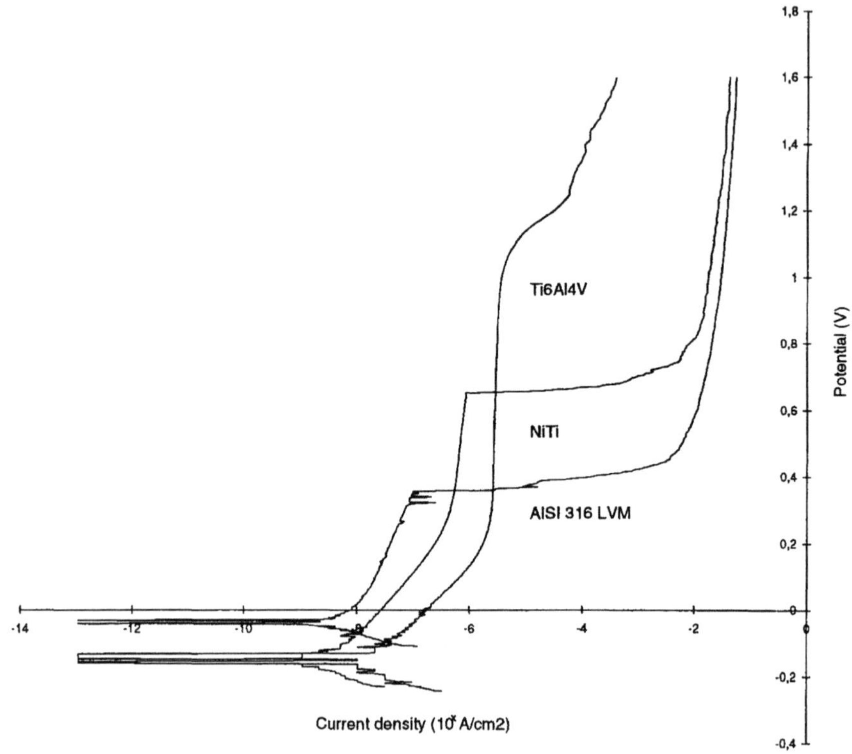

Fig. 4. Anodic polarization curves in a deaerated Hanks' solution of the shape memory metal (NiTi), Ti6Al4V and AISI 316 LVM samples

rence between the corrosion and breakdown potential, i.e., a better corrosion resistance was found for the shape memory alloy, compared to AISI 316 LVM stainless steel. As expected, the Ti6Al4V alloy demonstrated a superior corrosion resistance without pitting (Figure 4).

c) A Varian atomic absorption spectrophotometer (model SpectrAA 300 plus) with graphite furnace GTA (Zeeman 96 plus) was used for the determination of the passive release rate of nickel out of the samples in Hanks' solution. The passive release rate of nickel from the shape memory metal samples reduced in time from an initial release of 14.5×10^{-7} mg/cm^2 to a nickel release that could not be detected anymore after 10 days (Figure 5). In contrast to the shape memory metal samples the nickel release of the AISI 316 LVM samples remained under the detection limit (2.5 µg/l) during the measuring period.

d) 1. An extract test was chosen to determine the cytotoxic potential of the samples. The samples were extracted under sterile conditions for 24 h at 37°C in minimal essential medium. A monolayer of human skin fibroblasts were then exposed for 72 h to four different concentrations of the sample extracts.

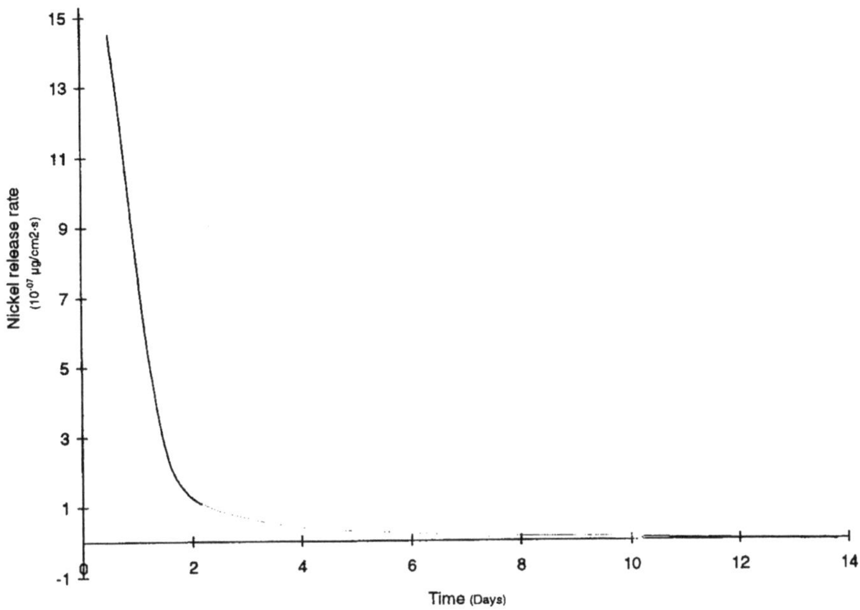

Fig. 5. Mean nickel release rate (mg/cm² s) versus time (days) of the shape memory metal (NiTi) samples in Hanks' solution, determined by atomic absorption spectrophotometry

At 24, 48 and 72 h, the monolayer was examined microscopically for cytotoxic effects: There was no effect on the cellular monolayer: no signs of cellular lysis, intracellular granulation's or cell morphologic changes were observed. The presence of the shape memory metal extracts had also no inhibitory effect on cell growth.

d) 2. The potential to produce an allergic response of the samples was assayed in the Guinea-pig skin-sensitization test. The samples were extracted in 0.9% NaCl at 50°C for 72 h under gentle movement. For the induction phase, soaked patches were applied to a clipped area of the dorsal side of the test animals under an occlusive dressing for 6 h. The induction procedure was repeated twice, at weekly intervals. After 2 weeks rest, the test animals were challenged at the untested pre-shaved skin sides with the extract of the samples. 24 h and 48 h after removal of the challenge patches the test sites were scored for erythema and edema formation: No dermal sensitizing response was observed at the test sites of the animals induced and challenged with the extracts of the shape memory alloy. Also, the test animals induced with AISI 316 LVM and the control animals showed no dermal sensitizing response.

The potential mutagenic toxicity was assessed using two assay systems: the Salmonella reverse mutation test (3) and the Chromosomal aberration test (4).

d)3. The Salmonella reverse mutation test is designed to establish the potential of a test material to induce reverse mutations in different Salmonella typhimurium strains. The shape memory metal samples were extracted as described in the sensitization test. Five concentrations of the sample extracts (20, 40, 60, 80 and 100%) were tested. Four Salmonella typhimurium strains were used to differentiate between base pair and frame shift mutations. Each strain and dose level, including the controls were tested in triplicate. After incubation at 37°C for 48 h the revertant colonies were counted. None of the assays of the Salmonella reverse mutation test showed a mutation factor higher than two in any of the four Salmonella strains (a mutation index of 2.0 was considered to be a sign of mutagenicity).

d)4. The in vitro chromosomal aberration test enables the detection of chromosomal mutations in mammalian cells. The Chinese hamster fibroblast cell line, V79 (GSF Neuerberg, Dld) was chosen to detect chromosomal aberrations after exposure of the sample extracts. The shape memory metal samples were also extracted as described in the sensitization test. The test was carried out with three extract concentrations: 10, 8 and 6%. Two cultures were used for each concentration of the sample extracts and controls. One hundred-well spread metaphases per culture were microscopically scored for chromosomal aberrations: Chromosomal breaks, fragments, deletions, exchanges, chromosomal disintegration's and gaps: no significant increase in number of cells with chromosomal aberrations was observed after exposure of the shape memory alloy extracts as compared to the solvent control.

3.2 Animal Experience with Shape-Memory Metal Scoliosis Correction Device

In clinical applications, especially in orthopaedic implant conditions, localized mechanical stress or fretting may damage the surface layer and can cause an increase in ion release or even a release of wear particles. It is known that spinal implants are especially prone to corrosion due to their construction, in particular the connection between pedicle screw, hooks and sub-laminar wires to the rod [60, 61]. Therefore, new spinal implants should be examined in vivo to ascertain whether corrosion and adjacent tissue reaction occur. The functionality of a new device should also be tested before it is introduced in clinical practice.

For the in vivo tests of the new device the action of the shape-memory metal rod was reversed. It was decided to *induce* a scoliotic curve in an animal model, instead of *correcting* it (Figure 6). Six immature pigs of ±6 months old, weighing between 70 kg and 90 kg were used in these experiments. The size and shape of the spines of these test animals resembled the human spine. The operations took place under standard sterile conditions and were performed under general anaesthesia. Permission for the protocol followed was obtained from the ethical committee for experiments on animals.

Preoperatively a blood sample was taken to determine the serum nickel concentration. Determining the serum nickel concentration was achieved using atomic absorption spectrometry, as described earlier where the measurements of

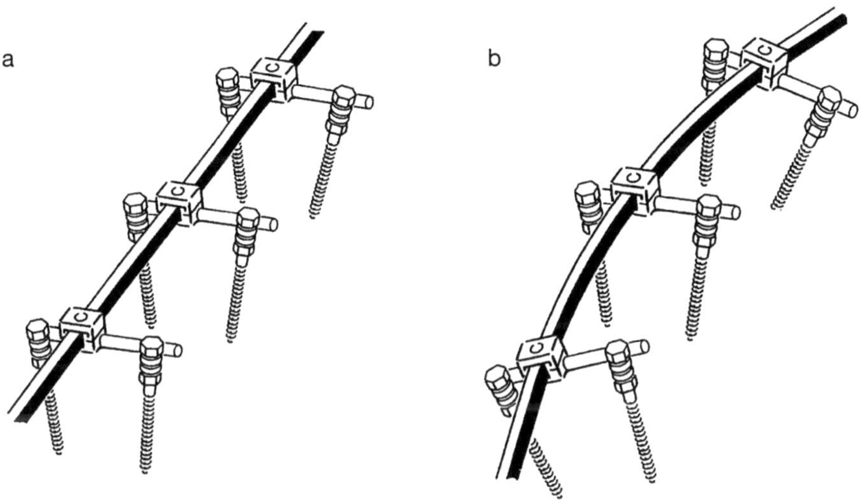

Fig. 6a, b. Dorsal view of the used instrumentation in the animal experiments: the straightened rod before heat treatment (**a**) and the recovered, original curve of the rod after heat treatment (**b**). The system is anchored to the spine of the animal with pedicle screws on three levels

the in vitro nickel release of memory metal were discussed (Sect. 3.1c). After being positioned the animals were shaved, disinfected and covered with a sterile cloth. A longitudinal incision was made along the median line from Th11 to L5. The vertebrae were identified prior to surgery using x-rays. The paravertebral musculature was shifted to the processus transversus using the usual method. Then the pedicle insertion points were identified on both sides of Th12, L2 and L4. Access to the pedicle was obtained with the pedicle probe and after tapping, the pedicle screws were placed. Care was taken to damage as little periosteum or bone as possible during this procedure. The pedicle screws were connected to a bridge at three levels.

In order to initiate the scoliotic curve a square shape memory metal rod with rounded edges was used with a cross section of 6.35 mm^2 and an original curve of 40° Cobb angle. The transition temperature of the rod, i.e. the temperature at which low-temperature phase martensite changes to the high-temperature phase austenite, was 25°C. The rod had been stored in sterile conditions at –18°C. The

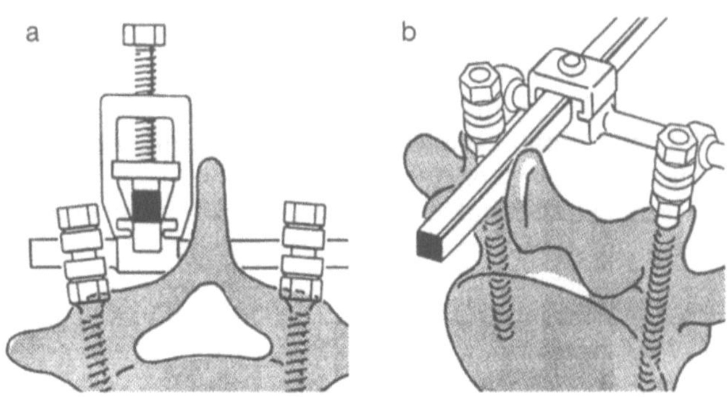

Fig. 7a,b. Transversal view of the approximeter (**a**), which is used for positioning of the rod on the rod-bridge interface and a three-dimensional view of the definitive fixation of the rod on the bridge (**b**)

bent rod was easily straightened in this cold condition using the "bender" prior to its application to the animal spine. Then it was positioned on the bridge with the approximeter and fixated (Figure 7). Next, the rod was heated with high frequency, low voltage electric current using the heating device. Two thermocouples continuously monitored the rod's temperature during this procedure. After being heated, the rod was definitively fixated to the bridge and the wound was closed layer by layer.

Directly after the surgical procedure anteroposterior and lateral radiographs were made and a second intravenous nickel sample was taken. Follow up took place after 1 week, after 6 weeks and after 3 months. After 3 months, three animals were sacrificed with T61 (euthanasia solution). After 6 months, the remaining three animals were sacrificed using the same procedure. An extensive histological analysis was performed on these animals. Biopsies were taken from the tissue around the device and from the lungs, spleen, liver and kidneys.

In all test animals a significant scoliosis was induced by heating the rod during the surgical procedure. Macroscopically, no signs were found of tissue damage when the rod was heated to 50°C. The first radiographs made immediately after surgery showed a curve of about 40° Cobb angle, the original curve of the shape memory metal rod. There were no indications that the pedicle screws had broken loose or of any other failure of the device. Post-operatively the test animals recovered quickly and often mobilized a few hours after surgery. One of the test animals developed a wound infection, caused by a rod, which extended too far distally. This animal was sacrificed prematurely and excluded from the study. The radiographs taken after one week of the other animals showed a loosened connection between the rod and the bridge in one animal. This caused no further complications and therefore this animal remained included in the 3 months evaluation.

The animals, which were sacrificed after 3 months reached a weight of about 120 kg. Macroscopic inspection of the scoliotic segment showed that the device was almost overgrown with newly formed bone. Corrosion, fretting, tissue discoloration or accumulation of black granular material due to wears processes around the rod or near the rod-bridge interface was not observed.

In two of the three animals which were sacrificed after 6 months (they weighed about 160 kg each) the radiograph showed a rod breakage. In both cases the breakage had occurred near the rod–bridge connection. The breakage had not been observed in the radiological follow up after 3 months. Moreover, the induced scoliotic curve remained unchanged. Macroscopic inspection of the instrumented spine segment of all cases showed a completely fused spine. The rod had been covered almost completely with newly formed bone (Figure 8). The two animals with the broken rod showed no indication of pseudoarthrosis. In addition, a gross inspection of the instrumented spine of these animals shows signs of local stress or fretting corrosion.

Histologic examination of the sections of the surrounding tissues and sections of the lung, liver, spleen and kidney showed no evidence of a foreign body response. The serum nickel measurements until six months postoperatively showed values hovering around the detection limit of 2.5 μg/l (max: 4.8 μg/l) and were not significantly higher in the case of the loosened rod-bridge connection or in the animals with the rod breakage.

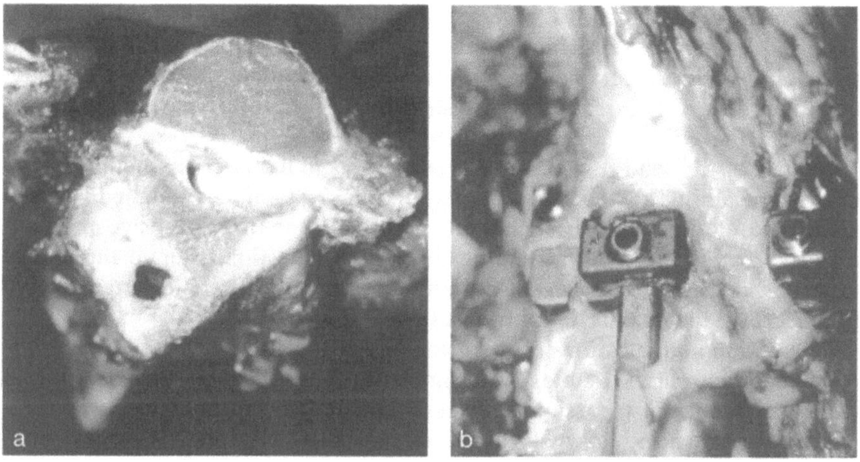

Fig. 8a,b. Transverse (a) and dorsal view (b) of the shape memory metal correction system covered almost completely with newly formed bone after 6 months follow up

4 Conclusions

The preliminary biocompatibility and biofunctionality evaluation of a new scoliosis correction device, based on the intriguing properties of the shape memory, nickel-titanium alloy was discussed in this chapter. With this innovative device the shape recovery forces of a shape memory metal rod are used to achieve a gradual three-dimensional scoliosis correction during and after the operation.

Clinically, the shape memory alloy is already used in several orthodontic, orthopaedic and cardiovascular applications [1, 5, 15, 29, 47, 55]. Two successful shape memory metal devices are the Simon Nitinol filter and the Mitec bone suture anchor. No adverse tissue reactions or allergic reactions through these implants have been described since their introduction. In spite of this satisfactory clinical use, the biocompatibility of nickel-titanium devices is still being discussed. The relatively large content of nickel in the alloy gives rise to considerable concern, especially when such nickel-titanium device is implanted into young patients. The possible toxic effects of the shape memory alloy used in the new scoliosis correction device were studied by performing short-term biocompatibility tests according to ISO regulations. The shape memory metal samples provoked no cytotoxic, allergic or genotoxic responses [63]. These good biological responses are due to a minimal release of ions and in that way a reflection of the good corrosion resistance of the shape memory alloy. These good corrosion properties of the shape memory alloy may be ascribed to the presence of a stable, mainly TiO_2 based, surface layer [64]. Animal experiments were performed to determine the in vivo functionality and biological safety of the shape-memory metal scoliosis correction device. In these experiments the action of the shape memory metal rod was reversed: a scoliotic curve was induced instead of corrected. The rod was capable of inducing significant scoliosis in the animals. In all cases, the original curve of the rod was obtained without the rod breaking out or without failure of the rod-anchor system. By far the greatest part of the correction was achieved within the first hour.

In spite of the solid fusion, a rod breakage was found after six months in two of the three animals. It should be noted that the test animals weighed between 120 kg and 160 kg in the phase that the rod broke. Moreover, extensive mechanical in vitro tests were performed on the rods in question prior to the animal experiments. Nevertheless, extensive fatigue testing of the whole system should be performed before the clinical trials can be initiated.

There were no indications of corrosion or fretting around the rod or near the rod-anchor interface. Histological examination showed good tissue responses without evidence of a foreign body response. Moreover, the device was completely overgrown with newly formed bone, which is a clear indication of a spontaneous fusion. Given these good tissue reactions and the good responses of the NiTi alloy in the short term biocompatibility tests, it is expected that the NiTi scoliosis correction device will show good biocompatibility in clinical applications.

References

1. Andreasen G (1980) A clinical trial of alignment of teeth using 0.019-inch thermal nitinol wire with transitional temperature range between 31°C and 45°C. Am J Orthod 78:528–537
2. Arkin AM (1950) The mechanism of rotation in combination with lateral deviation in the normal spine. J Bone Joint Surg A 32:180–188
3. Asher MA, Strippgen WE, Heinig CF, Carson WL (1994) Isola instrumentation. In: Weinstein SL (ed) The pediatric spine: principles and practice. Raven, New York, pp 1619–1659
4. Baumgart F, Bensmann G, Haasters J, Nolker A, Schlegel KF (1978) Zur Dwyerschen Scoliosenoperation mittels drähten aus memory-legierungen. Eine experimentelle studie. Arch Orthop Trauma Surg 10:67–75
5. Bensman G, Baumgart F, Haaster J (1982) Osteosyntheseklammern aus nickel titan: Herstellung, versuche und klinischer einsatz. Tech Mitt Forsch Berl 40:123–134
6. Bischoff R, Bennett JT, Stuecker R, Davis JM, Whitecloud TS (1993) The use of Texas Scottish-Rite instrumentation in idiopathic scoliosis. A preliminary report. Spine 18:2452–2456
7. Brooks HL, Azen SP, Gerberg E, Brooks R, Chan L (1975) Scoliosis: a prospective epidemiological study. J Bone Joint Surg A 57:968–972
8. Buehler WJ, Wang FE (1968) A summary of recent research on nitinol alloys and their potential application. Ocean Eng 1:105
9. Bunnell WP (1988) The natural history of idiopathic scoliosis. Clin Orthop 229:20–25
10. Callen BW, Lowenberg BF, Lugowski S, Sodhi RN, Davies JE (1995) Nitric acid passivation of Ti6Al4V reduces thickness of surface oxide layer and increases trace element release. J Biomed Mater Res 29:279–290
11. Castleman LS, Motzkin SM (1981) The biocompatibility of nitinol. In: Williams DF (ed) Biocompatibility of clinical implant materials. CRC, Boca Raton, pp 129–154
12. Coillard C, Rivard CH (1996) Vertebral deformities in scoliosis. Eur Spine J 5:91–100
13. Cotrel Y, Dubousset J (1984) Nouvelle technique d'osteosynthese rachidienne segmentaire par voie posterieure. Rev Chir Orthop 70:489–495
14. Cotrel Y, Dubousset J (1988) New universal instrumentation in spinal surgery. Clin Orthop 227:10–23
15. Cragg AH, De Jong SC, Barnhart WH, Landas SK, Smith TP (1993) Nitinol intra vascular stent: results of preclinical evaluation. Radiology 189:775–778
16. Deacon P, Archer IA, Dickson RA (1987) The anatomy of spinal deformity: a biomechanical analysis. Orthopedics 10:897–903

17. Dubousset J, Herring JA, Shufflebarger H (1989) The crankshaft phenomenon. J Pediatr Orthop 9:541–550
18. Dwyer AF, Newton NC, Sherwood AA (1969) An anterior approach to scoliosis. A preliminary report. Clin Orthop 62:192–202
19. Ecker ML, Betz RR, Trent PS, Mahboubi S, Mesgarzadeh M, Bonakdapour A, Drummond DS, Clancy M (1988) Computer tomography evaluation of Cotrel Dubousset instrumentation in idiopathic scoliosis. Spine 13:1141–1144
20. Hanawa T, Ota M (1992) Characterization of surface film formed on titanium in electrolyte using XPS. Appl Surface Sci 55:269–276
21. Harrington PR (1962) Treatment of scoliosis: correction and internal fixation by spine instrumentation. J Bone Joint Surg A 44:591–610
22. Healy KE, Ducheyne P (1992) The mechanisms of passive dissolution of titanium in a model physiological environment. J Biomed Mater Res 26:319–338
23. Herzenberg JE, Waanders NA, Closkey RF, Schultz AB, Hensinger RN (1989) Cobb angle versus spinous process angle in adolescent idiopathic scoliosis. The relationship of the anterior and posterior deformities. Spine 15:874–879
24. ISO 10993/EN 30993 (1992) Biological evaluation of medical devices. Part 1: guidance on selection of tests 1992. International Organization for Standardization, Geneva
25. ISO 10993/EN 30993 (1992) Biological evaluation of medical devices. Part 3: tests for genotoxicity, carcinogenicity and reproductive toxicity. International Organization for Standardization, Geneva
26. ISO 10993/EN 30993 (1992) Biological evaluation of medical devices. Part 5: tests for cytotoxicity: in vitro methods. International Organization for Standardization, Geneva
27. ISO 10993/EN 30993 (1995) Biological evaluation of medical devices. Part 10: tests for irritation and sensitization. International Organization for Standardization, Geneva
28. Jarvis JG, Ashman RB, Johnston CE, Herring JA (1987) The posterior tether in scoliosis. Clin Orthop 227:126–134
29. Kambic H, Sutton C, Oku T, Sugita Y, Murabayashi S, Harasaki H, Kasick J, Shirey E, Nose Y (1988) Biological performance of TiNi shape memory alloy vascular ring protheses: a two-year study. Int J Artif Organs 11:487–492
30. Kaneda K, Shono Y, Satoh S, Abumi K (1997) Anterior correction of thoracic scoliosis with Kaneda anterior spinal system. A preliminary report. Spine 22:1358–1368
31. Klein JA, Hukins DWL (1983) Functional differentiation in the spinal column. Eng Med 12:83–85
32. Labelle H, Dansereau J, Bellefleur C, Poitras B, Rivard CH, Stokes IA, de Guise J (1995) Comparison between preoperative and postoperative three-dimensional reconstructions of idiopathic scoliosis with the Contrel-Dubousset procedure. Spine 20:2487–2492
33. Langenskiöld A, Michelsson JE (1961) Experimental progressive scoliosis in the rabbit. J Bone Joint Surg B 43:116–120
34. Lausmaa J, Kasemo B (1990) Surface spectroscopic characterization of titanium implant materials. Appl Surface Sci 44:133–146
35. Lenke LG, Bridwell KH, Blanke K, Baldus C, Weston J (1998) Radiographic results of arthrodesis with Cotrel–Dubousset instrumentation for the treatment of adolescent idiopathic scoliosis. A five- to ten-year follow-up study. J Bone Joint Surg A 80:807–814
36. Lonstein JE, Carlson JM (1984) The prediction of curve progression in untreated idiopathic scoliosis during growth. J Bone Joint Surg A 66:1061–1071
37. Lu S (1990) Medical applications of Ni-Ti alloys in China. In: Duerig TW, Melton KN, Stöckel CM, Wayman CM (eds) Engineering aspects of shape memory alloys. Butterworth-Heineman, London, pp 445–451
38. Matsumoto K, Tajima N, Kuwahara S (1993) Correction of scoliosis with shape-memory alloy. J Jpn Orthop Assoc 67:267–274
39. Meyer GH (1966) Die Mechanik der Skoliose. Arch Pathol Anat Physiol Klin Med 35:15–253
40. Murray DW, Bulstrode CJ (1996) The development of idiopathic scoliosis. Eur Spine J 5:251–257
41. Nachemson AL (1966) The load on lumbar discs in different positions of the body. Clin Orthop 45:107
42. Nachemson AL (1981) Disc pressure measurements. Spine 6:93–97
43. Nachemson AL, Elfström G (1971) Intravital wireless telemetry of axial forces in Harrington rods in patients with idiopathic scoliosis. J Bone Joint Surg A 53:445–465
44. Nijenbanning G (1998) Scoliosis redress. Design of a force controlled orthosis. Thesis. University of Twente, Twente
45. Oda I, Abumi K, Lü D, Shono Y, Kaneda K (1996) Biomechanical role of the posterior elements, costovertebral joints, and rib cage in the stability of the thoracic spine. Spine 21:1423–1429
46. Panjabi MM (1992) The stabilizing system of the spine. Part 1. Function, dysfunction, adaptation and enhancement. J Spinal Disord 5:383–389

47. Prince MR, Salzman EW, Schoen FJ, Palestrant AM, Simon M (1988) Local intravascular effects of the nitinol blood clot filter. Invest Radiol 23:249–300
48. Roaf R (1958) Rotation movements of the spine with special reference to scoliosis. J Bone Joint Surg B 40:312–332
49. Roaf R (1960) Vertebral growth and its mechanical control. J Bone Joint Surg B 42:40–59
50. Rogala EJ, Drummond DS, Gurr J (1978) Scoliosis: incidence and natural history. A prospective epidemiological study. J Bone Joint Surg A 60:173–176
51. Sanders JO, Little DG, Richards BS (1997) Prediction of the crankshaft phenomenon by peak height velocity. Spine 22:1352–1355
52. Sanders JO, Sanders AE, More R, Ashman RB (1993) A preliminary investigation of shape memory alloys in the surgical correction of scoliosis. Spine 18:1640–1646
53. Sanders MM (1993) A memory metal based scoliosis correction system. Ph.D. thesis. University of Twente, Twente
54. Schmerling MA, Wilkow MA, Sanders AE, Woosley JE (1976) Using the shape recovery of Nitinol in the Harrington rod treatment of scoliosis. J Biomed Mater Res 10:879–892
55. Simon M, Athanasoulis CA, Kim D, Steinberg FL, Porter DH, Byse BH, Kleshinsky S, Geller S, Orron DE, Waltman AC (1989) Simon nitinol inferior vena cava filter: initial clinical experience. Work in progress. Radiology 172:99–103
56. Stokes IA, Ronchetti PJ, Aronsson DD (1994) Changes in shape of the adolescent idiopathic scoliosis curve after surgical correction. Spine 19:1032–1037
57. Thompson GH, Wilber RG, Shaffer JW, Scoles PV, Nash CL Jr (1985) Segmental spinal instrumentation in idiopathic scoliosis. A preliminary report. Spine 10:623–630
58. Thulbourne T, Gillespie R (1976) The rib hump in idiopathic scoliosis. Measurement, analysis and response to treatment. J Bone Joint Surg B 58:64–71
59. Veldhuizen AG, Sanders MM, Cool JC (1997) A scoliosis correction device based on memory metal. Med Eng Phys 19:171–179
60. Vieweg U, Van Roost D, Wolf HK, Schyma CA, Schramm J (1999) Corrosion on an internal spinal fixator system. Spine 24:946–951
61. Wang JC, Yu WD, Sandhu HS, Betts F, Bhuta S, Delamarter RB (1999) Metal debris from titanium spinal implants. Spine 24:899–903
62. Wever DJ, Veldhuizen AG, Klein JP, Webb PJ, Nijenbanning G, Cool JC, Van Horn JR (1999) A biomechanical analysis of the vertebral and rib deformities in structural scoliosis. Eur Spine J 8:252–260
63. Wever DJ, Veldhuizen AG, Sanders MM, Schakenraad JM, Van Horn JR (1997) Cytotoxic, allergic and genotoxic activity of a nickel-titanium alloy. Biomaterials 18:1115–1120
64. Wever DJ, Veldhuizen AG, De Vries J, Busscher HJ, Uges DRA, Van Horn JR (1998) Electrochemical and surface characterization of a nickel-titanium alloy. Biomaterials 19:761–769
65. White AA, Panjabi MM (1990) Clinical biomechanics of the spine. Lippincott, Philadelphia
66. Williams DF (1982) Corrosion of orthopaedic implants. In: Williams DF (ed) Biocompatibility of orthopaedic implants. CRC, Boca Raton, pp 197–229
67. Zielke K, Stunkat R, Beeaujean F (1976) Ventrae derotations spondylodese. Arch Orthop Unfallchirurg 85:257–277

Shape-Memory Implants in Spinal Surgery: Long-Term Results (Experimental and Clinical Studies)

Boris M. Silberstein, Victor Gunter

1 TiNi Device for the Anterior Fusion of the Spine

1.1 Introduction

Anterior fusion of the spine is the most common surgical technique for the severe fractures and diseases of the spine. It provides correction of the deformity and fixation of the injured spinal segment. It is generally accepted that anterior fusion of spine should be followed by the internal fixation of the bone graft to prevent its dislocation and bony block formation between adjacent vertebral bodies. It allows early ambulation and rehabilitation so that many severe complications can be avoided.

Method of bone graft fixation by screws and plates are well known [1, 2]. However, their deployment may be technically difficult and there is also a significant risk of loosening and migration within and outside the vertebral bodies.

We developed a new device for the graft fixation, which is made of shape memory alloy (TiNi-10). It is simple and provides a strong fixation of the spine in patients with some kinds of injuries, but particularly in those with stable fractures as classified by F. Holdsworth [3].

1.2 Material, Method and Experimental Results

The device is a stirrup with two hooks, which are directed towards each other. The initial shape of the device is obtained under the temperature of +500°C. After resection of the injured vertebral body and implantation of the bone graft, the stirrup is cooled in chlorethyl and the hooks are straightened at the right angles to the base of the stirrup. The hooks are then inserted into the adjacent vertebral bodies. Once the close contact between the fixative device and the vertebral bodies has been achieved, the device is warmed using hot saline. A few seconds later the hooks begin to bend towards each other and compress the vertebral bodies providing a firm and strong fixation of the graft.

We conducted a study to compare two types of fixation rigid (with anterior plate) and elastic (by TiNi stirrup) in terms of the biomechanical properties and their influence on the formation of the bony block. In order to investigate the loads, which are experienced by the graft being fixed by superelastic TiNi stirrup, we carried out a series of experiments on cadaveric spine. Organic glass of the same shape and size was used in the place of the bone graft and photoelastic method was employed for assessment and measurement of these loads. It was demonstrated that TiNi shape memorizing stirrup produced a significant sustained compression on the anterior part of the graft while its posterior part became unloaded. However, when the plate was used, the graft did not experience any gradual compression (as it was previously found by other investigators). Thus, our device created a unique elastic system (vertebral bodies, bone graft and a stirrup) which provided a dynamic fixation of the graft. As will be shown later, this fact has a major effect on the duration of the bony block formation.

To establish the relationship between the biomechanical properties of fixation and the length of consolidation period, we performed an experimental study on morphogenesis of the ventral bony block after anterior fixation with Ti-Ni stirrup (TiNi-10 alloy). Grafting and anterior fixation of the graft was carried out on 24 mongrel dogs.

Morphogenesis of artificial bony block was investigated during the period up to 1 year and the results were as follows. In a week after the operation there was a gap between the graft and its bed. By the end of second week, this gap was filled in with granulations and osteogenous tissue. In a month after the operation resorption and formation of the new bone was observed within the graft. The proximal and the distal part of the graft adjacent to the vertebral bodies were tethered to them by fibrous tissue. At 3 months, the ossification of primary osteogenous tissue between posterior wall of the graft and its bed continued, while the tissue between the graft and adjacent vertebral bodies remained fibrous. At 6 months, the process of the bony block formation was completed and observed in 96.7% of cases. In two cases, the bony block was not obtained.

A similar experimental study was performed to assess the quality of the bony block and rapidity of consolidation after fixation with anterior plate. It was shown that solid bony block was obtained only in 75% of cases and within 9–12 months after the operation, which is much later than in the series with shape memory fixative device. Thus, our experimental results were very promising and we decided to conduct a clinical trial.

1.3 Clinical Results

From 1984 to 1995 a total of 88 patients with wedge-shaped and burst fractures of cervical, thoracic and lumbar spine were treated with anterior fusion using above-mentioned device. Neurological deficit was in 59 patients, out of them 11 patients died within 4–36 days after the operation. All of them had severe lesions (Frankel grade A) of cervical spine. Out of remaining 77 patients, 74 (96.1%) had high quality fusion and correction of the posttraumatic kyphotic deformity. Solid bony block was formed within 3–4 months after operation in cervical spine and within 6 months in thoracal and lumbar spine. In 64 (87.1%) patients the achieved correction was maintained and in 13 patients there was some loss (the mean of 4.5°) due to protrusion of the graft into the vertebral bodies or resorb-tion of one of its ends. Recurrence of kyphotic deformity was observed only in three cases. Patients with cervical fusion were mobilized the next day and patients with lumbar fusion within 2 weeks after surgery.

Postoperatively, in patients with cervical fusion the neck was immobilised by semi-soft collar for 2–3 weeks if the neurological deficit was absent. In patients with thoracic and lumbar fusion the bed rest was prescribed for 2–3 weeks after the operation. No external immobilization was applied. Device was not removed after the formation of the bony block (except for complications).

1.4 Complications

Complications were observed in four cases. In one case, the hooks were placed too close to the endplates of adjacent vertebral bodies. This resulted into penetration of the hooks through the endplates and ventral dislocation of the graft. In two cases the middle part of the stirrup and the hooks were overextended (more than 8%) and memory of TiNi alloy was partly loose. Fixation of the graft was therefore loose, which led to the migration of the device. In the last case, one of hooks was inserted into the adjacent disc and bony block included this disc. In three out of four cases the device was removed.

1.5 Discussion

In experiment the process of bone block formation with the device of shape memory alloy completed by the end of the 6 months after bone graft fixation and was observed in 96.7% cases. A similar experimental study was performed to assess the quality of the bony block and rapidity of consolidation after fixation with anterior plate. Solid bony block was obtained only in 75% of cases and within 9–12 month after the operation that is much later than in cases of bone graft fixation with shape memory device. This can be explained by the anatomical features of the spine in the quadrypedal animals where compression between the implant and vertebral bodies following fixation with a plate is not sufficient.

However, it is more likely to result from the biomechanical properties of TiNi superelastic stirrup as it produces a sustained compression of the anterior part and distraction of the posterior part of the bone graft. This principle is also used in Ilizarov's method and has been proven to be responsible for acceleration of the bone growth. We try to create the compression similar to the physiological load on the spine and by this way to accelerate reparative process in operated zone of the spine.

1.6 Conclusion

Thus, experimental and clinical results have demonstrated the advantages of fixation with a TiNi stirrup over fixation with a plate. The unique biomechanical properties, shape memory and superelasticity of the TiNi fixative device give a basis for its future use in spinal surgery.

2
Porous TiNi Implants

2.1
Introduction

One of the major advances in spinal surgery has been the introduction of synthetic materials, which can be used for anterior fusion instead of the autologous bone graft. Currently hydroxyapatite [4], coral [5], porous ceramics [6, 7], biocompatible osteoinductive polymer block (BOP-B) [8, 9] are widely used for this purpose.

We have developed and successfully used an implant made of porous TiNi alloy. It has the biomechanical properties, which are very similar to those of bone tissue of the vertebral body. We would like to present our clinical results in patients who underwent surgery for fractures, tumors of the vertebral bodies and degenerative disk disease. Our conception was following: to use metal porous implant as a prosthesis of vertebral body for retention of spine deformity correction (for example, kyphotic) after removal of whole or part of the injured vertebral body or the disc, providing stabilization in the operated zone by ingrowth of bone tissue into pores of the implant). Autologous bone tissue was not used alone as a support material for this purpose, because their mechanical characteristics are not sufficient and was used only for osteoinductive purposes.

2.2
Material, Method and Experimental Results

TiNi Implants made of TiNi-1P alloy were used. Their mechanical and physical characteristics are shown in Tables 1 and 2. In animal experiments, anterior fusion of the spine by porous TiNi implants alone was conducted in 23 rabbits.

Table 1. Physical characteristics of porous TiNi alloy (porosity 8–60%)

Tensile strength (MPa)	200–1000
Yield strength (MPa)	5–200
Relative elongation (%)	1–20
Relative deformity during shape recovery (%)	1–10
Step of shape recovery (%)	40–90
Stress developed during shape recovery (MPa)	Up to 400
Temperature interval for the shape changes	Up to 150

Table 2. Physical characteristics of TiNi-1P (porosity) and TiNi-10 alloys

Technical name of alloy	Composition	Temperature range of shape memory effect (°C)	Volume of deformity (%)	Shape recovery (%)	Shape recovery efforts (kg/mm)
TiNi-1P	TiNi (C, O, N, Fe, Cu)	−180 to +190	1–2	60	20
TiNi-10	TiNi (Mo, Ag, Fe, C)	+10 to +25	6–8	95–99	60–80

The operation was performed via the anterior extraperitoneal approach. Following the exposure of two vertebral bodies in the lumbar spine, the intervertebral disc was removed and the defect was filled in with a porous TiNi implant.

Radiological and histological studies were conducted at 1, 3 and 6 months following anterior fusion with the porous TiNi implant. At 1 month, the implants were unchanged and surrounded by a connective tissue penetrating through the pores of the implant. At 3 months, the TiNi implant was surrounded by the solid fibrous capsule. At 6 months, a bony block around the implant was observed. It is important to note that a solid bony block was formed in all animals. There were no osteoporotic changes around the TiNi implant in the adjacent vertebral bodies.

2.3 Clinical Results

Between 1988 and 1998, a total of 401 patients underwent spinal surgery with porous TiNi implants. Out of them two patients had tumorous of the vertebral bodies, osteomielytis of vertebral body-2 (cervical and lumbar level), fractures of the spine-172 (87 in cervical, 53 in thoracic and 32 in lumbar), 210 patients had degenerative disc disease (56 cervical, 10 thoracal, 144 lumbar).

Besides, in 15 patients only porous TiNi implants were used for filling up the posterior bone defects formed after laminoplasty in the cervical and lumbar spine in patients suffering from spinal canal stenosis. Anterior fusion with only porous TiNi implant was performed in 142 patients and in combination with autobone graft taken either from the rib, endplates of the adjacent vertebral bodies or the iliac crest in 244 patients. Patients with cervical fusion were mobilized the next day and patients with lumbar fusion – within 2 weeks after surgery. Plain X-ray, CT-scan and MRI were used to monitor the formation of the bone-metal block.

In group of patients with only TiNi implants and in combination with autologous bone graft it was demonstrated within 3 months in cervical spine and 6 months in lumbar spine. 184 patients were followed up for a period of 1–3 years, 136 patients 4 years, 44 patients 7 years and two patients 9 years. In 87.4% cases the results of treatment were excellent, and in 12.6% good and satisfactory. Recurrence of kyphotic deformity was observed only in five cases. This can be attributed to the technical fault during the operation, inaccurate estimation of osteoporosis of the vertebral body prior to surgery or its subsequent progression.

2.4 Conclusions

Thus, our preliminary results suggest that porous TiNi alloy can be successfully used for the vertebral body replacement since its biomechanical properties are similar to those of the vertebral body. Further studies are required to establish the place of TiNi porous implants in spinal surgery.

References

1. Caspar W (1987) Anterior Stabilization with the Trapezial Osteosynthetic Plate. Technique in Cervical Spine Injuiries. In: Cervical Spine I (Edited by Kerh P, Weinder A), Springer-Verlag Wien New York, p 198–209
2. Zucherman J, Hsu K, White A, Wynne G (1988) Early results of spinal fusion using variable spine plating system. Spine 13:570–579
3. Holdsworth FW (1963) Fractures, Dislocations, and Fracture-Dislocations Of The Spine. J Bone Joint Surg 45-B p 6–20
4. Cook SD, Whitecloud TS, Reynolds MC, Harding AF, Routman AS, Kay JF, Jarcho M (1987) Hydroxylapatite Graft Materials for Cervical Spine Fusions In: Cervical Spine I, (Edited by Kerh P, Weinder A) Springer-Verlag Wien New York, p 257–262
5. Privat JM (1988) Matériaux biorésorbables. In: Ostéosynthèse Rachidienne. Sauramps Medical, Montpellier, p 71
6. Tomita K, Toribatake Y, Kawahara N, et al. (1994) Total en bloc spondylectomy and circumspinal decompression for solitary spinal metastasis. Paraplegia 32:36–46
7. Kiwerski JE, Ogonowski A, Bieniek J, Krasuski M (1994) The use of porous corundum ceramics in spinal surgery. Int Orthop 18:10–13
8. Lozes G, Fawaz A, Cama A, et al. (1989) Discectomies of the lower cervical spine using interbody biopolymer (B.O.P.) implants. Advantages in the treatment of complicated cervical arthrosis. A review of 150 cases. Acta Neurochir (Wien) 96:88–93
9. Brotchi J, Levivier M, Raftopoulos C, Baleriaux D, Noterman J (1989) Use of synthetic graft biocopolymer BOP in anterior cervical spine surgery J. Rachis 1, p 367

The Use of a Memory-Shape Staple in Cervical Anterior Fusion (about 100 Human Implantations)

Olivier Ricart

1 Introduction

This study will present one application of nickel–titanium shape-memory staple in spinal surgery. Clinical and radiological results based on 100 human anterior cervical spine implantations will be reported. These cervical fusions have been performed as described by Smith and Robinson [25], and were stabilized with a NiTi shape memory staple producing a dynamic compression effect on the bone graft.

We will try to demonstrate that NiTi shape memory staple provides excellent functional capabilities for implantation in the cervical spine, improving fusion time, implant and graft stability, safety of the surgical procedure and finally allowing an easier implantation technique compared to other anterior cervical surgical techniques.

2 Nitinol: Properties, Biocompatibility

In the early 1960s, Buehler [5] discovered the shape memory effect in an equiatomic alloy of nickel and titanium (Nitinol, chemical symbol NiTi). Since that time, intensive metallurgical investigations have been made to explore its basic behavior. In the early 1970s, NiTi was first used biomedically, in particular for vascular and orthodontic applications [9, 10]. It was only in the mid-1990s, that the first widespread commercial stent and orthopedic applications made heir breakthrough in medicine [23, 24].

NiTi has unique properties as compared to ordinary implant metals. Using its thermal shape memory property, this material sensing a change in external temperature is enable to convert to a preprogrammed shape. Nitinol is an intermetallic compound made up of 50% titanium and 50% nickel, the atoms of which are closely linked in a crystalline framework which prevents release of ions and corrosion after an appropriate surface treatment.

Nitinol has the property of being soft and easily deformable under the influence of temperature (martensite form). This so-called martensitic transformation is a solid state phase change, which leads to homogeneous deformation by shearing of the crystalline framework. The return to its original shape and rigidity after being heated to a higher temperature (austenite form) is known as "one-way memory effect phenomenon". Nitinol provides a possibility to make self-locking, self-expanding, and self-compressing implants activating at body temperature.

Because of the high nickel content of NiTi, it is theoretically possible that ions of nickel may release from the material due to corrosion, and may have deleterious effects and pathological consequences. Therefore, the biocompatibility of NiTi must be very well confirmed before it can be safely used as an implant material. At present time, there are enough conclusive biocompatibility data available on NiTi [1, 2, 7, 19, 22, 27, 28]. However, some questions remain about the appropriate surface condition of Nitinol in order to decrease the low initial release of nickel ions.

NiTi has good in vitro biocompatibility with human osteoblasts and fibroblasts. There were no toxic effects or inhibition of cells proliferation related to NiTi contact. The neutral and perineural responses to implanted NiTi were clearly non-toxic, and no qualitative differences in histology could be seen as compared with stainless steel [22].

At the beginning, NiTi with a water-sanded surface released more nickel in the cell culture medium than electrolytically polished stainless steel. After two days, however, the concentration decreased to about the same level. Some special treatments have been reported to improve NiTi's corrosion behavior. It has been demonstrated that thermal oxidation and titanium nitride coatings improve the corrosion resistance of NiTi [11, 26]. Electrochemical passivating treatment can also been used to have a very good biocompatibility and an excellent potential for clinical use.

3 Device Description

The most widely used shape memory staple measures 24 mm along its long axis and is slightly convex to provide a better fit with the inferior cervical lordosis [20]. The feet measure 10 mm, so as to avoid any intra-canal penetration and are at an angle of 70° in relation to the main part in its martensite shape. The wire has a rectangular section of 3 mm by 2 mm.

4 Surgical Technique

The surgical technique is simple, which is its major advantage. To complete the procedure a staple-holder forceps, a bradawl, a tube-key enabling screwing of threaded screws from the Caspar retractor as well as spreading of the feet of the staple, and a modified Caspar type retractor are necessary (Fig. 1). The Smith-Robinson technique via a left presternomastoid approach is used, the first step being placement of the modified Caspar type retractor directly to the vertebral bodies, on either side of the disk to be treated.

Two 2.7-mm holes are drilled in the vertebral bodies using a drill-bit through the holes of the retractor. Threaded wires are inserted to hold the retractor, in place during the diskectomy and decompression of the anterior surface of the spinal cord. The modified Caspar retractor enables distraction of the inter-somatic space for insertion of the graft. Anti-cortical iliac graft was used in all cases. The amount of distraction between the feet is shown in mm directly on the retractor. Distraction between 24 mm and 26 mm was considered adequate.

Fitting of the staple can then start. The same holes drilled to hold the retractor are used for the feet of the staple. After removing the threaded wires, the holes are carefully identified on the interior surface of the vertebral bodies and the staple, which has been kept in a freezer below 0°C before the operation, is then deformed by spreading the feet at a right angle in relation to the main part. This remains stable at room temperature and enables handling without any risk of the staple changing shape spontaneously.

Heating the alloy above 25°C is necessary for transformation to begin. Once the staple has been inserted into the holes in the vertebral bodies over the bone graft, warm physiologic saline solution at a temperature of 30°C is poured on it, leading in a few seconds to closure of the feet of the staple. Direct compression of the bone and clamping of the staple is then observed. After the operation, the patients are all immobilized using a soft cervical collar for 1 month.

5 Materials and Methods

Clinical results were evaluated using Odom's classification [18]. Radiological results were assessed on standard and flexion-extension dynamic views (Fig. 2). Fusion was confirmed by the presence of a continuous bridge of bone joining the vertebral bodies, and graft collapse was defined by a decrease in height of more than 30%.

6 Results

Between May 1992 and June 1998, 100 interbody anterior cervical fusions have been performed using this NiTi shape-memory staple. The series include 52 men and 48 women, with a mean age of 52 years (33 to 80). Mean follow-up is 48 months (12–72 months). Indications for surgery are presented in Table 1. Eighty-three (83) mono-segmental arthrodesis (Fig. 3), 12 multi-segmental arthrodesis (Fig. 4) and five corpectomies were performed. The level most often involved was C5–C6.

Table 1. Indications for surgery

Soft disk prolapse	47
Spondy losis	39
Metastatic disease	5
Post-traumatic instability	7
Infection	1
Correction of kyphosis	1

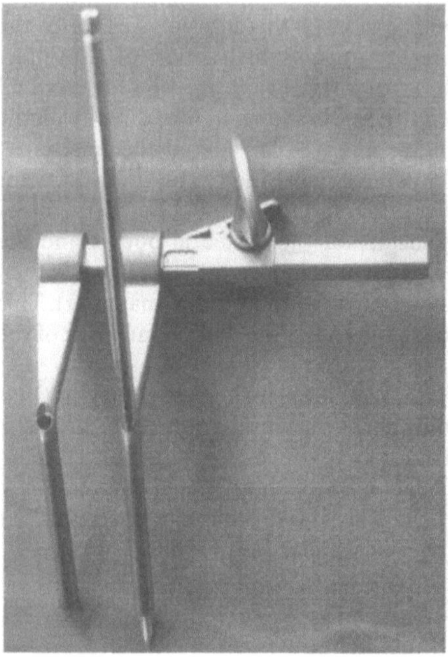

Fig. 1. Modified caspar retractor

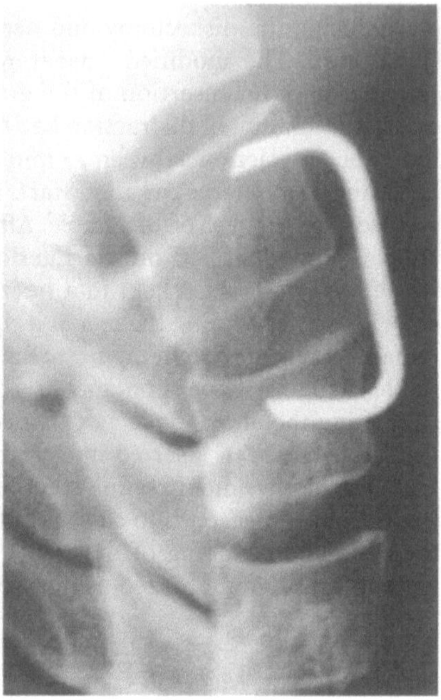

Fig. 2. Immediate post-operative X-ray

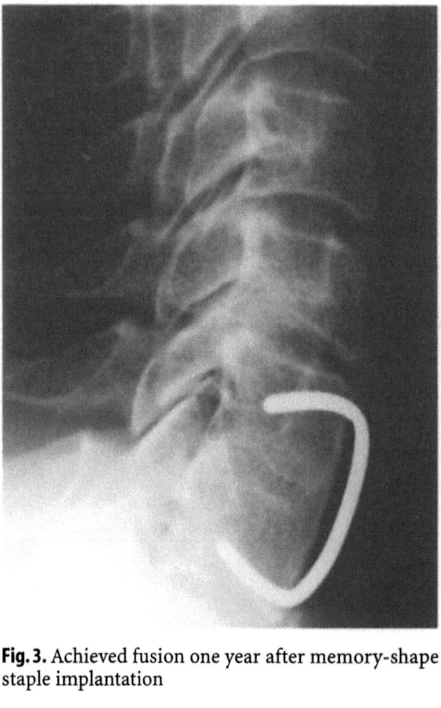

Fig. 3. Achieved fusion one year after memory-shape staple implantation

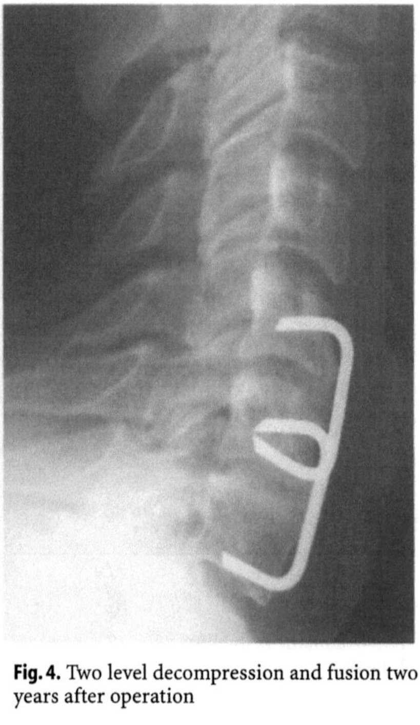

Fig. 4. Two level decompression and fusion two years after operation

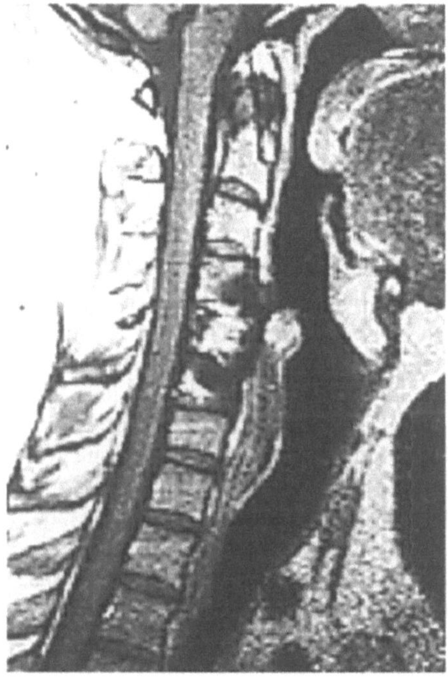

Fig. 5. MRI after NiTi staple implantation (note the absence of artefacts)

Good and very good clinical results were observed in 90% of the cases, while fair result was seen in 10%, the latter nevertheless involving an improvement for the patient in relation to the preoperative status. There were no poor results.

Radiologically, 99 fusions were observed in a relatively short time (mean 7 weeks, range from 5 weeks to 15 weeks). One non-union was found but this patient did not complain of any pain. There were three graft collapses (two early and one late at 8 months).

There were three partial shifts of a staple foot. This was attributed to technical errors, the foot having been inserted into a disk space, above or below the affected one. This had no clinical consequences in two patients, while one staple was removed because of dysphagia.

Dislocation of bone graft, loosening of the staple or kyphotic deformity did not occur. No complication related to the donor site was observed, except for short-lived post-operative pain (Fig. 5).

7 Other Complications

The usual complications of this type of surgery and unrelated to osteofixation were encountered: one Horner's syndrome, which regressed, one tumor recurrence following corpectomy for metastasis which required subsequent radiotherapy with a good result 5 years after surgery and one case of insufficient spinal-cord decompression requiring secondary laminectomy with a good final result. In three others cases, the staple were removed for dysphagia, after the fusion has been achieved.

8 Discussion

Three years ago, we reported our experience of ten cases of an arthrodesis using hydroxypatite (HA) interbody shaped blocks and the staple, this technique having been abandoned with an almost 50% hydroxyapatite fracture rate or worrisome lipping. The low resistance under loading of the HA explains the frequent incidence of implant fracture and the subsequent bad radiographic results such as the disc-height loss and the subsidence of the implants. However, a new generation of HA implants more resistant seem promising. In our opinion, for the present, the tri-cortical iliac graft appears to be the most reliable solution.

Since its description by Smith and Robinson in 1954, anterior cervical fusion has become a reliable procedure [3, 6, 15, 25]. Its use in single level degenerative disease is widespread with predictable good results in 63–81% of cases. However, graft migration (2–9%) or graft collapse (28%) has led to an unacceptable rate of progressive kyphosis and non union [6, 12–14, 31].

Bailey [3] and Cloward [8] tried to solve the problem by improving bone graft shape and fusion technique. Nevertheless, there was still a high rate of graft extrusion (2–9%), more specifically in multilevel decompression, tumor resection, cervical fracture or dislocation, and kyphotic deformity.

In fact, post-operative stability following anterior cervical fusion is not related to the load-bearing capacity of the bone graft. As reported in the literature, all types of grafts could bear a great compressive load. For instance, Smith and Robinson's graft could bear a compressive load of 50.9 kPa/cm^2. Cloward's, with the highest rate of collapse, could bear 41.6 kPa/cm^2, Bailey's could bear 35.2 kPa/cm^2 [29]. The major cause of graft migration or collapse at early post-operative stage seems to be the anterior shear at the interface between the graft and the adjacent end plates of the vertebrae which is mainly due to cervical motion. Therefore, long term external post-operative bracing (10–12 weeks) is required in cases without internal fixation.

At late post-operative stage (after 2 months), the risk of graft extrusion is very low, but graft collapse may occur leading to progressive kyphosis and non union. At this stage the major cause of graft collapse seems to be the avascular necrosis of the graft [6]. In order to prevent these complications, many authors have reported the use of an additional cervical anterior plate.

Bohler [4] was the first to describe the use of a standard plate for anterior fixation of the cervical spine. Since its introduction, the anterior cervical plate has not gained universal acceptance largely because of potential risks of spinal cord injury during bicortical screw placement or soft tissue injury secondary to hardware migration (5%) [6, 12–14].

In 1986, Morscher [17] introduced the use of a porous ingrowth expandable screw that would lock into the anterior cervical plate. This system was developed to prevent the migration and loosening of screws. A secondary advantage is that with this system it is not necessary to purchase the posterior cortex of the vertebral body to obtain excellent fixation. Kostuik [14] reported good results using the instrumentation with a mean fusion time of 13 weeks.

Mei [16] describes in 1997 the use of a NiTi shape memory expansion clamp for anterior cervical fusion. The advantages in terms of fusion time, biocompatibility, and hardware loosening were clearly demonstrated. However, in our opinion, the radiological assessment for fusion was difficult because the clamp is implanted to the disc space.

NiTi is an equiatomic alloy of nickel and titanium, which provides a possibility to make self-locking, self-expanding, and self-compressing, implants activating at body temperature. In our series, the indication for surgery was most often monosegmental degenerative disease (83 cases). We used the shape memory anterior staple to obtain immediate stable fixation. The patients did not require any major external stabilization device after surgery. The arthrodesis rate was 99% as determined by plain radiographs. The average time for fusion was 7 weeks.

Three patients had staple feet placed in a disc space, which led to partial hardware migration. There were three graft collapses quite likely due to an avascular necrosis. The incidence of this complication is not higher than in other large series. The self compressing effect of the shape memory staple does not increase the risk of graft crushing or collapse because the compression force exerted by the staple is very low (3 kg/cm^2).

The staple permits to avoid the risk of graft extrusion in a very simple way, as the procedure does not length the operative time compared with a non instrumented fusion. The shape memory property of the implant provides a self-locking effect on the bone graft, promoting intervertebral fusion.

In other large series the incidence of significant hardware loosening is approximately 5% (4.3–15.4%). In our opinion, the reason for this rate of loosening is the excessive axial stiffness of the plate and screws. Excessive stiffness may also cause a stress shielding effect on bone graft leading to delayed graft incorporation. The axial stiffness of the staple is significantly lower than that of a plate. That explains the absence of hardware loosening in our series if the staple is properly inserted, as well as quicker fusion time compared with fusions using other means of internal fixation.

We did not find any additional complication when staples were used for multi-level fusions or corpectomies (Fig. 3). However, use of staples for multi-segmental athrodeses in degenerative disease processes is open to more concern since it is preferable to use one staple at each level, the insertion of which is more difficult. In addition, a screw-plate restores lordosis better in these situations.

Use of the shape-memory staple in traumatic or tumoral pathology must be considered case by case in relation to spinal stability and the surgeon's experience. More rigid osteofixation methods are nevertheless to be recommended.

9 Conclusions

The advantages of this osteofixation technique compared to a screwed-plate are as follows: minimal space taken up (low profile), easy fitting, no risk to the spinal cord (in comparison with bi-cortical screwing). Time required for fusion seems shorter (7 weeks) in comparison to instrumented fusions using a plate (13 weeks).

The risk of post-operative shifting of the graft is excluded in comparison to non instrumented fusions without internal fixation. The shape-memory effect helps to hold the staple in place and prevents its migration, if it is properly inserted. Axial compression of the graft enhances its incorporation, but without crushing or kyphotic effect (the compression force exerted by the staple is 3 kg/cm^2). The stability provided by the implant offers the possibility to limit cervical immobilization to a simple soft collar for one month. Nickel–titanium allows the possibility of post-operative magnetic resonance imaging.

In our opinion, the ideal indication for this type of implant is mono-segmental cervical arthrodesis for degenerative disease processes, where it is sought to prevent possible anterior shift of the graft. The shape-memory staple fully meets this criterion in a very simple way. The time required for the operation is not increased since holes have been drilled previously for insertion of the modified Caspar type retractor. Additional risk related to vertebral screwing is avoided. Finally, the shape-memory staple is safe and effective, for this reason it should be considered for use when treating patients who need monosegmental anterior cervical decompression and fusion.

References

1. Assad M, Lombardi S, Berneches S, Desrosier EA, Yahia LH, Rivard CH (1994) Assays of cytotoxicity of the nickel–titanium shape memory alloy. Ann Chir 48:731–736
2. Assad M, Yahia LH, Rivard CH, Lemieux N (1998) In vitro biocompatibility assessment of a nickel-titanium alloy using electron microscopy in situ end-labeling (EM-ISEL). J Biomed Mater Res 41:154–161
3. Bailey RW, Badyley CE (1960) Stabilization of the cervical spine by anterior fusion. J Bone Joint Surg Am 42:565–594
4. Bohler J, Gaudermerk (1980) Anterior plate stabilization for fracture dislocations of the lower cervical spine. J Trauma 20:203–205
5. Buehler WJ, Wang FE (1967) A summary of recent research on the Nitinol alloys and their potential application in ocean engineering. Ocean Eng 1:851–859
6. Caspar W, Barbier DD, Klara PM (1989) Anterior cervical fusion and Caspar plate stabilization for cervical trauma. Neurosurgery 25:491–502
7. Castelman LS, Motzkin SM (1981) The biocompatibility of Nitinol. In: Williams DF (ed) Biocompatibility of clinical implant materials, vol 1. CRC Press, Inc., Boca Raton, Florida pp 129–154
8. Cloward RB (1958) The anterior approach for removal of ruptured cervical discs. J Neurosurg 15:602–617
9. Dotter CT, Buschmann RW, McKinney MK, Rosch J (1983) Transluminal expandable nitinol coil stent grafting: prelimenary report. Radiology 147:259–260
10. Drugacz J, Lekston Z, Morawiec H, Januszewski K (1995) Use of TiNiCo shape-memory clamps in the surgical treatment of mandibular fractures. J Oral Maxillofac Surg 53:665–671
11. Endo K, Sachdeva R, Araki Y, Ohno H (1994) Effects of titanium nitride coatings on surface and corrosion characteristics of NiTi alloy. Dent Mater J 13:228–239
12. Gassman J, Seligson D (1981) The anterior cervical plate. Spine 7:700–707
13. Goodman J, Seligson D (1983) The anterior cervical plate. Spine 8:700–7067
14. Kostuick JP, Conolly PJ, Esses SI, Suh P (1993) Anterior cervical plate fixation with the titanium hollow screw plate system. Spine 18:1273–1278
15. McAfee PC, Bohlman HH (1989) One stage anterior cervical decompression and posterior stabilization with circumferential arthrodesis. J Bone Joint Surg A 71:78–88
16. Mei F, Ren X, Wang W (1997) The biomechanical effect and clinical application of a NiTi shape memory expansion clamp. Spine 22:2083–2088

17. Morscher E, Sutter F, Jennis M, Olerud S (1986) Die Vordere Verplattung der halswirbelsaule mit dem hohischrauben plattensystem. Chirurgie 57:702–707
18. Odom GL, Finney W, Woodhall B (1958) Cervical disc lesions JAMA 166:23–28
19. Otsuka K, Wayman CM (1998) Mechanism of shape memory effect and superelasticity. In: Otsuka K, Wayman CM (eds) Shape-memory materials. Cambridge University, Cambridge, pp 27–48
20. Ricart O (1997) The use of a memory shape staple in cervical anterior fusion. In: Pelton AR, Hodgson D, Russell SM, Duerig TW (eds) Proceedings of SMST 1997. Shape Memory and Superelastic Technologies, Pacific Grove, pp 623–626
21. Ryhänen J, Raatikainen T, Kaarela O (1998) Complete acromioclavicular dislocation repair with a new shape memory AC-hook implant and operative technique and prospective pilot study. In: Proceedings of 7th international congress on surgery of the shoulder, Sydney, Australia, p 292
22. Ryhänen J (1999) Biocompatibility evaluation of nickel-titanium shape memory metal alloy. Acta Univ Ouluensis
23. Sanders JO, Sanders AE, More R, Ashman RB (1993) A preliminary investigation of shape memory alloys in the surgical correction of scoliosis. Spine 18:1640–1646
24. Silberstein B (1997) Subtotal and total vertebral body replacement and interbody fusion with porous Ti-Ni implants. In: Pelton AR, Hodgson D, Russell SM, Duerig TW (eds) Proceedings of SMST 1997. Shape Memory and Superelastic Technologies, Pacific Grove, pp 617–621
25. Smith GW, Robinson RA (1958) The treatment of cervical spine disease by anterior removal of the intervertebral disc and interbody fusion. J Bone Joint Surg Am 40:604–624
26. Trepanier C, Tabrizian M, Yahia LH, Bilodeau L, Piron DL (1998) Effect of modification of oxyde layer on NiTi stent corrosion resistance. J Biomed Mater Res 43:433–440
27. Van Humbeeck J, Stalmans R (1998) Characyeristics of shape memory alloys. In: Otsuka K, Wayman CM (eds) Shape memory materials. Cambridge University, Cambridge, pp 49–83
28. Villermaux F, Tabrizian M, Rhalmi S, Rivard C, Meunier M, Czeremuszkin G, Piron DL, Yahia LH (1997) Cytocompatibility of NiTi shape memory alloy biomaterials. In: Pelton AR, Hodgson D, Russell SM, Duerig TW (eds) Proceedings of SMST 1997. Shape Memory and Superelastic Technologies, Pacific Grove, pp 417–422
29. White AA III, Jupiter J, Southwick WO, Panjabi MM (1973) An experimental study of the immediate load bearing capacity of three surgical constructions for anterior spine fusions. Clin Orthop 91:21–28
30. Younger WM, Chapman MW (1989) Morbidity at bone graft donor site. J Orthop Trauma 3:192–195
31. Zdeblick TA, Cooke ME, Wilson D, Kunz DN, McCabe R (1993) Anterior cervical discetomy, fusion and plating. A comparative animal study. Spine 18:1974–1983

The Double Compressive Nickel–Titanium Shape-Memory Staple in Foot Surgery

Louis Samuel Barouk

1 Introduction

Since 1973, we were looking for a reliable and easy osteosynthesis system for the different osteotomies we perform in the forefoot. These osteotomies are the great toe first phalanx osteotomy, the arthrodesis of the first metatarso-phalangeal joint, and the calcaneo-cuboïd and the talo-navicular fusion. Screws, Kirschner wires or staples were used before, but were not provided a very reliable osteosynthesis, mostly in cases needing a strong fixation, as for example in shaft shortening of the great toe first phalanx or arthrodeses.

Since 1986, we have been using a double compressive nickel titanium staple invented by M. Bertholet. From 1986 until now, we have modified this staple in order to improve not only its reliability but also the easiness of implantation. In this chapter we will describe: the staple, its application in different useful locations of the foot, the results including the long term results and the local tolerance.

2 The Doubly Compressive Nickel–Titanium Shape-Memory Staple

2.1 Description

It is composed of two prongs joined by an oval part, as described in the Figure 1. The two prongs have the following particularities:

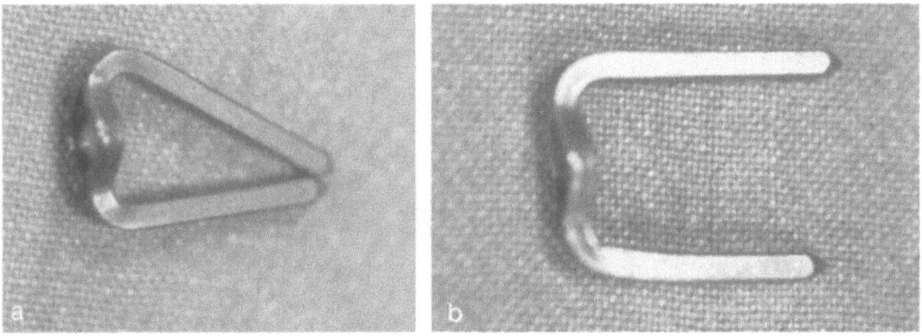

Fig. 1a,b. The shape memory staple: **a** at body temperature, the two prongs are closer and the oval part is shortened, providing a double compression effect; **b** on a temperature less than 30°C, the staple can be shaped so the prongs become parallel, and the oval part is lengthened (**a** and **b** are two different pictures of the same staple)

1. The section diameter is rectangular, in order to present a larger surface where the compression is applied.
2. The end is smoothed because the staple is inserted after the bone has been drilled, not much strength is then needed to place the staple into the bone; the prongs can also go beyond the bone surface without resulting in soft tissue trauma.
3. The oval part is composed of two diverging branches, oval-shaped.

2.2 Working Principles

The nickel–titanium alloy has received a previous treatment, which allows to modify the shape of the staple in a predetermined temperature: in this case, from 10°C to 25°C. The shaping is performed as followed:

1. The two prongs are pulled to be parallel.
2. The oval part is elongated, thus the length of the middle part joining the prongs is increased and this is the main interest of this staple.
3. At body temperature, after placement into the bone, the two prongs become closer to each other, resulting in a compression at this location.
4. The two branches of the oval part are diverging, the oval thus becomes wider, resulting in a diminution of the length of the middle part, providing also a compression at this location.
5. Therefore, there is a double (bicortical) compression: located both at the prongs and at the oval part (Fig. 2).
6. Furthermore, the oval part being in a perpendicular plane to the prongs, the stability of the osteosynthesis is notably increased.

The compression provided by this staple is not decreasing with time. This constant compression effect is very useful for the osteosynthesis, first to provide solid fixation despite sub-optimal reduction of bone fragments (Fig. 3a), secondly to maintain a continuous compression force when bone resorption of the

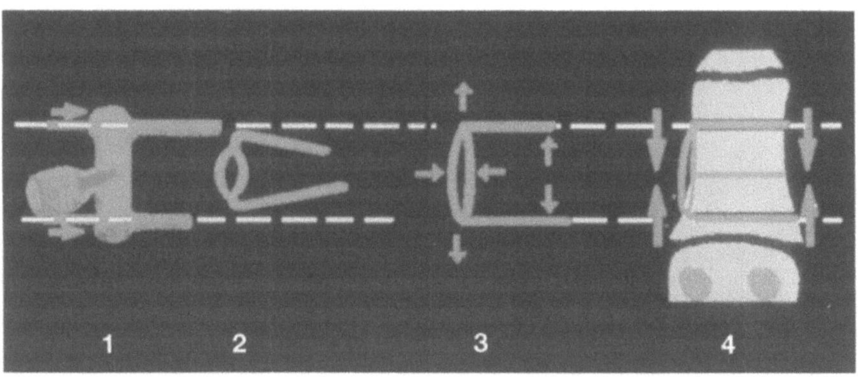

Fig. 2. Working principles. *1*, The drill guide. *2*, The staple natural shape at 35°C or more. *3*, The shape of the staple can be modified at less than 30°C, the prongs becoming parallel and the oval part lengthened. *4*, The double compression effect, one on the oval part, one on the prongs

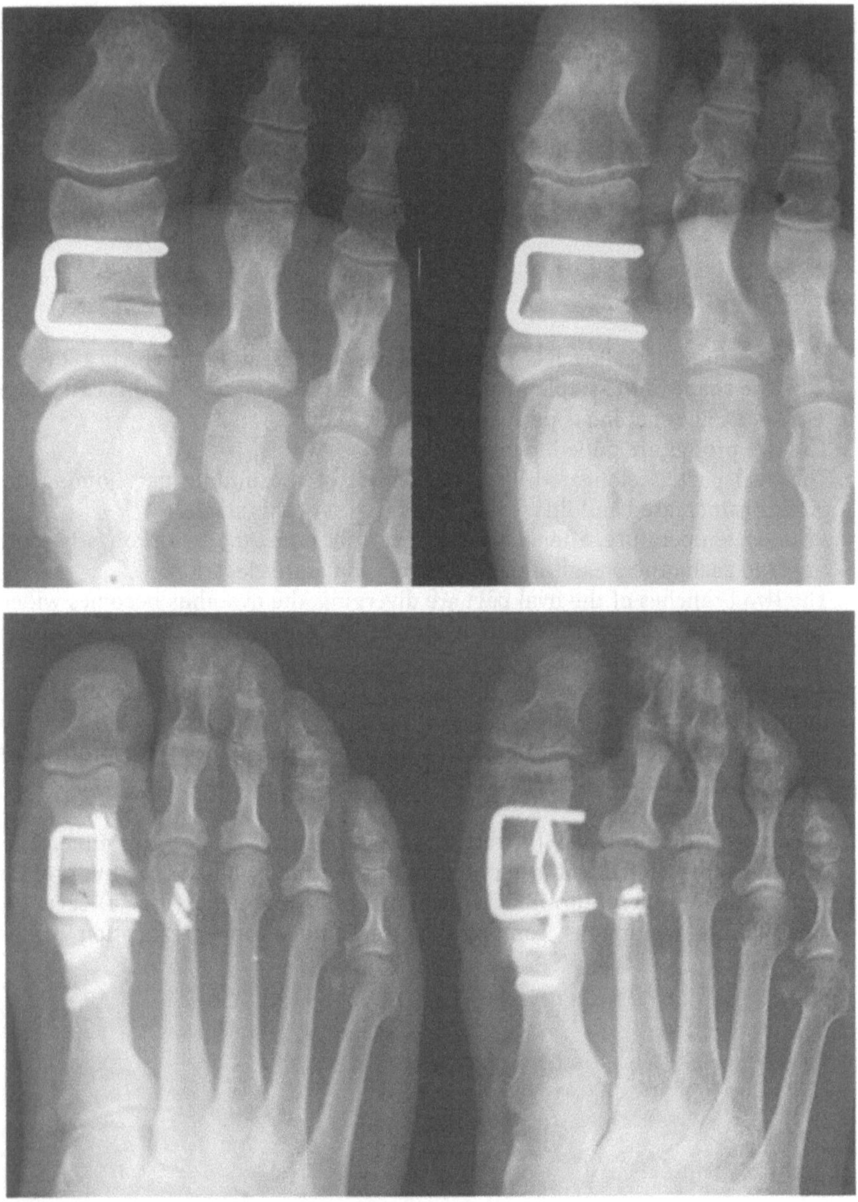

Fig. 3a, b. Remaining compression effect of the staple. **a** Great toe shortening: on the left, the osteosynthesis immediately after surgery; on the right, one month later. **b** Treatment of a pseudarthrodesis on a previous fusion tentative of the first MTP joint, by only removing the previous staple and setting the memory staple without any other procedures on the fragment extremities or addition of bone graft fragment ends occurs. One more effect of this permanent compression is given by the possibility to treat pseudarthrosis. The surgery consists of removing the previous osteosynthesis implants and placing the shape memory staple. No any additional surgical procedure at the fragment extremities or bone grafting is necessary (Fig. 3b).

3 Materials and Methods

From 1986 to 1999, 1850 cases (feet) have been operated on (1180 patients) with the use of shape memory staple, for the following indications:

1. Shaft osteotomy of the great toe first phalanx: 1530 cases, combined in 1380 cases with the scarf osteotomy of the 1st metatarsal (hallux valgus correction) and in 150 cases with scarf osteotomy and Weil osteotomy of the lesser metatarsals
2. Arthrodesis of the first metatarso-phalangeal (MTP) joint: 190 cases
3. Arthrodesis of the Lisfranc joints, osteosynthesis or arthrodesis of the hindfoot: 130 cases

   Follow up:

   1. 1–12 years
   2. Average: 4.5 years

4 Contraindications

The only contra indication using this staple is due to the strength of the compression effect, specially at the prongs extremities: in fact, in very osteoporotic bones we observed pinching of the prongs extremities. One solution was then to increase the distance between the two drill holes. However, we still consider the two following situations as contradiction:

1. In the great toe, first-phalanx osteotomy with very osteoporotic bones
2. In the first MTP joint, arthrodesis needing a large resection for patients presenting very osteoporotic bones

These contra indications represent less than 5% of the total cases we have performed.

5 Clinical Results

5.1 Shaft Osteotomy of the Great Toe First Phalanx

In most cases, a resection of part of the shaft is necessary, in order to shorten the great toe; advantages of great toe shortening are emphasized by many authors, specially the decrease lever arm in both transverse and sagittal plans. However,

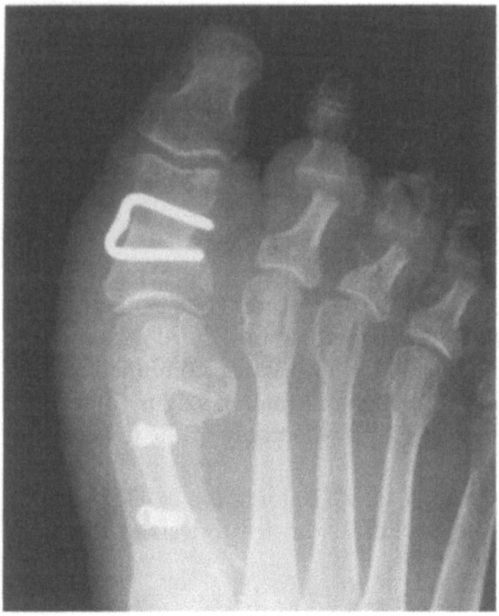

Fig. 4. One case with osteoporotic bone (on the right) secondary pinching of the prong extremities is observed. This problem does not occur anymore since modification of the drill guide and since we avoid to use the shape memory staple in extremely osteoporotic bones

this resection results in more instability, needing a solid fixation. The different osteosynthesis performed from 1973 to 1986 used staples, K-wires or screws. They were considered unsatisfactory because of the insufficient strength of fixation. On the contrary, since we have been using the shape memory staple, the bicortical compression (medial and lateral cortex) appears to provide a much solid fixation [1–4, 5].

During our learning experience with shape memory staples, some technical problems were encountered. Too much pinching of the prongs, particularly in osteoporotic bones, and secondary displacement of the staple were observed and caused by the strong compressive force of the staples (Fig. 4).

We have brought the necessary solution to this drawbacks. In fact, since 1992, the osteotomy technique has been modified (Fig. 5), and this complication is almost not observed anymore: first, a K-wire and a cannulated drill are implanted proximally (Fig. 5a). The osteotomy is performed in the right place using the drill guide (Fig. 5b, c), axial K wire (Fig. 5d) allows the control of the correction deformity as well as avoids the displacement of the fragments while setting the staple (Fig. 4e–g). A radiographic study was followed after the osteotomy (Fig. 4h)

Actually, the shaft first phalanx osteotomy of the great toe using the shape memory staple is extremely reliable and easy. The strong fixation allows immediate functional recovery. Furthermore, no healing delay neither no pseudoarthrosis were observed, except in our very early experience (1986–1988).

The isolated great toe osteotomy (150 cases) had less indications than combined with the scarf osteotomy (1380 cases) as used in the hallux valgus surgical treatment (Fig. 6a) [1, 2]. We have been using this combination since 1991; the results are so good that actually we use this combination in any case of hallux valgus deformity: the great toe osteotomy is performed after the three previous steps which are:

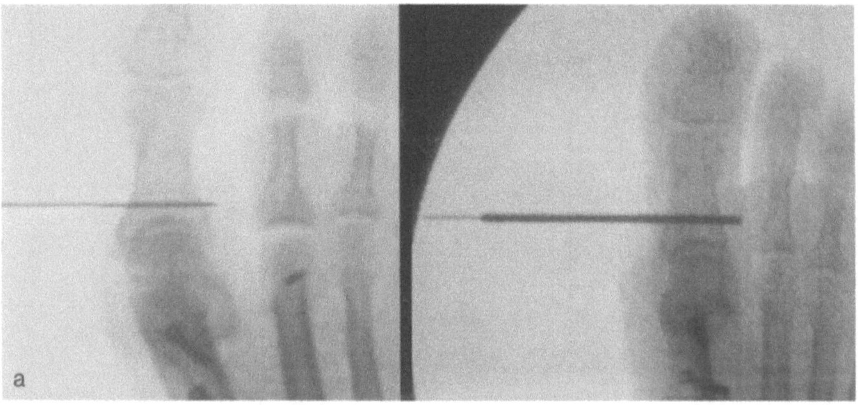

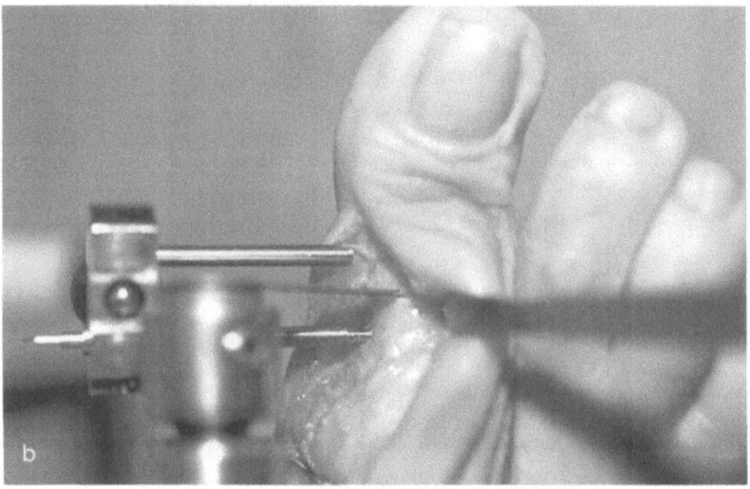

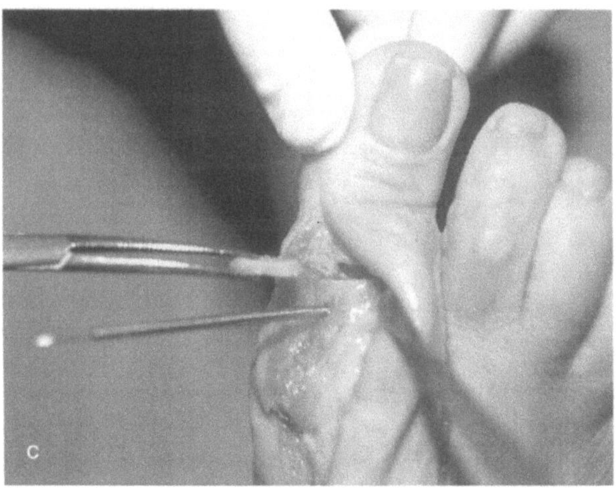

Fig. 5a–h. The different steps of the shortening of a great toe first phalanx using a shape memory staple. **a** Setting the proximal K-wire and the cannulated drill. **b, c** Resection using the drill guide. **d** The temporary axial wiring is obligatory. **e–g** The set staple. **h** X-ray 1 year following the osteotomy

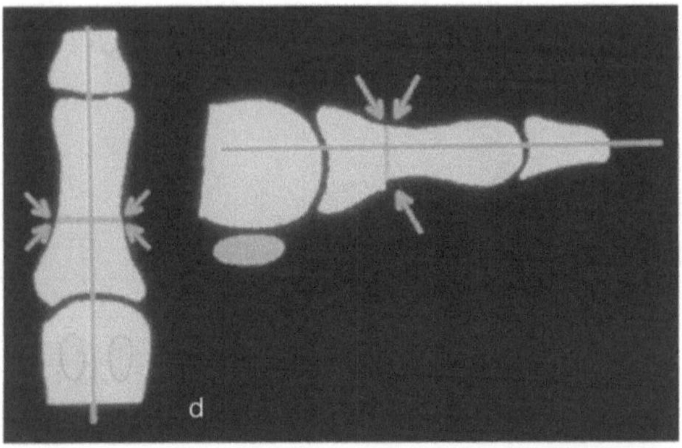

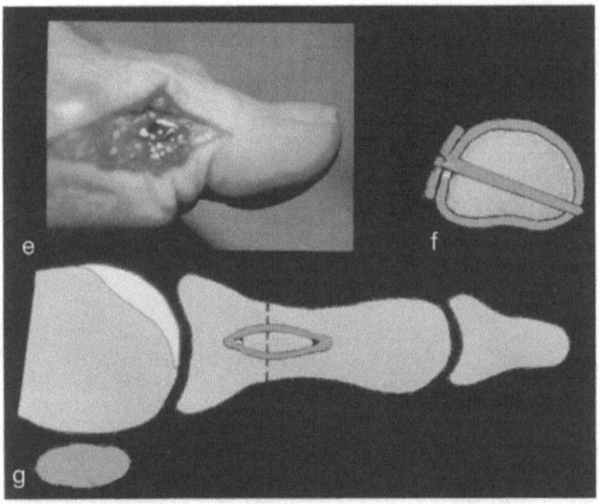

Figures 5d–h.

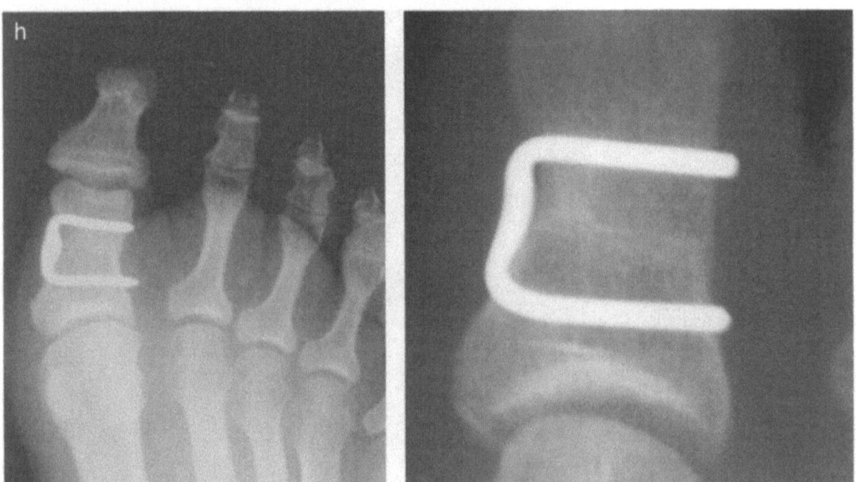

1. lateral released of the MTP joint,
2. scarf osteotomy of the first metatarsal,
3. medial capsulloraphy.

The great toe osteotomy is the fourth and last surgical step, which completes accurately and efficiously the deformity correction, the aim being also to result in a square type toe formula (equally of length of the first and second toes; Fig. 6b). If necessary this procedure is combined with lesser rays surgery (Weil osteotomy).

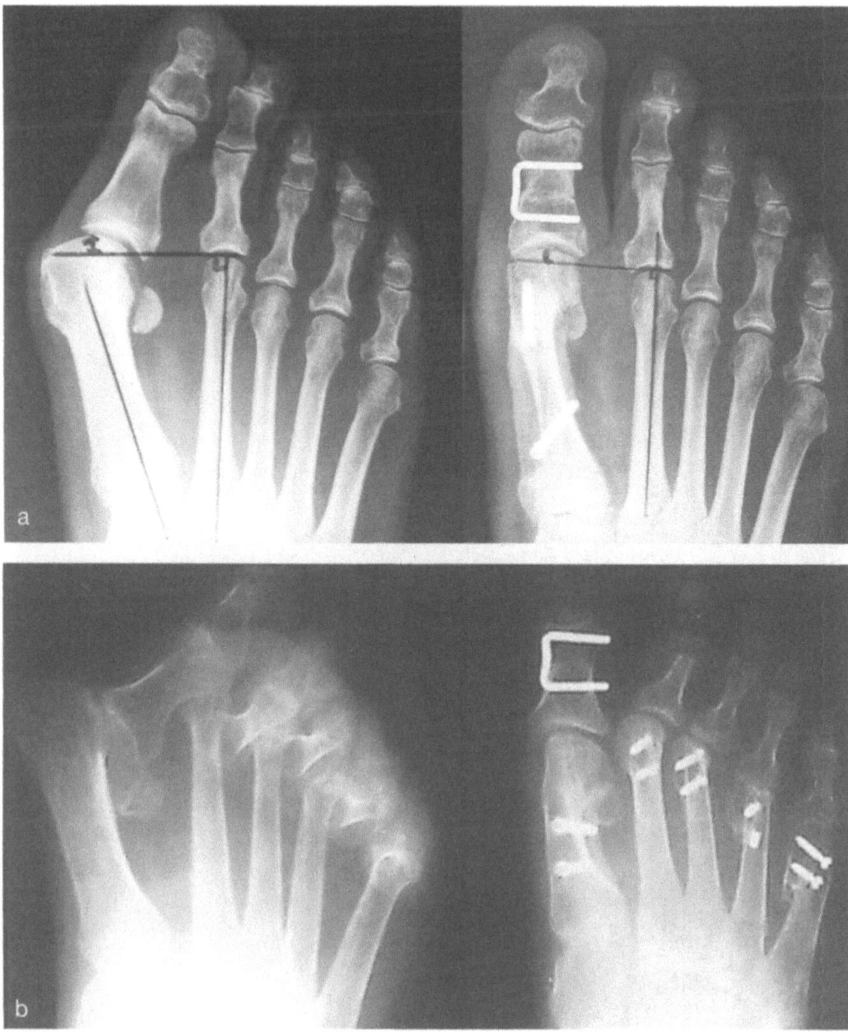

Fig. 6a, b. Two exmples of great toe osteotomy combined of the metatarsal. **a** Correction of hallux valgus deformity by scarf M1 osteotomy combined with great toe osteotomy. **b** Rheumatoid forefoot, three years follow up: the osteotomies of the metatarsal (shortening) combined with great toe osteotomy allows both, the preservation of the joints and the stability of the correction

5.2
Arthrodesis of the First Metatarso-Phalangeal Joint

Indications: hallux rigidus or important alteration of the first MTP joint (Fig. 7a). Two problems were observed before in these cases. The first one is the resulting angulation of the great toe, in fact the metatarso-phalangeal joint can easily be angulated too much in a dorsal or plantar direction and it is difficult to adjust the angulation in the sagittal plane. The major problem seen, however, with this arthrodesis is the healing delay caused by the difficulty to ensure a strong fixation using staples, K-wires or screws.

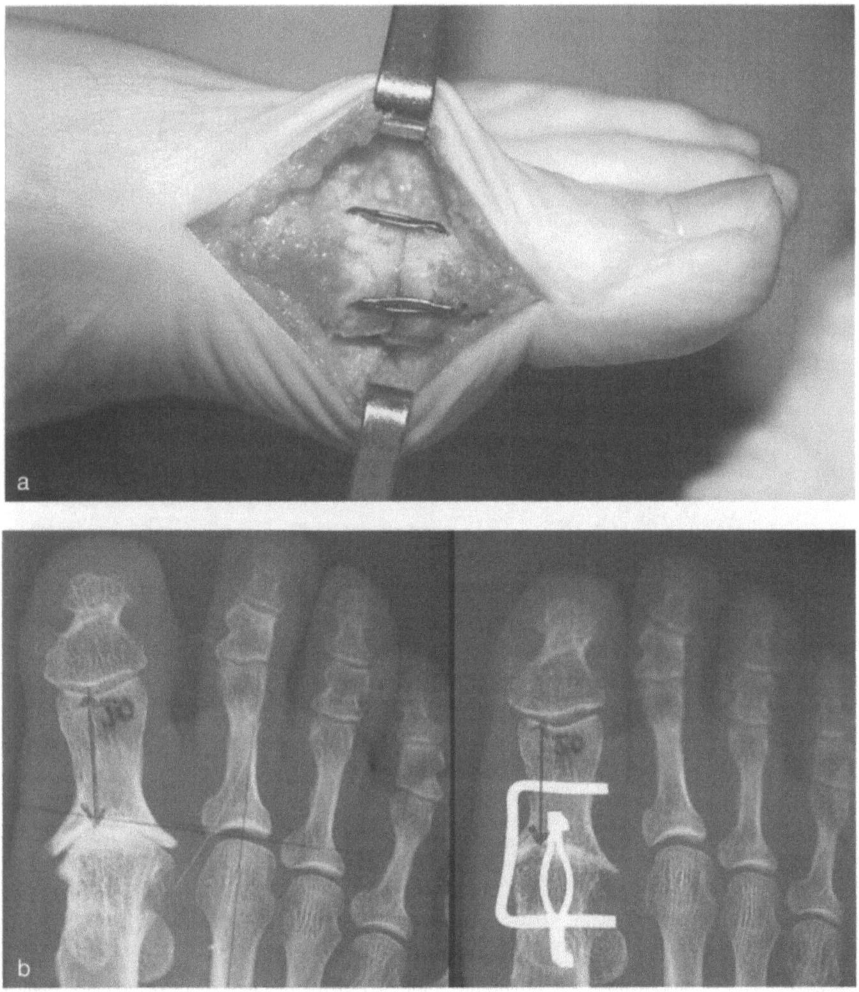

Fig. 7a,b. a Operative view. **b** The permanent and dynamic compression provided by the staples ensure the healing of the arthrodesis of the first MTP joint with a minimal resection, thus preserving the length of the great toe

The use of shape memory staples results in a notable improvement by providing strong fixation. The technique is easy: after resection of the remaining cartilage, exposing the subchondral bone, the position of the two fragments is carefully determined and fixed with two small temporary K-wires. Two shape memory staples are then implanted.

The staples used for this indication are larger and stronger than the staples used for the shortening of the first phalanx. The oval part is 20-mm long.

The first staple is introduced in a transversal plan, as plantar as possible and the second staple is placed on the dorsal surface, near the lateral border. Thus, the two staples are located in two perpendicular planes, ensuring an extremely strong fixation, not only immediately after surgery but long term due to the continuous compression of the two fragments.

This strong fixation permits minimal bone resection for shortening of the great toe without risking some instability (Fig. 7b). All patients wore a heel support shoe during 40 days, then the full weight bearing was permitted on the forefoot. No healing delay and no pseudarthrosis were observed. In five cases, a previous pseudarthrosis was treated by using this method (Fig. 3b).

The only contra indication appears to be cases where it is necessary to make a large resection of the fragments in order to shorten the great toe. In these cases, the resection is performed as far as the cancellous bone is reached, but the remaining bone may not be strong enough to resist the compression force of the staple. In these case, other fixations, with screw in particular, are recommended.

5.3 Arthrodesis of the Lisfranc Joints, Osteosynthesis or Arthrodesis of the Hindfoot

The major indication in the middle foot is arthrodesis of the Lisfranc joint. In such indication, the healing was difficult to obtain; in these cases, the shape memory staple appears to provide reliable and fast healing (Fig. 8).

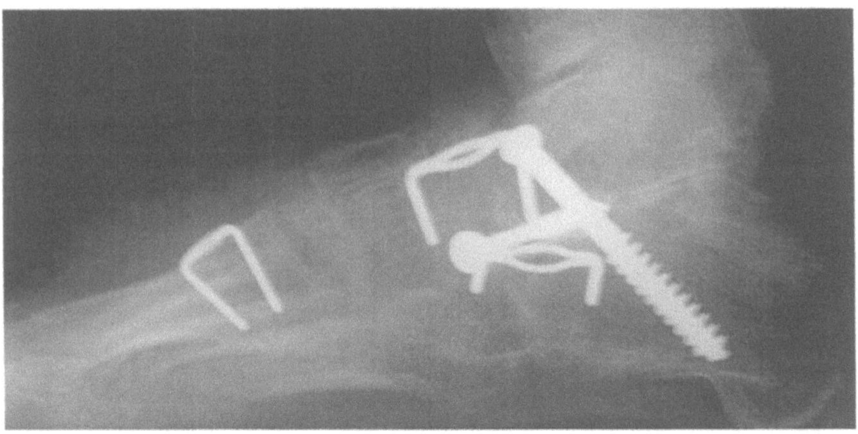

Fig. 8. The use of shape memory staple in arthrodesis of the Lisfranc's joint, in this case the first cuneo-metatarsal joint. The healing on 2.5 post-operative months

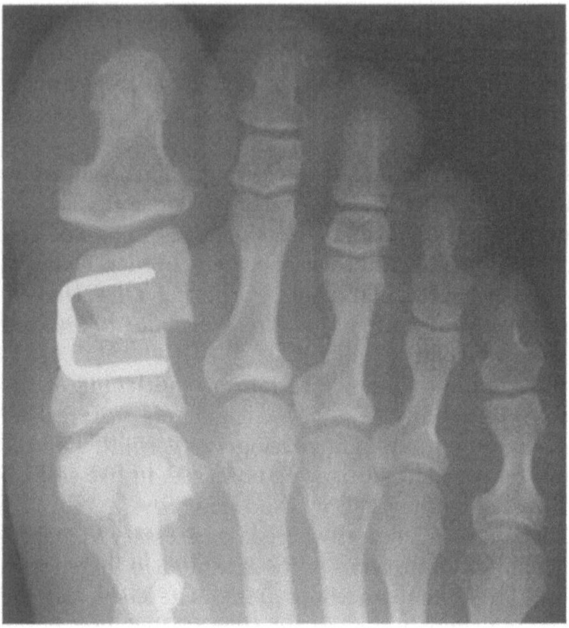

Fig. 9. Use of shape memory staple in hindfoot arthrodesis: on 2.5 post-operative months, we observed a complete fusion of the talo-navicular and the calcaneo-cuboid joints

In hindfoot arthrodesis, the use of shape memory staples is recommended for the fusion of the talo-navicular joint and for the calcaneo-cuboid joint fusion (Fig. 9). In each case, we observed an extremely fast and reliable fusion, notably in the talo-navicular joint where the fusion were usually more difficult to obtain.

Tolerance: in 55 patients re-evaluated clinically and radiolographically at or after five post operative years, no problem of local tolerance of the nickel titanium alloy has been observed, even in the longest follow up. Among these

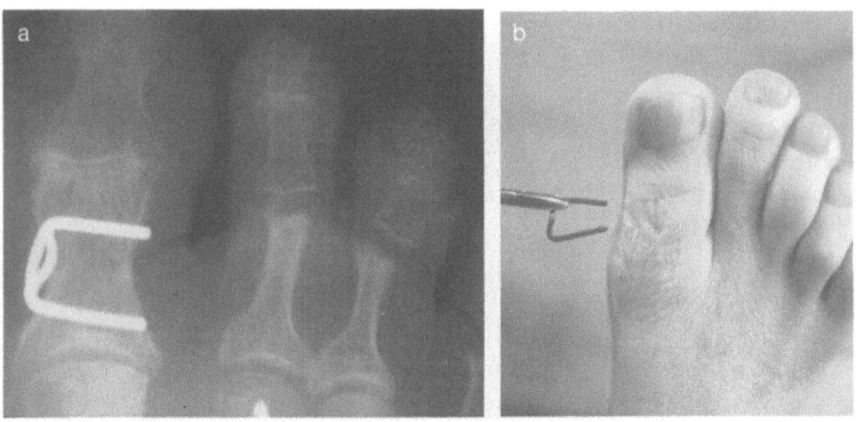

Fig. 10a, b. Long term results. **a** Seven post-operative years: no resorption around the prongs are observed. **b** Five post operative years: removing of the staple; we note the remaining elasticity of the staple

55 patients, staples were removed in eight cases. We observed that removal was considerably more difficult after 5 years than at 1 year post-operatively. No local tissue reaction or deposit of alloy particles were noticed. Radiographically, no resorption around the prongs were observed (Fig. 10a).

Permanence of elasticity: when the staples were removed, even after long-term implantation, their elasticity appears to remain intact (Fig. 10b). This is a particularity of this type of alloy.

6
Conclusions

The shape-memory staple with an oval part joining the two prongs provides not only a double-compression permanent effect, in the prongs and in the oval part, but also provides better stability of the osteosynthesis. Its application in the foot has been studied in 1850 cases since 1986. The contra indications are few (5%). They are associated to the compression strength of the staple in very osteoporotic bones and in the arthrodesis with important resection of the subchandral and cancellous bones. They have been used successfully for shaft osteotomies of the great toe first phalanx, but also for first MTP joint arthrodesis, and for middle or hind foot arthrodesis. In each of these indications the use of shape memory staples results in a great improvement not only concerning the reliability of the osteosynthesis but also in healing time due to the double compression effect.

References

1. Barouk L-S (1994) Scarf osteotomies in the hallux valgus. Personal experience. Therapeutic proposition. Foot Dis 1:79–89
2. Barouk L-S (1993) Le raccourcissement du gros orteil: intérêt de l'agrafe à mémoire spécifique. In: Actualités en médecine et chirurgie du pied 8th série. Masson, Paris, pp 93–105
3. Barouk L-S (1992) Osteotomies of the great toe. J Foot Surg 31:388–399
4. Barouk L-S (1997) New osteotomies of the forefoot and their therapeutic role. Forefoot surgery. Cahiers d'enseignement de la SOFCOT. Expansion scientifique Française, Paris pp. 49–76
5. Viladot R, Rochera R, Alvarez G, Pasarin A (1996) Die resektionarthroplastik zur behandlung des hallux valgus. Orthopade 4:324–331

Orthodontic Applications

Corrosion Behavior of NiTi Alloys in a Physiological Saline Solution

Kazuhiko Endo, Hiroki Ohno

1 Introduction

NiTi alloys possess certain characteristics, such as super-elasticity and shape memory effect that render them useful as biomaterials [13]. Super-elastic NiTi alloy orthodontic wire has been widely used clinically since the 1970s because this wire produces more constant forces due to stress-induced martensitic transformation than do conventional orthodontic wires made of stainless steels or Co–Cr alloys [14]. The shape memory effect has been effectively utilized in the fixture of both blade-type and perio-root-type dental implants to provide a desirable firm fixation with bone [8, 22].

A high biocompatibility is required for NiTi alloys to fulfill these unique functions in the human body for a long period. It has been demonstrated that NiTi alloy does not inhibit the proliferation and growth of either human fibroblasts or osteoblasts [1, 15, 18]. No severe adverse reactions have been reported so far in patients treated with super-elastic NiTi orthodontic wires or NiTi alloy dental implants. At present, the greatest potential risk in the use of NiTi alloys is Ni hypersensitivity. An epidemiological study demonstrated that the incidence of Ni hypersensitivity is as high as 4% in the general population in Europe [9]. Contact hypersensitivity to Ni in austenitic stainless steels has also been reported under clinical conditions [17]. This adverse biological response may be triggered by a small amount of Ni ions released from a NiTi alloy device used in the human body.

It has long been recognized that the corrosion resistance of metallic implants is an essential characteristic that must be controlled to minimize adverse biological responses in vivo. In this chapter, we describe the passivity and selective Ni dissolution behavior of NiTi alloys under a simulated physiological condition and discuss the factors affecting the corrosion resistance of these alloys. Surface treatments for improving the corrosion resistance of the NiTi alloy are also reviewed.

2 Anodic Corrosion Behavior of the NiTi Alloy and Other Implant Alloys

Figure 1 shows the potential/current density curves for the equi-atomic NiTi alloy and other alloys used in orthopedic and dental implants that had been immersed

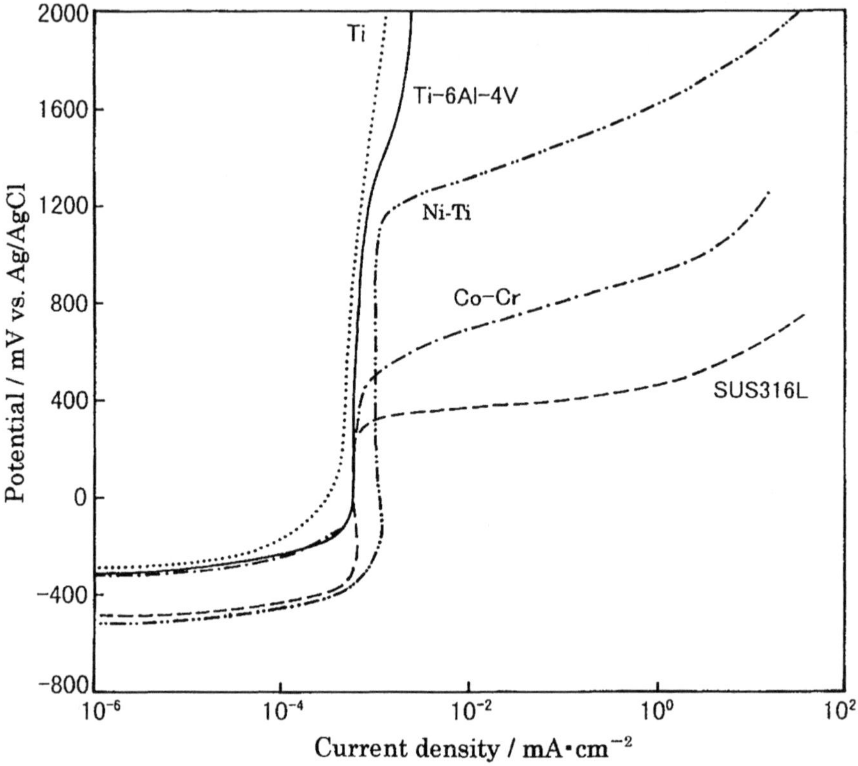

Fig. 1. Potential/current density curves for the Ni–Ti alloy and other alloys used for surgical implants in a saline solution at 37°C

in a saline solution (0.9% NaCl solution) at 37°C. The NiTi alloy exhibits passivity in the potential range from –400 mV to +1200 mV, where a passive current density (potential-independent current density) is observed. Above the breakdown potential of 1200 mV versus Ag/AgCl, the anodic current density increases steeply caused by the initiation of localized corrosion due to local breakdown of the passive film. In general, a highly passivated alloy is characterized by a low passive current density with a high breakdown potential in the potential/current density curve [23].

The breakdown potential for the NiTi alloy is approximately 700–800 mV higher than that of SUS316L and Co–Cr alloy (Vitallium) currently used for surgical implants such as artificial joints and bone plates. The passivity of the NiTi alloy is far more stable than that of SUS316L and Co–Cr alloy in a chloride solution, suggesting that the NiTi alloy exhibits a relatively low susceptibility to localized corrosion (crevice corrosion and pitting corrosion) when placed in a living body.

The passivity of commercially pure Ti (CPTi) and Ti-6Al-4V is stable from the free corrosion potential to +2000 mV in saline solution, and no localized corrosion takes place in this potential range. NiTi alloy contains approximately 50%

Ni, resulting in its inferior corrosion resistance as compared to those of CPTi and other Ti-based alloys.

3
Dissolution of Ni Ions from the NiTi Alloy

Table 1 shows the amount of Ni and Ti ions released into the saline solution as determined by graphite furnace atomic absorption spectrophotometry [7]. Ni ions preferentially dissolved into the solution, and total amount of Ni ions released during a 14-day immersion period was 0.43 μg/cm². Based on this data, the amount of Ni ions released from a single NiTi alloy orthodontic wire with a surface area of 3 cm² is calculated to be 0.1 μg/day. This Ni level is negligible compared with the daily intake of Ni in the diet, which is the main source of Ni for humans (300–600 μg/day) [20]. Although the amount of Ni ions released from the NiTi alloy devices is small, the local accumulation of Ni in surrounding tissues and their adverse reactions should be further investigated. This issue is of importance, especially for NiTi alloy devices embedded in the body, such as dental implants, craniofacial implants, and cardiovascular stents.

Table 1. The amount of Ni and Ti ions released from the polished Ni–Ti alloy into a saline solution. Data are expressed as means±SD

	Amount of metal ions released (μg/cm²/14 days)	
	Ni	Ti
Ni–Ti alloy (as-polished)	0.43±0.02	ND

ND, not detectable

4
Characterization of the Surface Oxide Film on a NiTi Alloy

The passivity of NiTi alloys is attributed to the protectiveness of the surface oxide film. This passive film formed on an equi-atomic NiTi alloy was characterized by X-ray photoelectron spectroscopy (XPS). Figure 2 shows Ti 2p, Ni 2p, and O1s spectra obtained from polished NiTi alloy surfaces before and after immersion in 0.9% NaCl solution or 1% lactic acid solution for 28 days [10]. A detailed description of the assignment of chemical states for the spectral peaks is found elsewhere [6], and the results of peak assignment are included in this figure.

The Ti 2p spectrum from the polished alloy surface consists of major peaks at 458.7 eV and 464.7 eV corresponding to TiO_2 and a small peak at 454.3 eV corresponding to metallic Ti under an oxide film. In contrast, the intensity of the Ni $2p_{3/2}$ peak at 856.2 eV corresponding to $Ni(OH)_2$ is very small compared with that of the peak corresponding to metallic Ni. These spectra indicate the presence of a Ti-rich oxide film with a small amount of Ni hydroxide on the polished alloy surface. After immersion in 0.9% NaCl solution or 1 % lactic acid solution, the inten-

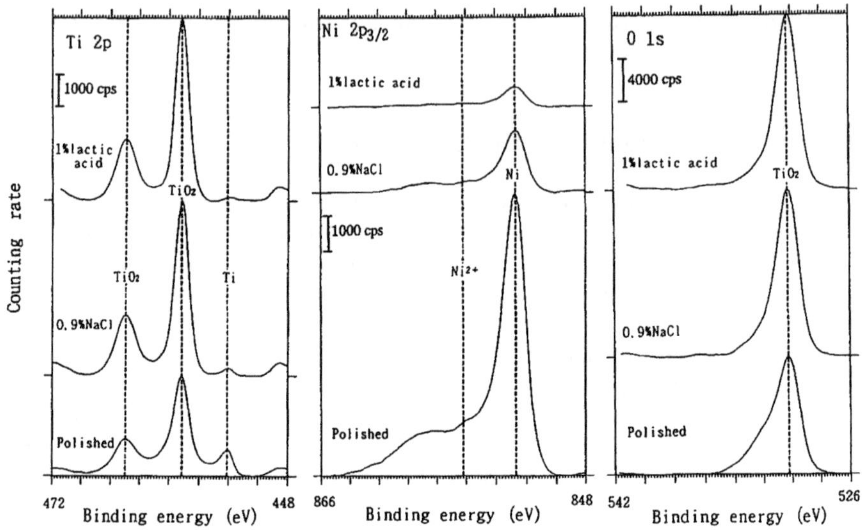

Fig. 2. XPS spectra obtained from polished Ni–Ti alloy surfaces before and after immersion in the solution for 28 days

sities of the Ti 2p peaks and the O1s peak corresponding to TiO_2 increased, while that of the Ti 2p peak corresponding to metallic Ti became smaller due to growth of the surface oxide film during immersion in these solutions. With the increase in the thickness of the oxide film, the intensity of the Ni $2p_{3/2}$ peak corresponding to both the metallic state and $Ni(OH)_2$ decreased, suggesting that the surface oxide film aged in the solution was mainly composed of TiO_2 with a trace amount of $Ni(OH)_2$. The decrease in the intensity of the Ni $2p_{3/2}$ corresponding to metallic Ni with the growth of the Ti-rich oxide film implies that metallic Ni does not disperse uniformly as clusters in the film but concentrates at the alloy/oxide film interface. Based on these findings, the surface structure of the NiTi alloy can be postulated as that shown in Figure 3. XPS analysis did not clearly reveal the position at which the $Ni(OH)_2$ was located. However, the intensity ratio of $I_{Ni(OH)2}$ to

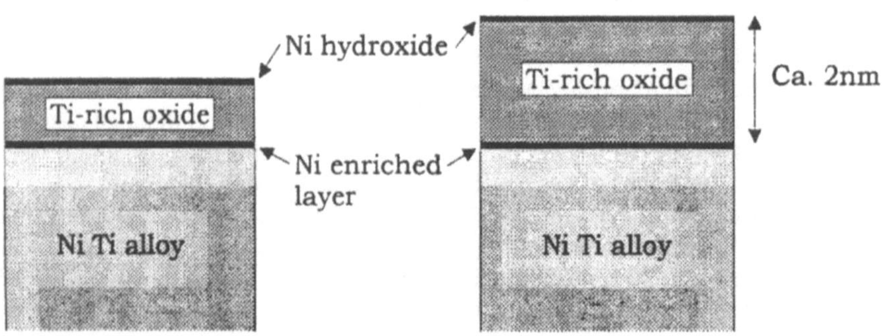

Fig. 3. The surface structure of the polished Ni–Ti alloy before and after immersion in a saline solution

I_{Ni} ($I_{Ni(OH)_2}/I_{Ni}$) increased sharply with a decrease in the take-off angle of the photoelectrons, indicating that $Ni(OH)_2$ is probably present in the outer region of the passive film.

5 Surface Structure and Corrosion Characteristics of the NiTi Alloy

The initial corrosion process of the NiTi alloy in a saline solution in the absence of fretting can be summarized as follows. After polishing, preferential oxidation of Ti occurred in air to form a Ti-rich oxide film, leaving a Ni enriched layer at the alloy/oxide film interface. As the preferential oxidation of Ti progressed in the solution, the Ti-rich oxide film grew with the dissolution of Ni ions from the alloy surface. However, the amount of released Ni ions was so small that the concentration of metallic Ni was maintained at a high level at the alloy/oxide film interface.

The formation of the Ti-rich oxide film is responsible for the markedly higher breakdown potential of the NiTi alloy as compared to those of SUS316L and Co–Cr alloy (Fig. 1). Polarization resistance measurements demonstrated that the corrosion rate of the NiTi alloy in a saline solution decreased with immersion time [7]. This behavior can be interpreted as the aging of the passive film in the solution. The formation of a thicker and more protective passive film in the solution may enhance the passivity of the alloy, resulting in a decrease in the corrosion rate. The breakdown potential of the NiTi alloy, however, was lower than those of CPTi and Ti6Al4V alloy. This lower breakdown potential for the NiTi alloy is due partly to the enriched Ni at the alloy/oxide film interface. Rondelli et al. also demonstrated that the pitting potential of NiTi alloy determined by the potentiostatic scratch test was much lower than that of Ti6Al4V [16]. The enriched Ni at the alloy surface under the passive film may retard the repassivation process when the passive film is locally broken under potentiostatic conditions. The composition of both the passive film and the alloy surface under the film is critical in determining the stability of passivity and the reaction rate of the repassivation process for NiTi alloys.

6 Factors Affecting the Corrosion Behavior of the NiTi Alloys

6.1 Effects of Alloying

Deformation behavior and super-elasticity characteristics can be improved by the addition of a third element to the equi-atomic NiTi alloy [13, 19]. In recent years, several types of super-elastic orthodontic wire containing small amounts of Cr and/or Cu have become commercially available, enabling a choice of the most appropriate wire for a variety of clinical cases. The addition of small amounts of other elements to NiTi alloys may influence not only the mechanical properties but also corrosion resistance. Figure 4 shows the potential/current density curves for Ni49.3Ti, Ni–48.4Ti–0.19Cr, and Ni–49.6Ti–4.97Cu–0.29Cr

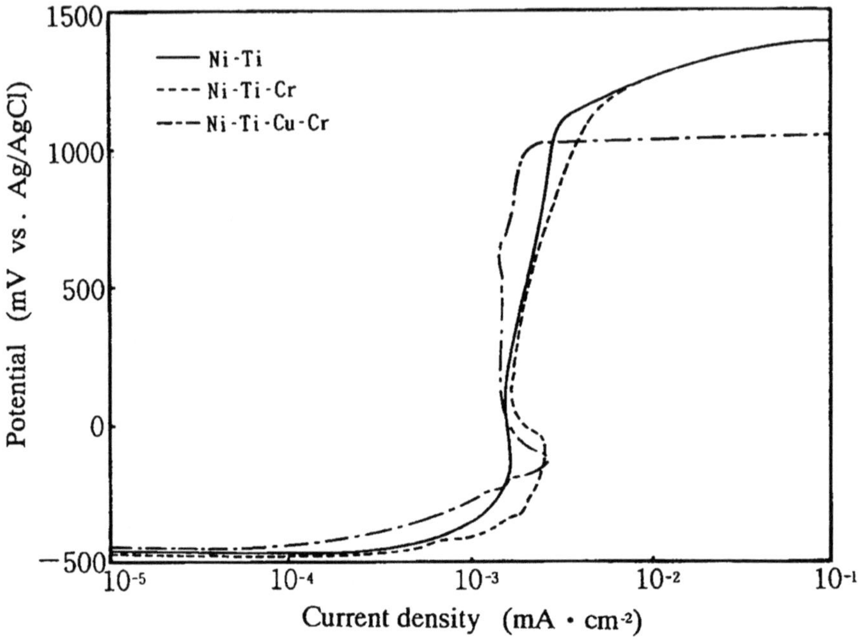

Fig. 4. Potential/current density curves of the three Ni–Ti alloys with different compositions in a saline solution

(elements in atomic %) alloys in a deaerated saline solution at 37°C [10]. There are no significant differences in corrosion potential and passive current density among the three alloys. The breakdown potential for the NiTi–Cu–Cr alloy is approximately +1000 mV, which is 200 mV lower than those for the other two alloys. For the NiTi–Cu–Cr alloy, corrosion progressed with highly localized pitting above the breakdown potential. Figure 5 shows typical pits, formed during anodic polarization measurement, propagating deep into the NiTi–Cu–Cr alloy with corrosion products surrounding them. This increased susceptibility of the NiTi–Cu–Cr alloy to localized corrosion in a chloride solution is associated with the enriched Cu as well as Ni at the alloy surface under the passive film.

Figure 6 shows the amounts of Ni and Cu ions released into the saline solution [10]. After immersion for 28 days, approximately 0.45 μg of Ni ions was released per unit area of all three-alloy specimens. The amount of Cu ions released from the NiTi–Cu–Cr alloy was less than one-fourth of that of released Ni ions. The Ti ion concentration in the saline solution was below the detection limit (0.012 $\mu g/cm^2$). These results indicate that the corrosion resistance of the three alloys under freely immersed conditions is identical. These results are consistent with those obtained by the anodic polarization measurements, where no significant differences in passive current density were observed among the three alloys (Fig. 4).

The free corrosion potential for the three alloys was from −495 mV to −280 mV during the immersion in saline solution for 28 days. It is also unlikely

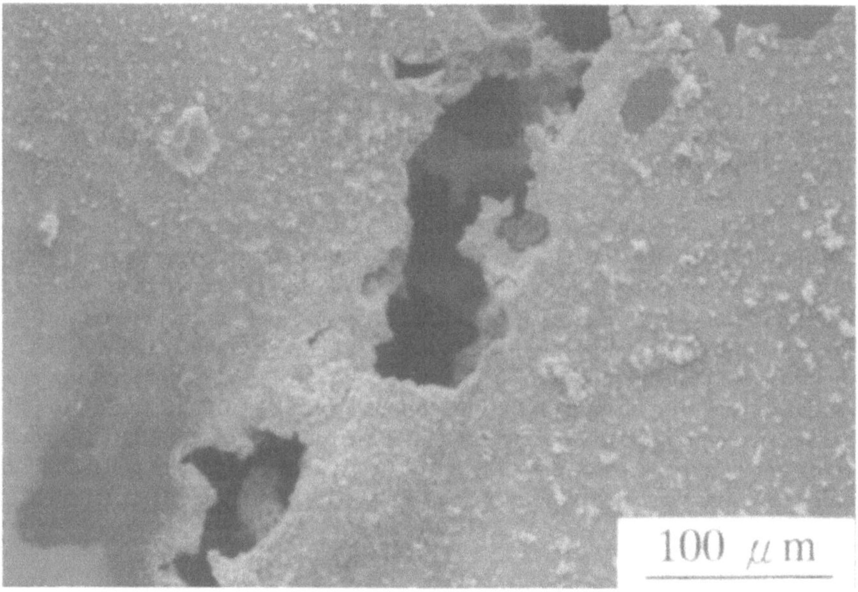

Fig. 5. SEM image of the NiTi–Cu–Cr alloy surface after anodic polarization measurement in a saline solution

that the free corrosion potential of the NiTi–Cu–Cr alloy in an oral environment exceeds +1000 mV. Although the localized corrosion susceptibility increased in the NiTi–Cu–Cr alloy to above +1000 mV, the corrosion resistance of this alloy in saline solution can be regarded as practically identical to those of the NiTi and NiTi–Cr alloys. In conclusion, small amounts of Cr and Cu added to change the super-elastic characteristics of NiTi alloy do not change the corrosion resistance of the alloy in a simulated physiological environment, unless the corrosion potential of the alloy exceeds +1000 mV.

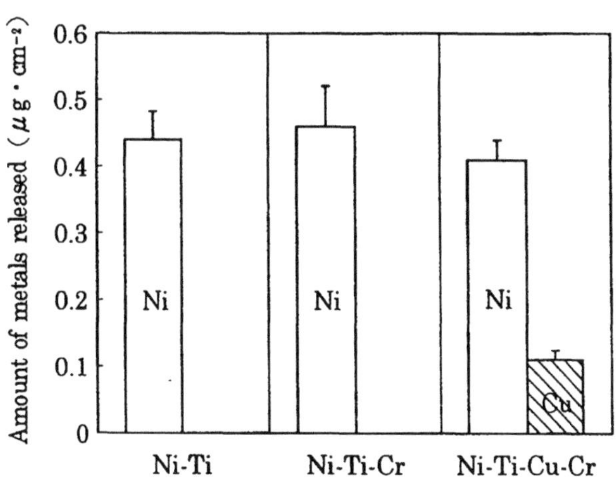

Fig. 6. Amounts of metal ions released from three NiTi alloys with different compositions after immersion in a saline solution for 28 days

6.2
Effects of Surface Texture

In order to examine the influence of surface texture on the corrosion resistance of NiTi alloys, anodic polarization measurements were carried out on the NiTi alloys with different surface textures: a polished plate specimen, a porous specimen made by powder sintering (mean pore diameter: 300 μm; mean void volume: 50%), and an amorphous thin- film specimen sputter-deposited on a silicon substrate [1]. Figure 7 shows the potential/current density curves for the three specimens in a saline solution at 37°C. The thin-film specimen exhibits a lower passive current density and a nobler breakdown potential by 200 mV than does the mirror-polished plate specimen. The enhancement of the passivity demonstrated for the thin-film specimen may be attributed to the uniform, protective passive film formed on the amorphous NiTi alloy film. In contrast to this, the passive region for the porous specimen is from +60 mV to +160 mV, which is extremely narrow as compared with the passive regions for the polished plate and thin film- specimens. The porous specimen is not readily passivated in

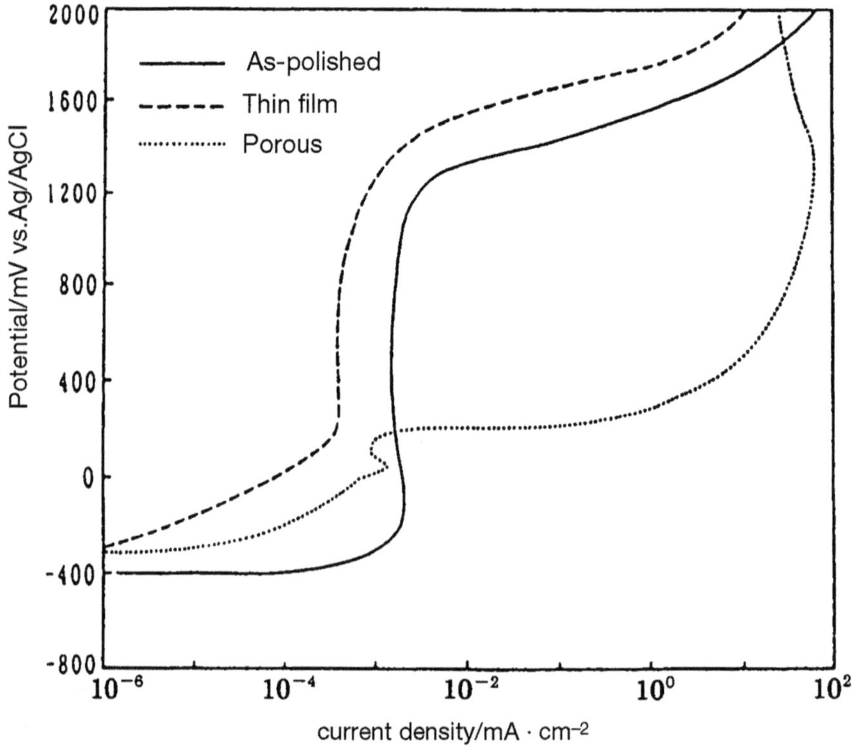

Fig. 7. Potential/current density curves of NiTi alloys with different surface textures in a saline solution

saline solution due probably to the marked irregularities on the surface, which may prevent the formation of a protective oxide film. Sohmura and Kimura demonstrated that the corrosion resistance of NiTi alloy was markedly reduced with increases in the surface roughness. They also reported that the edges of a NiTi alloy specimen were the preferred initiation sites for pitting [12, 21].

Bone implants with a porous surface are now being developed using Ti alloys including NiTi alloys. The porous structure is expected to improve the mechanical retention of the implant to the bone by stimulating new bone tissue to grow into the pores, but following problems have to be overcome before this new implant can be accepted:

1. the high susceptibility to localized corrosion due to surface inhomogeneity,
2. the increased amount of metal ions released due to increased effective surface area,
3. the delamination of metal fragments by stress at the bone/implant interface, and
4. the increased risk of bacterial infection [4].

6.3 Effect of Contact between Dissimilar Metals

When dissimilar metals are placed in contact with each other to form a galvanic cell, the corrosion rate of the more active (anodic) metal is accelerated, while that of the more noble (cathodic) metal is retarded. This type of corrosion is called dissimilar metal corrosion or bimetallic corrosion [11, 23].

Super-elastic NiTi alloy orthodontic wire has frequently been used in the oral environment in combination with metal brackets made of stainless steels. According to the free corrosion potentials for NiTi alloy and SUS316L in saline solution (Fig. 8a), the NiTi alloy, which exhibits a lower corrosion potential, will become the anode of the galvanic corrosion cell when these two alloys are placed in contact with each other. The galvanic current between the galvanic couple of NiTi alloy and SUS316L was measured in saline solution using a zero impedance ammeter, and the result is shown in Fig. 8b. The magnitude of the galvanic current density provides an indication of the degree of accelerated corrosion rate of the NiTi alloy produced by coupling with SUS316L. The galvanic current density is initially high and it decreases as time elapses to a value of 0.2 $\mu A/cm^2$ after 72 h of immersion. This value for galvanic current density is significantly higher than that of the uncoupled corrosion rate of approximately 0.001 $\mu A/cm^2$ for the NiTi alloy in saline solution. This high galvanic current density is due to the large potential difference between the NiTi alloy and SUS316L (Fig. 8a). The corrosion potential of CPTi is lower than that of SUS316L. Although the NiTi alloy becomes the anode and its corrosion is accelerated when coupled with CPTi, the galvanic current density is below 0.07 $\mu A/cm^2$ after 72 h of immersion. Recently, metal brackets made of CPTi have been commercially available, and the use of NiTi alloy orthodontic wire with this bracket is effective in reducing the amount of metal ions released into saliva.

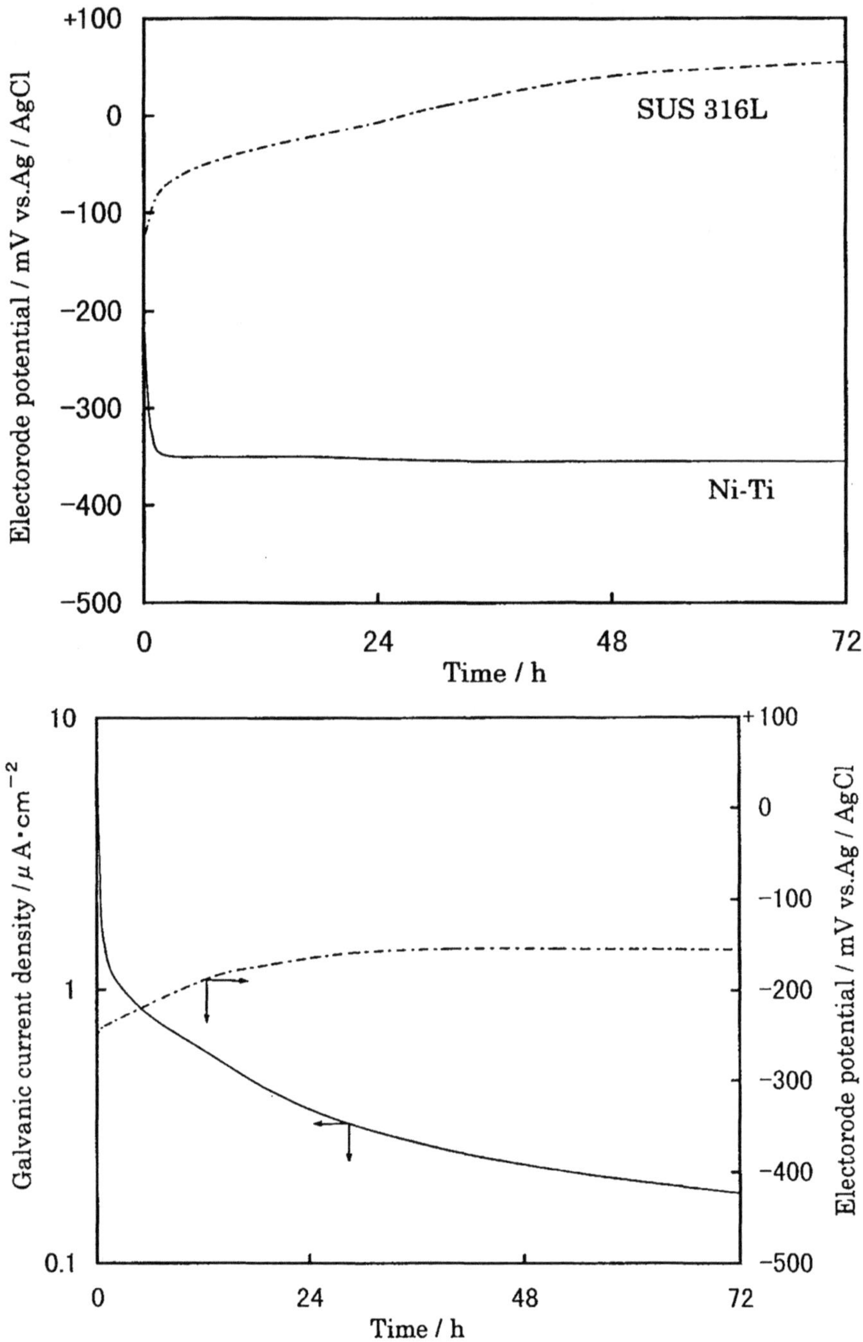

Fig. 8a, b. **a** Variations in corrosion potential with time for SUS316L and Ni–Ti alloy in a saline solution. **b** Variations in galvanic current density and mixed potential with time when the Ni-Ti alloy was coupled with SUS316L in a saline solution

6.4
Effect of Amino Acids and Serum Proteins

Human body fluid and saliva are electrolytic solutions rich in amino acids and proteins. Previous studies have demonstrated that proteins have some influence on the corrosion behavior of surgical alloys and their component pure metals [2, 3]. The influence of amino acids and serum proteins on the passivity of the NiTi alloy was studied by the potentiodynamic polarization measurements [7]. Figure 9 shows the potential/current density curves for the polished NiTi alloy in saline solution and a cell culture medium (Dulbecco's modified Eagle's medium with 18% fetal bovine serum). The passive current density in the cell culture medium is much higher than that in the saline solution, suggesting that the dissolution of the metal ions in the passive state is enhanced in the presence of amino acids and serum proteins. The increased susceptibility to localized corrosion in the presence of amino acids and serum proteins is also suggested by the lower breakdown potential (by 200 mV) in the cell culture medium than in the saline solution.

The corrosion rate of the NiTi alloy in both solutions was estimated by measuring the polarization resistance (R_p), which is a parameter inversely proportional to the corrosion rate. As shown in Figure 10, R_p increases with time in the saline solution, while it decreases with time in the cell-culture medium. The values of R_p

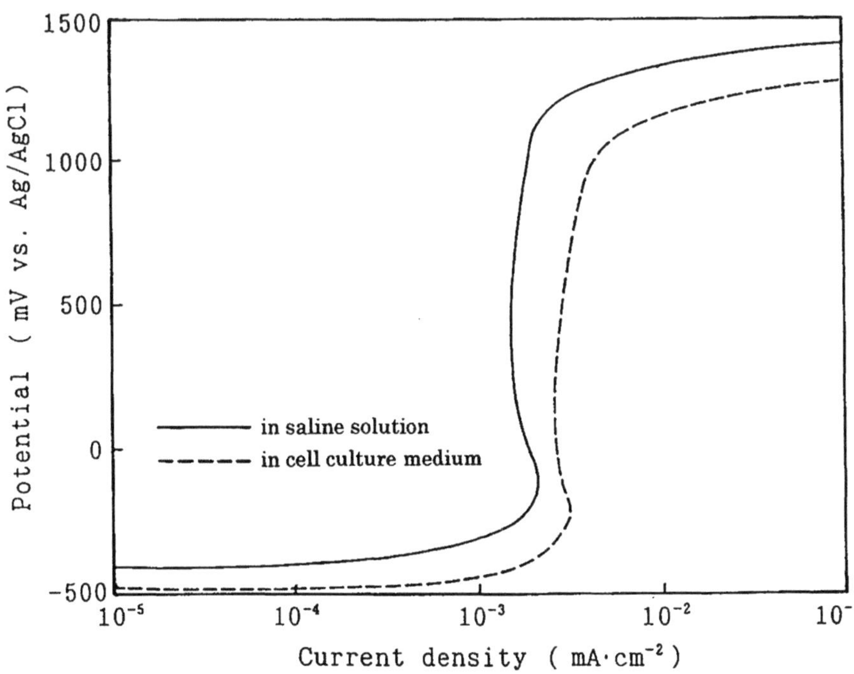

Fig. 9. Potential/current density curves for the polished alloy in a saline solution and a cell-culture medium

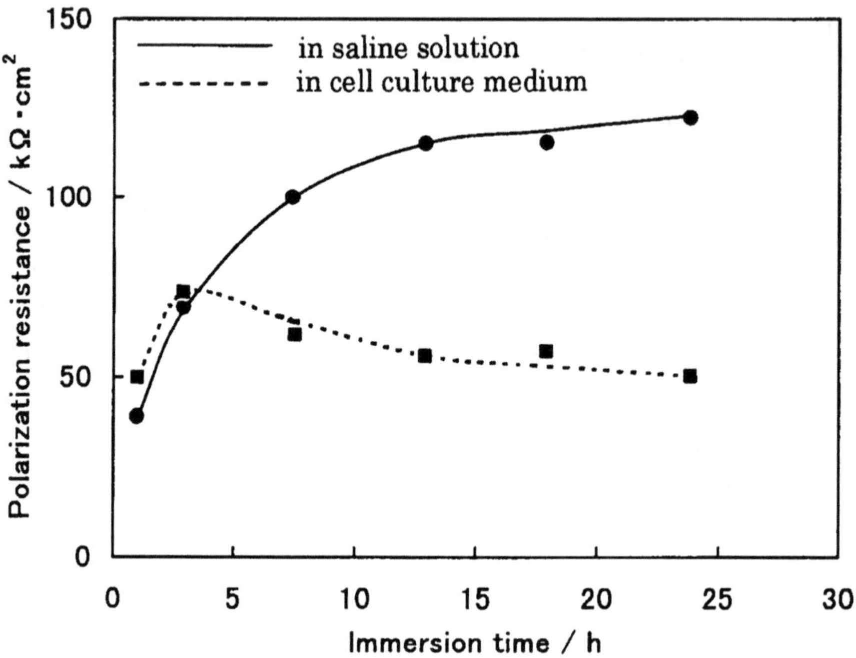

Fig. 10. Variations in the polarization resistance of the NiTi alloy in a saline solution and a cell culture medium

at 24 h of immersion indicate that the corrosion rate in the cell culture medium is approximately 2.5 times higher than that in the saline solution. It is apparent from these results that amino acids and serum proteins in the cell culture medium promote the corrosion of the NiTi alloy. However, the role of amino acids and serum proteins in enhancing the corrosion of the NiTi alloy is not yet fully understood. The characterization of the NiTi alloy surface by XPS after polarization at various potentials for 15 min in both solutions, demonstrated that the amino acids and serum proteins quickly adsorbed the alloy surface, and that the thickness of the passive film formed in the cell culture medium was less than that formed in the saline solution at all potentials [7]. These facts suggest that the adsorbed amino acids and serum proteins enhance corrosion either by inhibiting the formation of the protective passive film or by forming metal ion-protein complexes that facilitate film dissolution.

7 Surface Treatments for Improving the Corrosion Resistance of the NiTi Alloy

The NiTi alloy possesses a degree of corrosion resistance sufficient to enable it to be used in the human body. Nevertheless, several attempts have been made to form protective coating films on the NiTi alloy in dry or wet process in order to

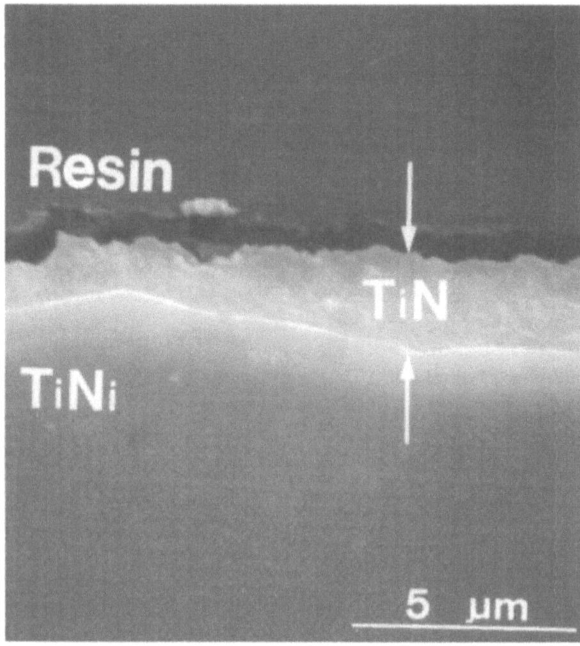

Fig. 11. Cross-section of the TiN film produced on the NiTi alloy by arc ion plating

reduce the risk of side effects including hypersensitivity to released Ni [5, 7, 12]. Of the numerous protective films, here we present the protective properties of titanium nitride and polysiloxane films.

Figure 11 shows a cross-sectional view of the TiNi film produced on the NiTi alloy by arc ion plating at 400°C under a N_2 atmosphere for 20 min [5]. The thickness of the film is approximately 1–2 μm. Figure 12 illustrates the molecular structure of the NiTi alloy surface after it was chemically modified with some biofunctional proteins using an aminosilane and a glutaraldehyde [7]. This surface modification was first introduced by the authors to provide an effective means to control the alloy/tissue interaction with specific proteins covalently immobilized onto the NiTi alloy surface [6]. The coupling layer composed of a highly cross-linked siloxane network also functions as a coating film and it prevents corrosion of the alloy.

Figure 13 shows the potential/current density curves for (a) the polished NiTi alloy, (b) the TiN-coated NiTi alloy and (c) the chemically modified NiTi alloy in a saline solution at 37°C. With TiN coating, the passive current density in the passive region (–50 mV to 500 mV) is reduced to 0.04 μA/cm^2, approximately two orders of magnitude lower than the value for the polished alloy. The breakdown potential for the TiN-coated alloy, however, is much lower than that obtained for the polished alloy. Mechanical cracking of the TiN film takes place at +500 mV, facilitating pit initiation and growth on the alloy surface (Fig. 14).

The passive current density for the chemically modified NiTi alloy is also significantly lower than that for the polished alloy. There is no significant difference

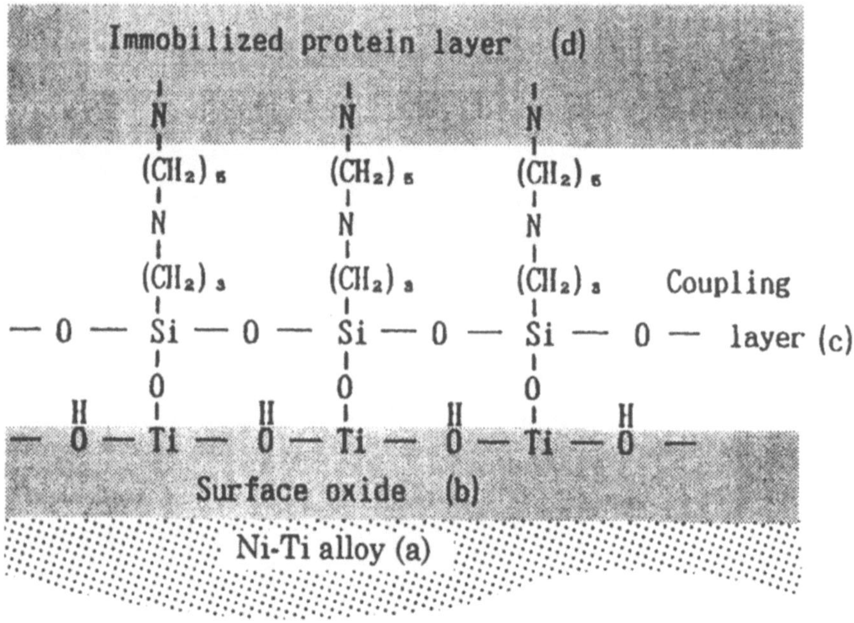

Fig. 12. Schematic illustration of the surface structure of the NiTi alloy chemically modified with biofunctional proteins using a aminosilane and a glutaraldehyde. The aminosilane is bound to the surface through Ti–O–Si bonds with a highly cross-linked siloxane network

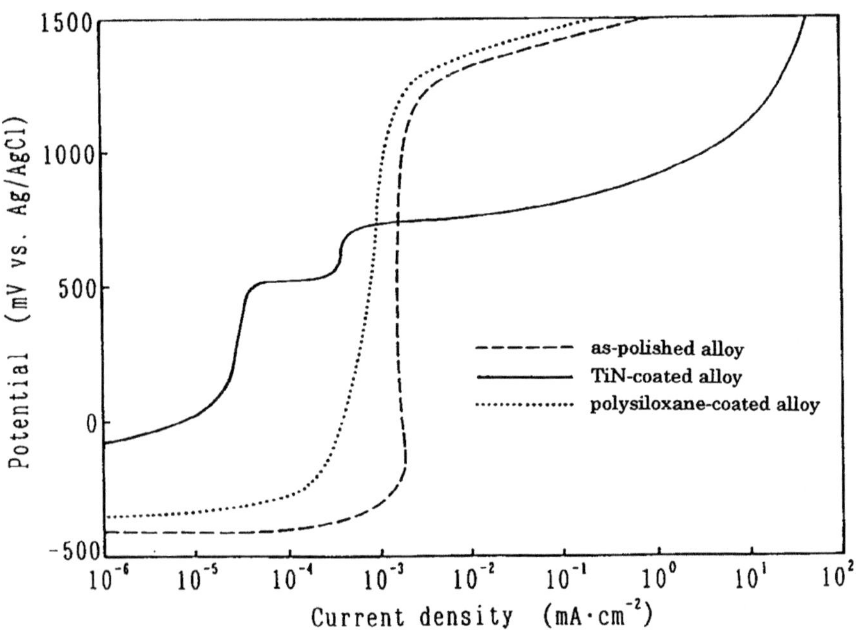

Fig. 13. Potential/current density curves for the polished NiTi alloys and the NiTi alloys with different surface treatments in a saline solution

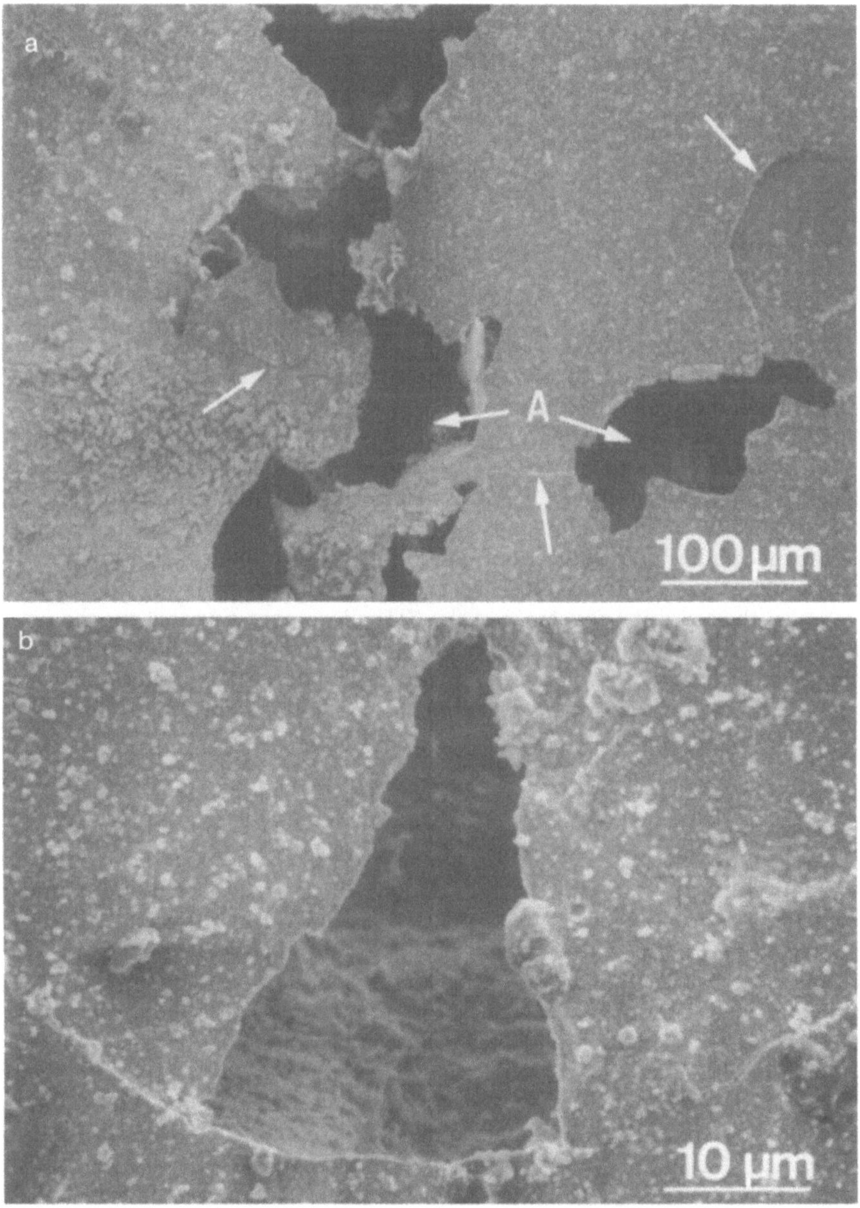

Fig. 14a, b. Secondary electron images of the TiN coated alloy surface after the anodic polarization measurement in a saline solution

in the breakdown potential between the chemically modified alloy and the polished alloy.

Table 2 shows the amount of Ni ions released from the NiTi alloys into a saline solution over a period of 14 days. With chemical modification, the amount of Ni

Table 2. Amount of Ni ions released from the NiTi alloys into a saline solution. Data are expressed as means±SD

Ni–Ti alloy	Amount of metal ions released (µg/cm²/14 days) Ni	Ti
As-polished	0.43±0.02	ND
Polysiloxane-coated	0.25±0.05	ND
TiN-coated	ND	ND

ND, not detectable

ion released is reduced by approximately 50%. The concentration of Ni ions was below the detection limit for the alloy coated with TiN film, suggesting that corrosion resistance in a saline solution is drastically increased by coating with the TiN film unless the corrosion potential exceeds +500 mV. The effectiveness of TiN coating in reducing the corrosion rate in the low potential region in simulated physiological environments was also suggested by Kimura and Sohmura [12, 21]. Before this coating technique is applied to metallic implants, the mechanism of the mechanical cracking of the coated film needs to be clarified in order to reduce the susceptibility to localized corrosion. The polysiloxane coating also improves the corrosion resistance of the NiTi alloy, suggesting that the chemical modification shown in Fig. 12 confers biological activity on the alloy surface and simultaneously improves corrosion resistance. This surface modification can provide an effective means to control metal/tissue interactions and can lead to bioactive metallic implants.

8
Summary

The breakdown potential of the NiTi alloy in a saline solution is significantly higher than that of the SUS316L and Co–Cr alloy currently used in surgical implants. The surface oxide film rich in Ti is responsible for this higher localized corrosion resistance. At present, preferential dissolution of Ni ions from the NiTi alloy devices is the subject of greatest importance in relation to Ni hypersensitivity. The immersion tests conducted using saline solution demonstrated that the amount of Ni released from a NiTi alloy device is 1/3000 to 1/6000 of the daily intake of Ni in the diet. Further study is needed, however, on the role of amino acids and serum proteins in enhancing the corrosion of the NiTi alloy in order to obtain a precise estimation of the amount of Ni ions released in vivo. The issues of the local accumulation of released Ni in the tissues around an embedded NiTi alloy device and the adverse reaction of these tissues also need to be addressed before this alloy can be widely employed as an implant material. The TiN and polysiloxane films are effective in reducing the dissolution of Ni ions from the Ti–Ni alloy. The development of more stable films, which prevent initiation of localized corrosion under the potentials exceeding +1200 mV, is desirable. Clinically, the use of a NiTi alloy device in combination with stainless steel or precious metal alloys should be avoided in order to prevent dissimilar metals corrosion.

References

1. Abiko Y, Endo K, Sachdeva R, Araki Y, Ohno H, Kaku T (1996) Effects of NiTi surface texture on corrosion resistance and cellular response. J Jpn Soc Dent Prod 9:9–17
2. Brown SA, Merritt K (1980) Electrochemical corrosion in saline and serum. J Biomed Mater Res 14:173–175
3. Clark GC, Williams DF (1982) The effects of proteins on metallic corrosion. J Biomed Mater Res 16:125–134
4. Cordero J, Munuera L, Folgueira MD (1994) Influence of metal implants on infection. J Bone Joint Surg B 76:717–720
5. Endo K, Sachdeva R, Araki Y, Ohno H (1994) Effects of titanium nitride coatings on surface and corrosion characteristics of NiTi alloy. Dent Mater J 13:228–239
6. Endo K (1995a) Chemical modification of metallic implant surfaces with biofunctional proteins. Part 1. Molecular structure and biological activity of a modified NiTi alloy surface. Dent Mater J 14:185–198
7. Endo K (1995b) Chemical modification of metallic implant surfaces with biofunctional proteins. Part 2. Corrosion resistance of a chemically modified NiTi alloy. Dent Mater J 14:199–210
8. Fukuyo S, Sachdeva R, Oshida Y, Sairenji E (1992) The perio root implant. In: Fukuyo S, Sachdeva R (eds) Perio root implant and medical application of shape memory alloy. Japan Medical Culture Center, Tokyo, pp 65–72
9. Hildebrand HF, Veron C, Martin P (1989) Nickel, chromium, cobalt dental alloys and allergic reactions: an overview. Biomaterials 10:545–548
10. Iijima M, Endo K, Ohno H, Mizoguchi I (1998) Effect of Cr and Cu addition on corrosion behavior of NiTi alloys. Dent Mater J 17:31–40
11. Iijima M, Endo K, Yonekura Y, Mizoguchi I, Yamada K, Ohno H (1999) Dissimilar metals corrosion of orthodontic metallic devices. Jpn J Dent Mater 18:45
12. Kimura H and Sohmura T (1987) Corrosion resistance in surface coated NiTi shape memory alloy. Jpn J Dent Mater 6:73–79
13. Miyazaki S, Shiota I, Otsuka K, Tamura H (1988) Effects of Cu addition on mechanical behavior of NiTi alloy. Mater Res Soc 9:153–158
14. Miura F, Mogi M, Ohura Y, Hamanaka H (1986) The super-elastic property of the Japanese NiTi alloy wire for use in orthodontics. Am J Orthod Dentofacial Orthop 90:1–10
15. Putters JLM, Sukul DMKSK, Zeeuw GR, Bijma A, Besselink PA (1992) Comparative cell culture effects of shape memory metal (Nitinol), nickel and titanium: A biocompatibility estimation. Eur Surg Res 24:378–382
16. Rondelli G, Vicentini B, Cigada A (1990) The corrosion behavior of nickel titanium shape memory alloys. Corrosion Sci 30:805–812
17. Rostoker G, Robin J, Binet O, Blamoutier J, Paupe J, Leibowitch ML, Bedouelle J, Sonneck PJM, Garrel JB, Millet P (1987) Dermatitis due to orthopaedic implants. J Bone Joint Surg A 69:1408–1412
18. Ryhanen J, Niemi E, Serlo W, Niemala E, Sandvik P, Pernu H, Salo T (1997) Biocompatibility of nickel-titanium shape memory metal and its corrosion behavior in human cell culture. J Biomed Mater Res 37:451–457
19. Saburi T (1989) Recent research on shape memory NiTi alloys. Kinzoku 8:11–18
20. Schroeder HA, Balassa JJ, Tipton IH (1962) Abnormal trace elements in man – nickel. J Chronic Dis 15:51–65
21. Sohmura T, Kimura H (1988) Evaluation of corrosion resistance for Ti-based alloy by anodic polarization. Jpn J Dent Mater 7:26–27
22. Tagawa K, Kim HK, Kuo YS, Fukuyo S (1992) Clinical application of 3-D shape memory implant. In: Fukuyo S, Sachdeva R (eds) Perio root implant and medical application of shape memory alloy. Japan Medical Culture Center, Tokyo, pp 61–64
23. Trethewey KR, Chamberlain J (1988) Corrosion for students of science and engineering. Wiley, New York

NiTi Alloys in Orthodontics

Andrea Wichelhaus

1 Introduction

In orthodontics, we are moving teeth within the bone in order to bring the teeth into a good functional position and alignment. Therefore, we are using a fixed appliance technique. In orthodontic-fixed appliance therapy, the physical properties of the materials play an important role in the application of force to the teeth. To move teeth, we are using different wire materials with different elasticity modules.

2 Conventional Wires and their Problems

Conventional wires are made out of stainless steel. They are available in different shapes (round, square and rectangular) and sizes. There however are problems associated with conventional stainless steel wire materials.

2.1 The High-Elasticity Module

During loading, we can only use the linear part of the activation curve (Fig. 1). If we bend the wire, we can apply a relatively high force on the tooth yet in practice, there is a high risk of overloading the teeth.

2.2 The High Load/Deflection Rate

Because of the higher elasticity module of stainless steel wires, a small activation (deflection) can cause a higher force on the teeth. Depending on the precise manufacturing of the wire and cross section, as well as the precise activation or bend of the wire, for the clinician to apply the right amount of force on the tooth. When moving teeth over a defined area, especially in the initial phase of the orthodontic treatment, it would be advantageous to use a material with a low load/deflection ratio. When using stainless steel materials, you have to bend loops into the wire, in order to lower the ratio.

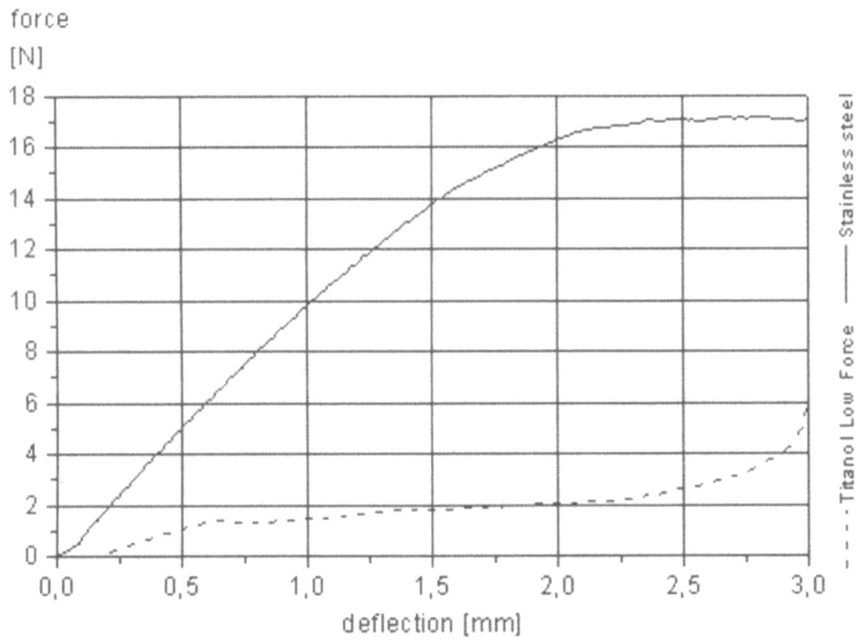

Fig. 1. Force/deflection diagram of a stainless steel and a NiTi wire with the same dimension. The difference in force emission on the tooth between these two wire materials is clearly visible

It is necessary to know the wire material stiffness, because of the problem associated with overloading the teeth. The result of overloading the teeth could be root resorption and irreversible damage to the hard and soft tissue. The wire stiffness is dependent upon:

1. The elasticity modulus
2. The wire dimension
3. The length of the wire
4. The interbracket distance

The gradient of the characteristic curve in the elastic range corresponds to the elasticity modulus of the applied wire material. Dependent upon the wire's cross-section, the stiffness of the wire material can be calculated over the elasticity module of the wire material [2, 3, 7–10].

When applying a steel wire material, it is necessary to bend it in loops in order to reduce the forces and to avoid an overloading of the teeth (see Fig. 11) [5]. Because of these loops, the wire's stiffness is reduced proportionally to the third power of the length (stiffness ratio $1{:}L^3$). At the same time, the force at work on the tooth is reduced by a relationship of 1:L. The disadvantage is that such loops are cumbersome, difficult to manufacture and can be deformed by the patient during mastication. This could change the direction of the forces considerably. Similar considerations come into play when other materials are used, if these materials are to be applied within the Hook's range (Fig. 1).

3 NiTi Wires

Compared to conventional wire materials, the nickel-titanium wires show characteristically different behaviours. Ever since they were described by Andreasen, NiTi alloys have been gaining importance in orthodontics [1]. The advantages of superelastic alloys lie in their special characteristics (Fig. 1), which can be described using tension tests. To look for the clinical behaviour and application of this material, we have to use bending tests. When viewing the cross-section, a strainless neutral phase can be detected in the middle of the wire during loading, whereas strain increases in the outer portions of the wire [16]. In orthodontics, pseudoelastic wires are used for the movement of teeth mainly by bending or torsion [19]. Stretching of a wire to use the energy stored in this stretching does not occur during treatment. The results of our own bending tests show the advantages of nickel-titanium wires in orthodontic applications and the differences between nickel-titanium alloys arising from manufacturing [18, 22]. The nickel-titanium wires shown in Figures 2 and 3 clearly differ regarding their behaviours. The tested material in Figure 2 shows no superelasticity in the area of the deactivation curve. This means for the clinical application in the patient that no definite constant force can be applied to the tooth. The activation becomes inexact – as with other conventional materials such as TMA or stainless steel. The tested material in Figure 1 shows smaller hysteresis behaviour. Consequently, the force emission of this wire is higher compared to the tested wire material in Figure 3. As observed, a different force application to the teeth is to be expected depending on the NiTi wire used. Therefore, the wire material shown in Figure 2, also known as "work-hardened", should not be used any more according to the actual scientific state-of-art. Materials with a corresponding pseudoelastic plateau in the area of the deactivation curve are better suited as archwire materials for orthodontic applications. Titanol[1], Sentalloy light and medium[2], Rematitan lite[3], NiTi[4], Nitinol-SE[5] and nickel–titanium[6] belong to this category.

Furthermore, the bending tests show that the force emission of the applied wire materials is dependant on the respective deflection. In the case of a small deflection of up to 2 mm, the clinician still cannot reckon on a constant force emission of the wire material. Here, only the low elasticity modulus and the good deflection behaviour of the wire material are exploited. The last-mentioned NiTi archwires have a good stability form, i. e. neither at mouth temperature nor at room temperature can the archwires be deformed permanently without pliers. This is achieved through a correspondingly low transformation temperature of the alloy. At room temperature, this alloy is in an austenitic state. Because of the low elasticity module and the good deflection in the case of corresponding stability form, these wires are especially suited for initial levelling tasks in the case of rotated teeth. Segmented archwires used to apply moments should also be made out of this NiTi material. Since the range of constant force emission of the tested wire materials increased with activation, these wires have been suited for large

[1] Forestadent, [2] GAC, [3] Dentaurum, [4] Ormco, [5] Unitek, [6] Technomed

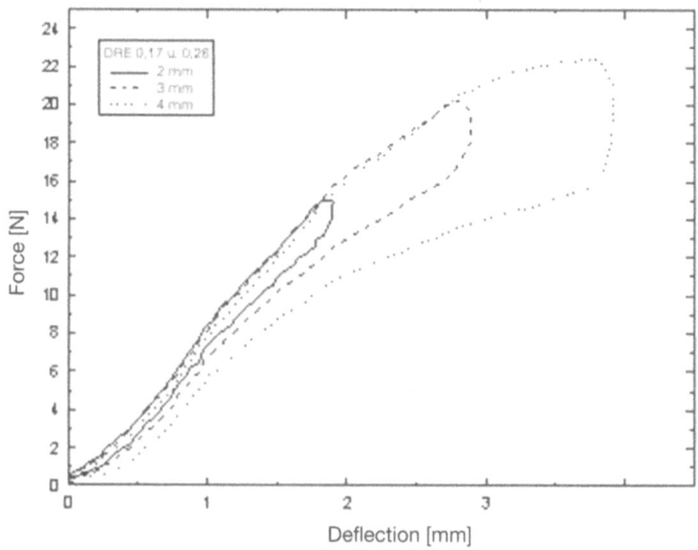

Fig. 2. Force/deflection diagram of a 0.017"x0.025" Rematitan (Dentaurum). The linear behaviour of the force emission in the case of different activations becomes apparent

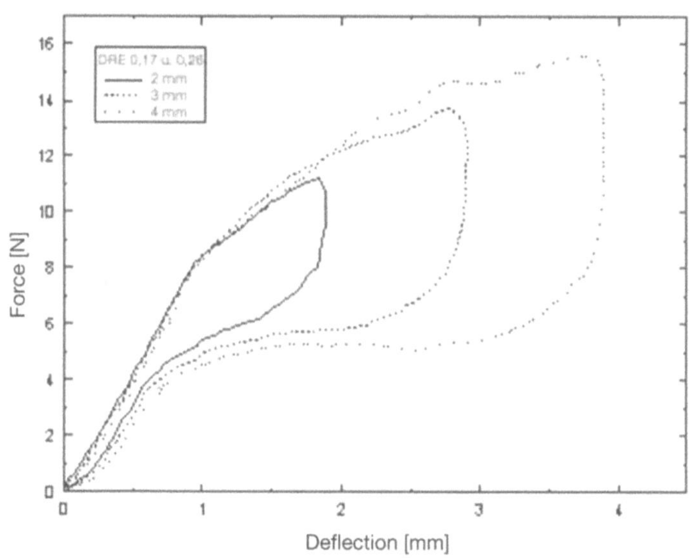

Fig. 3. Force/deflection diagram of a 0.017"x0.025" Rematitan lite (Dentaurum). An activation of 4 mm showed a pseudoelastic behaviour

deflection. Side effects on the adjacent teeth must be taken into consideration in the case of large deflections in order to avoid secondary effects. To reduce the side effects, especially during single tooth movements over a large range, the bypass and piggyback techniques proved to be efficient (Figs. 4, 5).

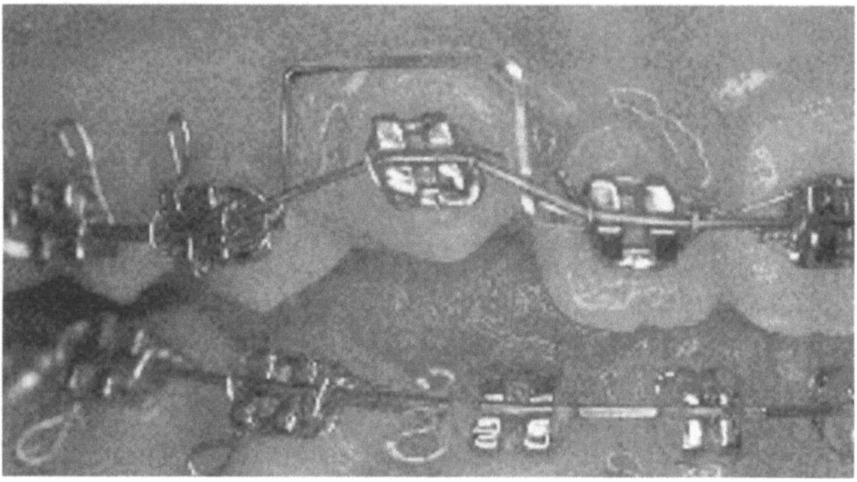

Fig. 4. Alignment of the canine in one step using a superelastic NiTi wire. Because of the good deflection of the material, the tooth was aligned without reactivation. Because of the side effects, an overlay-technique was used: a stainless steel archwire was inserted as a stabilization arch for the other teeth

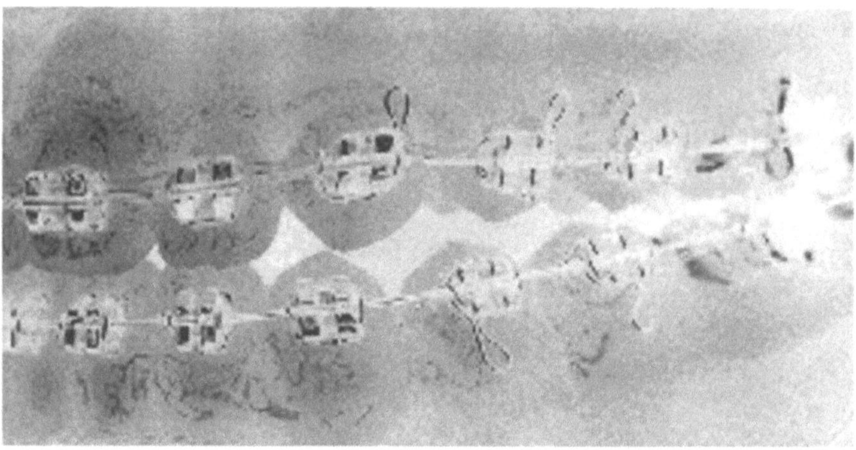

Fig. 5. Alignment of the canine in one step using a superelastic NiTi wire. Because of the good deflection of the material, the tooth was aligned without reactivation. Because of the side effects an overlay-technique was used: a stainless steel archwire was inserted as a stabilization arch for the other teeth

4 Thermal NiTi Wires

In the case of NiTi wires of the newer generation, the As temperature was changed through slight levelling differences or a corresponding manufacturing process. The transformation temperature (TTR) between martensite and austenite was adjusted, so that it could be exploited for the orthodontic application in the

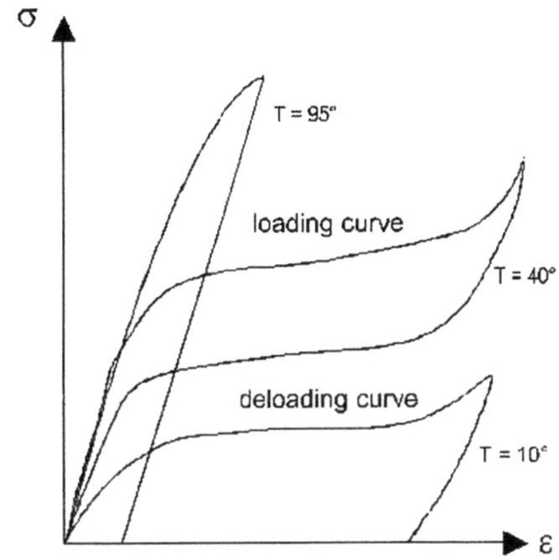

Fig. 6. Tension diagram of a NiTi wire. Depending on the temperature in the mouth it is possible to apply physiological low intermittant force to the teeth

form of inter mittent forces. In the case of this archwire material, the alloy is in martensitic condition at room temperature. Only when inserting the archwire into the mouth, does the alloy change from the martensite into the austenite phase. This means that a force is exerted on the tooth, only when the temperature is increased. Taking food and drinks of different temperatures can lead to varying force sizes of the wire. This is tantamount to an inter mitting force application. The NiTi archwires, as thermal archwires (Neosentalloy[1], copper–NiTi[2], Low-Force[3], etc.) exploit, besides the low elasticity module, the memory effect or thermoelasticity. Small intermittent forces are transferred to the teeth depending on the temperature in the patient's mouth (Fig. 6).

The results of our bending tests on NiTi archwires showed, through the marked hysteresis, the deactivation curve on a clearly lower level [22]. Consequently, the force level was the lowest in the case of this archwire material. However, large variations became apparent between the individual manufacturers. The force development of these NiTi archwires is represented by a moderate and defined area, as well as in the case of large archwire dimension (Fig. 7). When viewing Figure 7, it can be seen that the dimension of the archwire material has only a slight effect on the force emission. For clinical applications, this means that archwires with large cross-sections could be used at a very early stage in treatment without overloading the tooth. This enables a more physiological and more efficient tooth movement. The force developed by the wire is influenced only by the total deflection of the wire. It follows that the larger the deflection of the wire, the lower the pseudoelastic plateau and consequently the slighter the force emission of the wire. Therefore, a certain deflection of the wire is required to be able to exploit the superelastic properties for the treatment. In the case of a deflection of 2 mm, only a limited plateau behaviour appears, consequently, the superelasticity is not exploited.

[1] GAC, [2] Ormco, [3]Forestadent

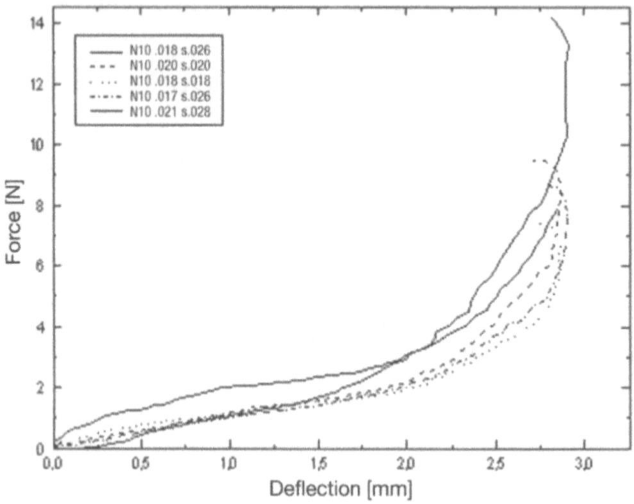

Fig. 7. Force/deflection diagram of a NiTi wire (Neosentalloy F100, GAC[14] Forestadent). Independent of the wire size, the force emission of the wire (deactivation curve) is nearly 1 N in the pseudoelastic plateau

5
Clinical Application of Thermal NiTi Wires

The thermal NiTi wires should be used for initial levelling and aligning of teeth due to the physiological force application to the teeth [4]. Figures 8 and 9 show thermal NiTi alloy clinical applications. The treatment can be resolved within a very short time by taking advantages of this alloy's properties. Consequently, the

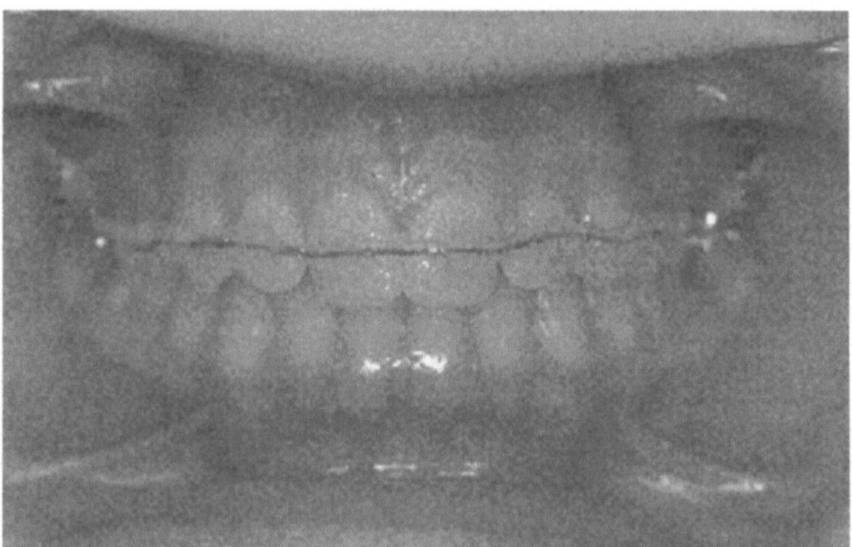

Fig. 8. Patient with an initially inserted thermal alloy as a first step in the orthodontic fixed appliance therapy. After eight weeks and the insertion of a second thermal NiTi wire the teeth were aligned

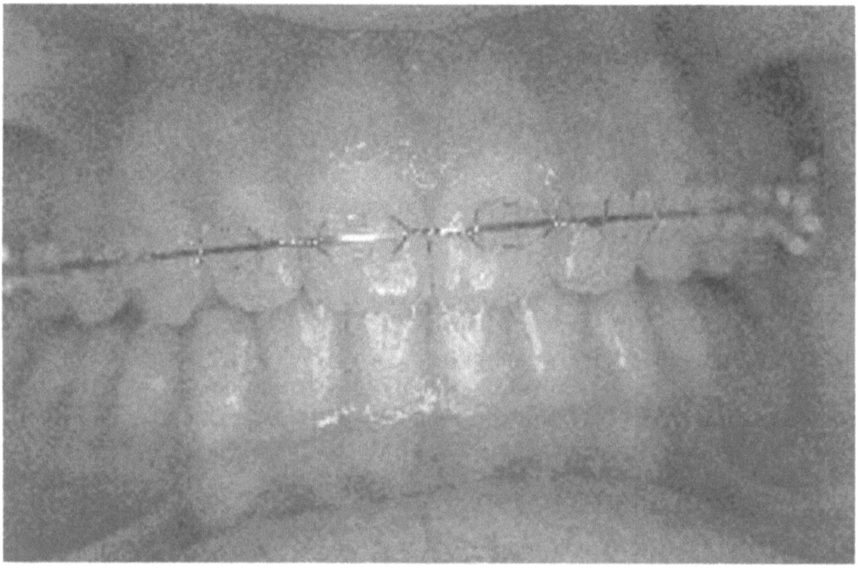

Fig. 9. Patient with an initially inserted thermal alloy as a first step in the orthodontic fixed appliance therapy. After 8 weeks and the insertion of a second thermal NiTi wire, the teeth were aligned

treatment time for the patient is considerably reduced. At the same time, it is possible for the clinician to achieve a good control of the tooth movement, because larger archwire dimensions can be used very early in the treatment. The insertion of a larger archwire with the same force emission enables to fill the bracket slot and therefore exploit the programmed values of the brackets when using the straight wire technique. Despite the application of larger archwire materials and better control, the force evoked and wire should be chosen wisely, such that, an overloading of the teeth and corresponding tissue damage can be avoided [4]. Table 1 shows the different application possibilities of NiTi alloys in orthodontics.

Table 1. Orthodontic application of NiTi wires

With derotation	
Teeth alignment	Superelastic NiTi wires; thermal NiTi wires
Without derotation	
Sectional arches	Superelastic NiTi wires
Overlay technique	Superelastic NiTi wires; thermal NiTi wires

6
Properties of NiTi Alloys in Orthodontics

The following properties of NiTi alloys are exploted in orthodontics for the movement of teeth.

6.1
The Great Ability to Deflect (Shape Memory)

The wire always returns to its initial form independently of masticatory forces. Therefore, the tooth movement can be carried out continuously. Readjustment is not necessary. The great ability to deflect is not only an advantage of the SE-NiTi alloys, but also an advantage of the so-called work-hardened alloys.

6.2
Small Load/Deflection Ratio

Through the small load/deflection ratio, the wire develops low forces throughout its deflection without overloading the tooth. The advantage of the small load/deflection ratio offers the clinician the safety of overactivating without overloading and possibly causing root resorptions.

6.3
Superelasticity

The term pseudoelasticity is often used instead of superelasticity. This term means that after stretching a wire, an elastic range of the stress/strain curve is followed by a plastic range, which does not lead to increase in strain despite further stretching. Independent upon the strain and temperature, this wire material shows a pseudoelastic plateau behaviour between As and Ms. Independent of the deflection of the wire when viewing the force/activation curve, continuous forces can be applied to the tooth. For the first time, the definite application of forces and moments is possible with the help of this wire material resulting in a more physiological tooth movement.

6.4
Memory

Martensitic materials are very special wire materials among superelastic wires. Depending upon the As-temperature, these wire materials show a temperature-dependent behaviour. In the low-temperature-phase, the wire can be formed and adapted to the patient's individual dental situation. After the wire material has exceeded the As-temperature, it turns into a high-temperature-phase and shows in the SEM-area the strain-induced martensite plateau with almost constant force emission. At the low-temperature-phase to high-temperature-phase transition, the wire remembers its initial form and shows the shape memory effect.

7
Advantages of NiTi Wires in Orthodontics

1. NiTi archwires are superior to conventional wire materials, especially in the levelling phase.
2. The exceptionally good deflection behaviour is achieved by no other alloy.

3. Use of definite force levels (superelasticity).
4. The archwire dimension only partly influences the force levels developed by the wire.
5. Greater control and efficient tooth movement in the early stages of treatment by the application of larger archwire dimensions at the beginning of orthodontic therapy.

8 Temperature Treatment of Orthodontic NiTi Wires

An ideal archwire would be one whose stiffness could be determined in advance in particular areas of the dental arch. Dependent on the tooth (canine or molar), the tooth position and the direction of the tooth movement, different force levels are necessary. In clinical practice, for extrusion of a canine, the wire should have a reduced force emission in that area of the wire. If too high force level is used, the risk of tissue damage increases. To reduce this pathological side effect, the force emission of the NiTi wire in that area is reduced, in order to treat the patient with a whole archwire. Loops were used to solve these problems using conventional techniques and materials (see Fig. 11). Besides their biomechanical negative effects, loops cause a lot of problems in oral hygiene, and therefore we have a higher risk in demineralisation of the teeth. By using NiTi alloys, it is possible to avoid loops. A change concerning the force behaviour of a superelastic NiTi wire in the plastic range is to be expected by additional heat treatment [11–13, 17]. The stiffness of the wire can be controlled correspondingly by heat treatment. The heat treatment of the NiTi wire can be carried out with an oven or with the Memory Maker [6, 16, 18, 21]. The last-mentioned method is much easier, because you can use it directly during treatment. The principle of the Memory Maker is to apply an electric current directly to the wire. The pulsing direct current that is carried over the wires produces heat treatment. The temperature at which the wire is brought is a determinant of its physical properties. In this fashion, it is possible to give the wire a reduction in the force emission in defined areas during heat treatment between 360°C to 520°C.

9 Memory Maker

According to Miura [12], another method to program the wire and to influence it in its superelasticity is the Archmade™. In both systems, Memory Maker and Archmade, the wire is heated by means of a source of electricity. However, the Memory Maker is better suited for the clinical handling: temperature is easier to control with a pulsing direct current, therefore overheating and irreversible damages of the wire can be avoided.

We carried out specific heat treatments at temperatures between 360°C and 520°C and the results in the bending test show a clear change in the pseudoelastic plateau of the desactivation curve. In the case of a temperature treatment with 360°C, a force of about 5 N is observed in the pseudoelastic plateau. When increasing temperature, the pseudoelastic plateau decreases its force level. The wire

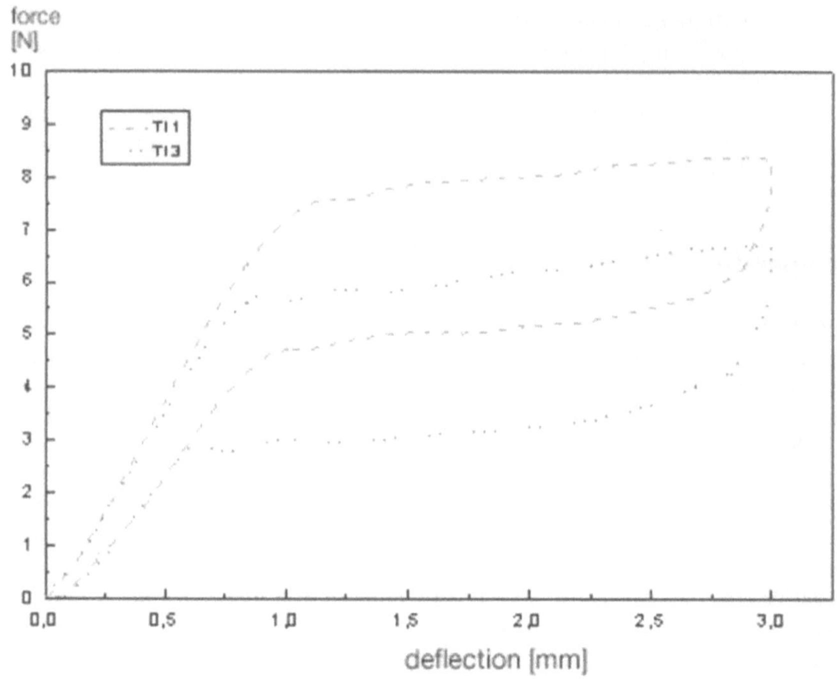

Fig. 10. Temperature treatment of a new 0.016"x0.022" Titanol wire. The reduction of the force emission is clearly visible when using a temperature of about 520°C (TI 1=360°C; TI 3=520°C)

dimension tested here reduces the force by about 2 N when the force is heated to 520°C instead of 360°C (Fig. 10); other investigations show the same trends [11–13, 16]. The higher the programming temperature, the lower the pseudoelastic plateau in the course of the deactivation curve. Furthermore, the wire material was not irreversibly damaged by changing the temperature. Former investigations have already shown that programmed and heat-treated wires can be reprogrammed [18, 21]. Wire materials that have been programmed with a low temperature, can be reprogrammed with a high temperature. This kind of transformation can change the behaviour of the wire. A wire material that has already been treated with a high programming temperature cannot be transferred to a higher force emission, in the pseudoelastic plateau, with a following low programming temperature. However, a wire that was initially treated with a low programming temperature of 360°C and a high force emission, can definitely be transformed with a programming temperature of 520°C with a low force emission in the pseudoelastic plateau.

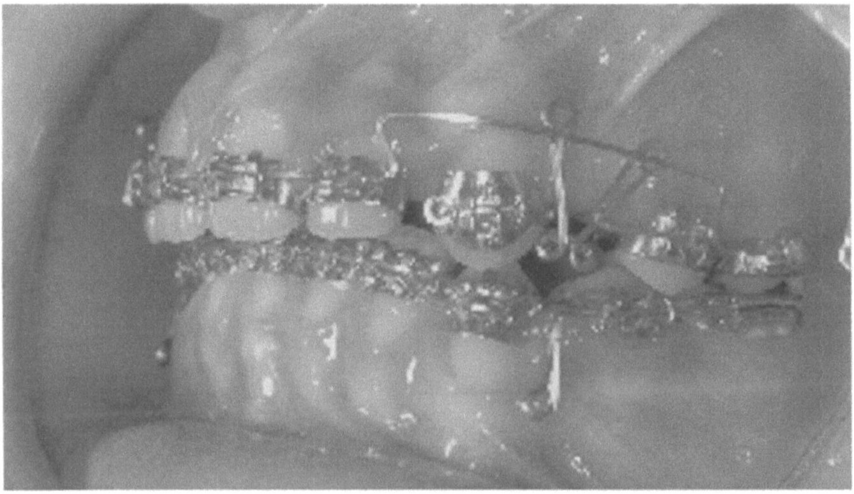

Fig. 11. Patient with a multi-loop archwire to reduce the stiffness of the stainless steel wire

10
Heat-Treated Archwires and Clinical Application

As clearly seen in the Figures 11 and 12, the oral hygiene is problematic in the case of conventional multiloop archwires, because they show prediliction to accumulate plaque. The consequence may include changes in the microflora of the oral cavity. Haematogenic dissemination of bacteria and its effect on the whole organism cannot be excluded. Furthermore, the bending of loops in conventional stainless steel archwires time-consuming. The programming of invisible loops in the wire, as effected to align the canine in the patient in Figure 12, is less disturbing for the patient's oral hygiene, less time-consuming regarding the fabrication, and does not require an additional reactivation of the archwire. A wire once programmed by means of heat treatment can be removed from the mouth when the corresponding tooth movement is finished. Furthermore, with the knowledge of the experimental results with heat-treated NiTi wires, an archwire appliance can be developed that demonstrates a low force in certain areas and a high force in other areas. In the case showed in Figure 12, the force developed by the wire was reduced by a high programming temperature. In a wire section where we use teeth as anchorage (for example the molars) the wire is treated with a lower programming temperature, which means a larger force emission of the wire in this area. Therefore, reactive forces and moments can be better intercepted at the anchoring teeth. Figure 12b shows the clinical situation of the patient after six weeks, where the alignment is finished. Because of the low force emission of the wire, the canine could be brought into the right position without additional reactivation.

The heat treatment of NiTi wires proves to be advantageous for a variety of treatment tasks. Besides the reduction of the force emission of the wire, it is pos-

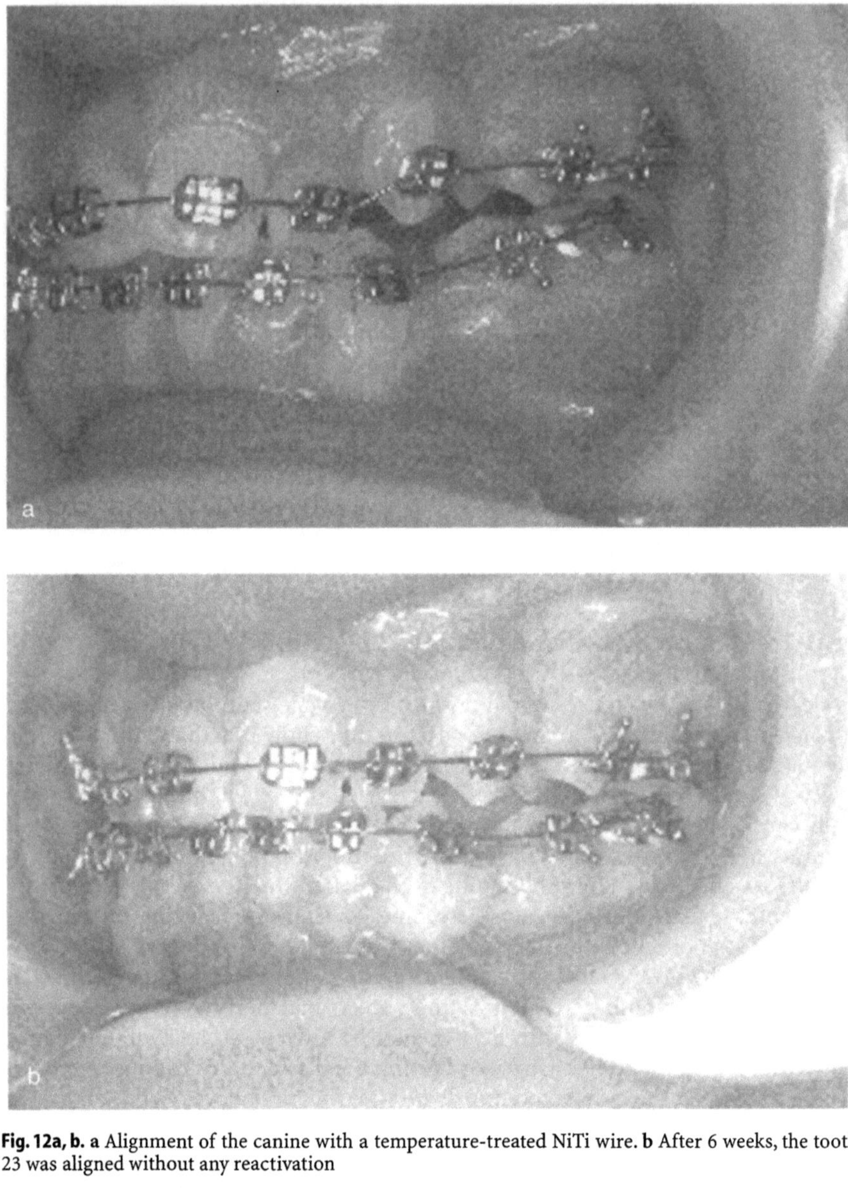

Fig. 12a, b. **a** Alignment of the canine with a temperature-treated NiTi wire. **b** After 6 weeks, the tooth 23 was aligned without any reactivation

sible to bring specific bends into the wire. For instance, to control the vertical position of the incisors during orthodontic treatment, you can incorporate second order bends by means of heat treatment.

The individual programming of a new memory in the form of first, second, or third order bend enables us to improve the treatment with NiTi archwires in the first phase of fixed appliance therapy. All in all, the heat treatment of NiTi

Table 2. Heat treatment with the memory maker

Vertical control of the incisors during levelling	Programming temperature 360°C; second order bends: curve of spee
Intrusion of the upper or lower incisors	Second-order bends: tip back
Transversal control of upper and lower arch during levelling	First-order bends: compression or expansion
Root movement in an early stage of treatment	Third-order bends: torque
Segmented archwires	First, second and third order bends
Invisible loop (reduction of the force emission)	Programming temperature: 520°C; no bending

archwires enables a more physiological, specific, and efficient tooth movement. Therefore, the treatment time can be considerably shortened. Table 2 shows the application possibilities of temperature treatment with NiTi alloys in orthodontics.

11 NiTi–Stainless Steel Combinations

The new development of NiTi-steel combinations [18–20, 22, 23] enables an additional clinical application of nickel-titanium alloys in orthodontics. The pseudoelastic plateau can be better exploited through a combination of the nickel-titanium alloy and the stiffer steel alloy. This way, specific and constant forces and moments can be applied to the teeth and the corresponding biomechanics can be fixed before the start of treatment. The superelastic material is combined with the steel wire material by means of a crimped connection offering completely new treatment perspectives. Moreover, this combination has a larger working-range, so that a readjustment is seldom necessary. Furthermore, the superelasticity is exploited for special application purposes, so that constant forces and moments can be expected in large areas. An extraordinarily favourable and physiological effect on the tooth movement is obtained. In addition, the archwire or segmented archwire can be adapted to the patient's individual situation by means of the flexible steel portion. As a result of our biomechanical investigations the following systems have been developed:

1. NiTi-SE–steel uprighting spring for the molars (Fig. 13)
2. Torque-segmented archwires for the root movement of the maxillary and mandibular anterior teeth (Fig. 14)
3. Compound retraction archwire for the bodily distal movement of incisors in extraction cases

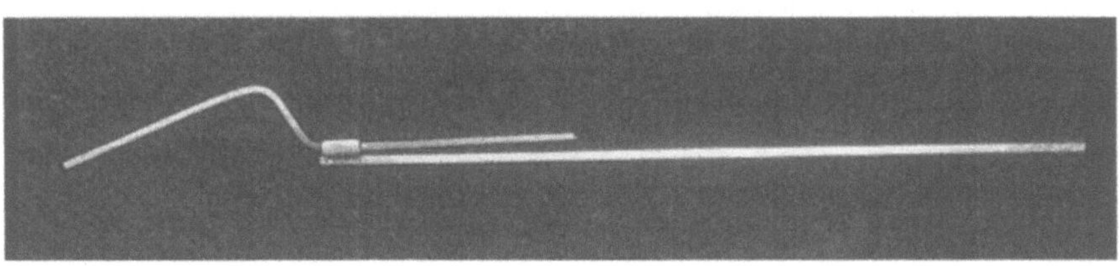

Fig. 13. NiTi-SE–steel uprighting spring

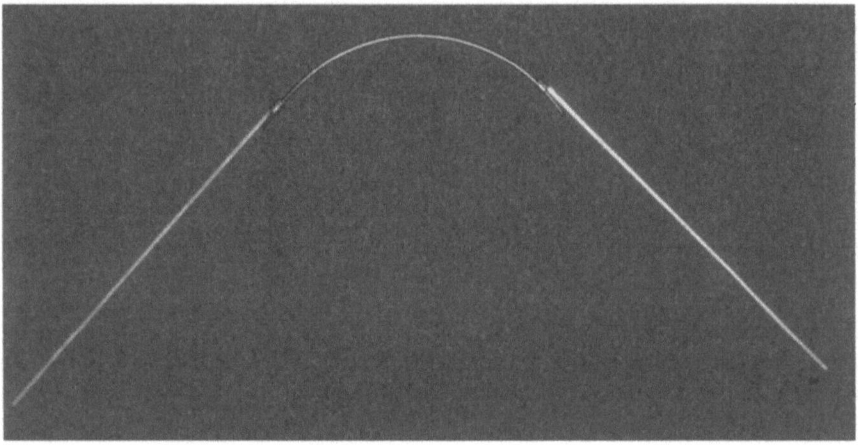

Fig. 14. Torque-segmented archwire. The torque-segmented archwire consists of a superelastic frontal segment with a pre-torque of 30° or 45°

4. Retraction spring for the canines
5. Intrusion mechanics for the bodily movement of the anterior teeth in the alveolar bone

The advantages of these treatment elements show a considerable improvement in all patients compared to conventional materials and spring systems [14, 15]. Further researches' activities in this field will certainly lead to the development of other treatment elements on this basis.

References

1. Andreasen GF, Hillemann TB (1971) An evaluation of 55 cobalt substituted nitinol wire for use in orthodontics. J Am Dent Assoc 82:1373–1375
2. Bantleon H-P, Droschl H (1985) Kraftabgabe von Loops bei Verwendung unterschiedlicher Loophöhen und Drahtqualitäten. Fortschr Kieferorthop 46:471–484
3. Bustone ChJ, Baldwin JJ, Lawless DT (1961) The application of continuous force to Orthodontics. Angle Orthod 31:1–14
4. Faltin RM, Arana-Chavez VE, Faltin K, Sander FG, Wichelhaus A (1998) Root resorptions in upper first premolars after application of continuous intrusive forces (intra-individual study). J Orofac Orthop 59:208–219
5. Jarabak JR (1960) Development of a treatment plan in the light of one´s concept of treatment objectives. Am J Orthod 46:481–514
6. Knox J, Jones M, Durning P (1993) An ideal preformed arch wire. Br J Orthod 20:65–70
7. Kusy RP (1981) Comparison of nickel-titanium and beta titanium wire sizes to conventional orthodontic archwire materials. Am J Orthod 79:625–629
8. Kusy RP, Greenberg AR (1982) Comparison of the elastic properties of nickel-titanium and b-titanium archwires. Am J Orthod 82:199–205
9. Kusy RP, Stevens LE (1987) Triple-stranded stainless steel wires – evaluation of mechanical properties and comparison with titanium alloy alternatives. Angle Orthod 48:18–32
10. Lane DE, Nikolai RJ (1980) Effects of stress relief of the mechanical properties of orthodontic wire loops. Angle Orthod 50:139–145

11. Miura F, Mogi M, Ohura Y, Hamanaka H (1986) The super-elastic property of the Japanese NiTi alloy wire for use in orthodontics. Am J Orthod 90:1–10
12. Miura F, Mogi M, Ohura Y (1988) Japanese NiTi alloy wire: use of the direct electric resistance heat treatment method. Am J Orthod 10:187–191
13. Okamoto Y, Hamanaka H, Miura F, Tamura H, Horikawa H (1988) Reversible changes in yield stress and transformation temperature of a NiTi alloy by alternate heat treatments. Scripta Metallurgica 22:517–520
14. Sander FG, Wichelhaus A (1995) Clinical experiences with the torque-segmented archwire (TSA). J Orofac Orthop 56:194–201
15. Sander FG, Wichelhaus A (1995) Clinical application of the new NiTi-SE-steel uprighting spring. J Orofac Orthop 56:296–308
16. Sander FG (1990) Eigenschaften superelastischer Drähte und deren Beeinflussung. Inf Orthod Kieferorthop 4:501–514
17. Thier M, Kubla G, Drescher D, Bourauel C (1991) NiTi wires for orthodontic application. J Phys IV France Vol 1, coll C4, suppl III n° 11:181–186
18. Wichelhaus A, Sander FG (1994) The behaviour of superelastic wires in the elastic and plastic range in dependence upon temperature-treatment. Kieferorthop Mitt 8:95–106
19. Wichelhaus A, Sander FG (1995) Biomechanical evaluation of the new torque-segmented archwire (TSA). J Orofac Orthop 56:224–235
20. Wichelhaus A, Sander FG (1995) Development and test of a new NiTi-SE–steel uprighting spring. J Orofac Orthop 56:283–295
21. Wichelhaus A, Sander FG, Hempowitz H (1997) The transformational behaviour of wires in the elastic and plastic range in dependence upon temperature-treatment. In: Pelton AR, Hodgson D, Russell SM, Duerig TW (eds) Proceedings of SMST 1997. Shape Memory and Superelastic Technologies, Pacific Grove, pp 449–454
22. Wichelhaus A (1996) Die Entwicklung und klinische Anwendung superelastischer Bögen und Teilbögen in der Kieferorthopädie. Habilitation
23. Wichelhaus A, Sander FG (1996) Anwendung des Compound-Retraktionsbogens. Inf Orthod Kieferorthop 3:407–424

Clinical Application of Shape-Memory Alloys in Orthodontics

Dietmar Siegner, Dagmar Ibe

1 Introduction

Among the different applications of NiTi materials in medicine, force delivery systems in orthodontics are among the most promising. To move teeth orthodontically, forces of a specific magnitude are necessary. Too little force and tooth movement is slow or there is no movement at all. Too much force and there might be undesired side effects such as pain, anchorage loss, and root resorption. With conventional elastic materials, force degradation due to movement of the teeth or time-related degradation is always present and requires frequent reactivations of the force delivery systems. The advent of superelastic NiTi alloys promised to almost eliminate the force degradation problem and to design force delivery systems with precisely tuned force magnitudes. Within the scope of this chapter we look into a number of aspects related to the use of NiTi materials in orthodontics.

2 History

Nickel titanium alloys were first developed by the Naval Ordinance Laboratory in the early 1960s. These materials were introduced for orthodontic wires by Andreasen [1, 2] with the name *Nitinol*, which is a near equi-atomic intermetallic compound that is in the martensitic phase at room temperature and also in the mouth. To achieve strength, elasticity, and springback, the wires were subjected to a work hardening process. Thereby internal stress resulted in a nearly linear stress-strain diagram with almost no hysteresis. The modulus of elasticity was measured to be around 57,000 MPa and thereby about one sixth that of stainless steel [3–5]. Advantages in orthodontic applications were the ability to reduce the forces of a wire with great cross-section and also the large deformation possible before permanent deformation took place. Prior to the advent of NiTi a reduction in force was achieved by decreasing wire size, by using multiple strands of wires, or by using loops to decrease the load defection and increase the amount of wire between brackets [6].

In 1978, Furukawa Electric Company of Japan produced a new wire with springback properties, shape memory effect and super-elasticity. In 1985, Burstone et al. [7] studied the mechanical properties of Chinese NiTi wire by means of a bending test. They described an unusual deactivation curve in Chinese NiTi wires which - unlike steel and opposed to Hooke's Law- produces a constant force over a long range of activation. The composition of the alloy of this material was rather similar to the classic nickel titanium alloys above minute differences as well as a combination of heat treatment and work hardening which resulted in a transition temperature that was slightly below mouth temperature. Miura et al. [8] introduced the so-called Japanese NiTi alloy wire for orthodontic applications in 1986 and described the transformation of the austenite phase in the passive material into the martensite phase through the application of stress. The idea behind this was to have almost constant force acting on the teeth from the very irregular situation at the beginning of treatment to the nearly perfect alignment at the end of the treatment. Thus the number of archwires used during the treatment could be reduced from about six per arch to two or even only one. Due to the special kind of elasticity that can be utilized in orthodontic applications these shape memory alloy wires are called *superelastic* in orthodontic terminology.

3 Basic Application Principles

Of the different types of mechanical deformation, tensional stress is almost nonexistent in orthodontics. The most common application is in bending mode (Figs. 1, 2). Any discrepancy between the actual position of a tooth as represented by the orientation of the bracket slot in three dimensions and the desired posi-

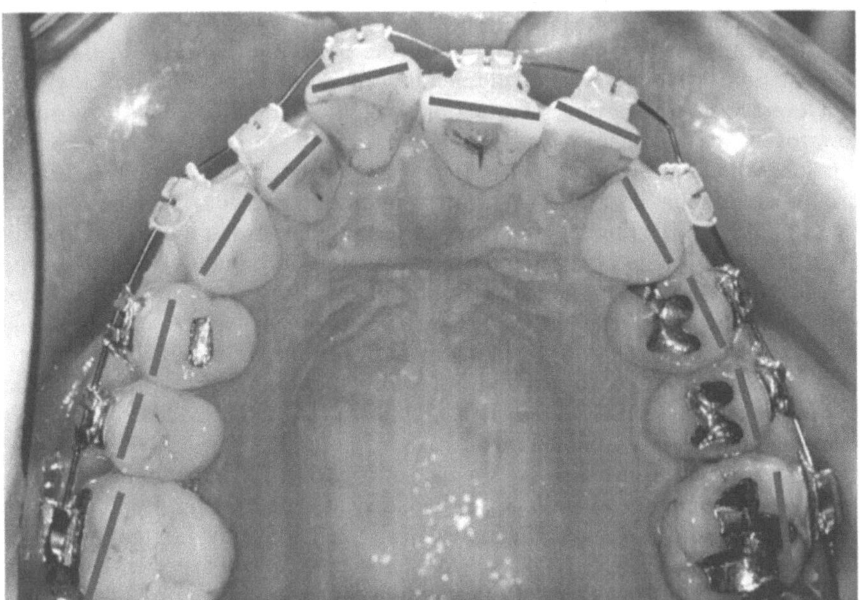

Fig. 1. Irregular teeth in the upper arch during the initial leveling stage. Please note the elastic deformation of the superelastic archwire in bending mode

Fig. 2. Test apparatus with three-point-bending test

tion as represented by the orientation of the passive archwire will result in a deformation of the archwire in the bending mode. The springback of the wire will exert the force or the force couple that moves and/or rotates the tooth towards the desired position and orientation.

Torsional mode in the wire is also used in two instances. A rotation of a tooth around an axis that is parallel to the archwire in the bracket slot can be achieved by using precisely calibrated archwires of rectangular cross-section and bracket slots that are grooves of rectangular cross-section only slightly wider than the inserted archwire. Any torsional activation of the archwire or an orientation of the bracket slot at an angle to the archwire plane will result in a torsional moment that is called a third-order moment or just *torque* among orthodontists. A second instance of torsional activation in orthodontic applications is in coil springs. When coil springs are activated, the wire in the individual windings is deformed mainly in a torsional mode.

The testing and comparison of shape memory wires for orthodontic applications creates a number of problems. The classic test in tension mode does not represent the clinical application and will lead to significantly different results. In bending mode different parts of the wire's cross section will experience different amounts of strain. The center of the wire will consist of the neutral fibre. Where no strain is present. Towards the outer curvature, the tension strain will increase and compression strain will be present towards the inner curvature (Fig. 3). As

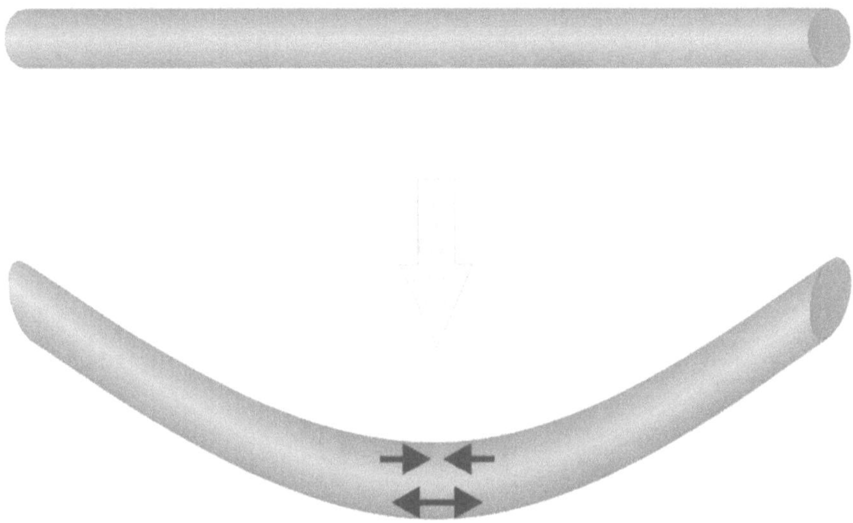

Fig. 3. Bending of a wire leads to compression (on the inside of the curvature) and tension (on the outside). In the center of the wire a «neutral fiber» can be found. The further away from this neutral fiber a particle of the wire is, the larger the internal stress will be. Often, a part of the wire cross-section will be in superelastic mode because the stress is sufficient to effect the martensitic transformation while other parts closer to the neutral fiber will remain in the austenitic state

the stress-strain relationship for shape memory alloys is not linear, different parts of the wire cross-section will have different relationships between martensitic and austenitic phases and at least some parts of the wire cross-section will not act in a superelastic fashion. In a tension test all parts of the wire cross-section will have the same strain and the wire will show a more consistent superelastic behavior. For the reasons shown, it is also important that the test set-up resemble the clinical situation as best as possible. Of special importance is the radius of the bending curvature and the wire span between the supports.

Three main methods exist, each of which have certain disadvantages (Fig. 4). The first is a simple three-point bending test where the test wire is bent by moving the middle one of three posts to deflect the wire from its passive state (Fig. 4a). At the same time, the force is measured and a force-deflection-diagram is created [9]. Advantages are the ease of use, relative cheapness, and the possibility to test curved wire segments. This is also the method envisioned for the new ANSI/ADA Specification 32 [10].

The second method uses three brackets mounted with the bracket slots in line where the center bracket is moved perpendicular to the long axis of the wire (Fig. 4b) [11–13]. The wire is held in the bracket slot using elastic ligature rings or steel ligature ties. This experimental set-up resembles the clinical situation best, but not optimally. Bending strain in this set-up is larger because the three bracket slots also introduce moments into the test wire. Problems with this method are the more complicated and time consuming procedure, a change in the geometric configuration between three-point contact to five-point contact depending on the

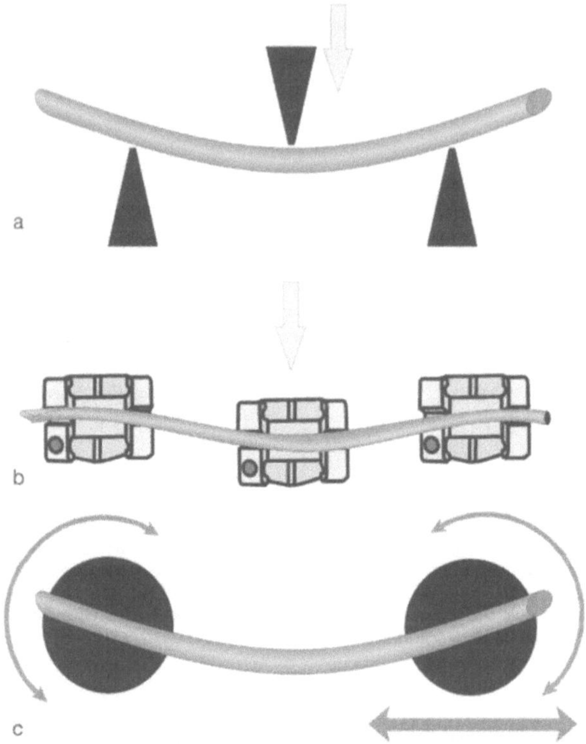

Fig. 4a–c. Schematic drawings of frequently used test methods in bending mode. **a** Three-point bending test. **b** Three-bracket test. **c** "Bonn test" (see text)

amount of deflection [11], and the influence of friction between wire and ligature and wire and bracket respectively. While extremely flexible and soft wires might not be able to achieve complete return to the passive state due to friction in the experimental set-up, the same friction might be overcome in the clinical situation due to the fact that mastication results in a situation where the teeth are in a vibrating state and static friction might be overcome.

The third method uses a more complex approach but is able to measure the stress-strain relationship in bending mode with hardly any side effects (Fig. 4c). This is achieved by having a short length of test wire held between two moment sensors that are mounted on an X–Y table so the distance between these sensors can be adjusted by feed back loop so that no horizontal forces act on the wire. Now one of the moment transducers is rotated incrementally while at the same time the moment is measured and the distance between the two transducers is adjusted so that no horizontal forces exist [14]. The advantage of precise measurements of the physical properties in bending mode is countered by an extremely complicated and expensive measurement apparatus and by the additional complication of transfering the resulting stress-strain data into meaningful force data for the orthodontic application.

When running tests of NiTi-wires that are supposedly superelastic it appears that a typical non-linear curve is present (Fig. 5). The curve has two distinctly dif-

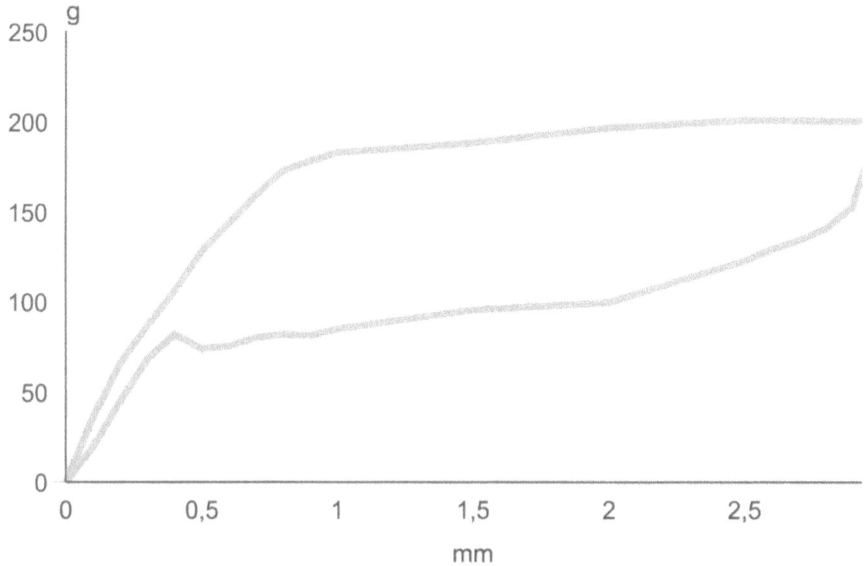

Fig. 5. Force deflection curve of a typical superelastic archwire (Forestadent Titanol low force; 0.016") tested in three-point bending mode with 12 mm between outer posts at 37°C

ferent branches, the activation curve and the deactivation curve, which is usually on a considerably lower level. The difference between the curves is mainly the result of hysteresis in the stress induced martensitic transformation and to a minor part the effect of friction. The activation curve consists of an initial almost linear part, which represents the elastic deformation of the austenitic material. At a certain amount of deformation the martensitic transformation is initiated and the curve becomes non-linear and the slope decreases. Usually the point where an elastic deformation of the martensitic material would result in a curve with increased slope will not be reached in orthodontic applications or bending tests. The beginning of the deactivation curve -after previous activation- is characterized by an initially rapid decrease of the force level before the curve shows a plateau with a relatively small slope. With further deactivation, a point is reached where the transformation of martensite to austenite is again completed and the curve becomes steeper and linear until it reaches zero.

When looking at different wire materials it becomes readily apparent that they can be categorized into three families called A, B, and C depending on their force-deflection-diagrams and the resulting clinical characteristics. The classic nickel-titanium wires consisting of work hardened martensite display a transition temperature clearly above the mouth temperature. Their force-deflection diagram shows only minor non-linearity, almost no hysteresis, and hardly any temperature dependence. The modulus of elasticity is about one sixth that of steel and the elastic working range is large. While more difficult to bend with pliers than steel it is nevertheless still possible. This type of NiTi wire is called *Family A*.

A wire of *Family B* is austenitic at mouth temperature but martensitic at room temperature or slightly below room temperature (A_f between 15°C and 30°C). They show a superelastic plateau in the force-deflection diagram. *Family C* wires are also austenitic at mouth temperature but as the transition temperature is only slightly lower (A_f between 30°C and 36°C), much smaller amounts of deformation or stress are needed to induce the martensitic transformation. Therefore they show a superelastic plateau on a much lower level. The forces developed by these wires are so low that it is possible to use a relatively thick wire or one of rectangular cross-section from the start of treatment, without the risk of harm to the teeth.

A typical application of these wires is leveling of irregularly positioned teeth during the beginning of treatment. The wire is pushed into the slot of a bracket on a malpositioned tooth by the orthodontist (activation) and tied in place using a ligature wire or an elastic ring. Through this activation/ligation, the wire is deformed. As it tries to return to the original shape (deactivation mode) it exerts a force and/or moment on the bracket and the corresponding tooth. When the tooth moves in the desired direction the amount of deflection of the wire decreases and eventually the deformation of the wire will be so small that it is will not operate on the superelastic plateau any more. Now the acting force will decrease rather fast and tooth movement will cease. The lower this point is (close to the passive state of the wire), the more complete the leveling with this wire will be. The wire in Figure 5, for example, has a good plateau that reaches down to 0.4 mm.

4
When are Orthodontic Wires Superelastic?

From the above it becomes clear that in orthodontic applications a wire does not react in a superelastic fashion per se, but rather that the presence of a response with superelastic characteristics depends on a combination of material properties and the geometry of the application. Of prime importance is the amount of stress-induced deformation of each minuscule part of the wire. If the lengthening of a part of the wire is sufficient to incur the stress-induced martensitic transformation, that part of the wire will react in a superelastic fashion. As orthodontic wires are almost exclusively stressed in bending, the amount of internal stress will depend on the amount of bending, which again depends on the amount of irregularity between the teeth. More precisely the strain in a particular part of the wire depends on the curve radius and the perpendicular distance of that part from the *neutral fiber* in the center of the wire. The strain will be proportional to the distance and inversely proportional to the curve radius:

$$\sigma = \frac{d}{r}$$

Two facts are obvious from this equation. One is that two parts of the same wire that have different distances from the neutral fibre will also be subjected to different amounts of strain. It is therefore the norm rather than the exception that some parts of the wire are in a superelastic state while others are not. And the second is that in a thicker wire it will be easier to achieve sufficient amounts of strain to have a superelastic reaction. Measurements of different orthodontic wires show that in a situation that resembles the clinical situation closely, a number of commercially available materials do not react in a superelastic fashion [12]. However, some of these materials can be made to react in a superelastic fashion if the inter-bracket distance is reduced by having smaller teeth (like in the mandibular anterior region) or by using wider brackets [13]. It is therefore not uncommon to have a wire react in a superelastic fashion in one region of the dentition and not in another. Because the inter-bracket distances tend to be greater in the upper jaw it is more of a problem to find a wire that will react sufficiently in a superelastic fashion in that situation. If the inter-bracket distance is further increased by extracting a tooth or by not bonding every single tooth it will be all but impossible to achieve superelasticity in that region with the wire materials available today. These problems will be even more relevant when the wire is used as a sectional archwire or as a spring. In these applications the curve radius is commonly so large that the strain in the wire will never be large enough to induce the martensitic transformation. In such cases the length of the elastic element usually allows the usage of conventional, linear wire materials. If shape memory alloys are to be used with benefit in such a situation the design of the elastic element must ensure that a part of the shape memory alloy wire achieves a curve radius that is small enough to induce the martensitic transformation in this type of alloy. In principle this can be effected through a very large overactivation or by combining stiff sections of wire with sections that are of lesser stiffness. Up to now such elements or archwires have been realized by combining sections of steel wire with very short sections of superelastic NiTi wire but in principle it should be possible to have the different stiffness properties in the same wire as will be seen later. If such an element is activated almost all the bending deformation will take place in the short wire section of lower stiffness. There the radius of the curvature will be large enough to induce superelasticity.

5
Limiting the Force

One of the main tasks during the biomechanical design of orthodontic applications is limiting the maximum force. Tooth movement is based on the biologic reaction of osteoblasts and osteoblasts in the periodontal ligament (PDL) to the stimulus of stress within the PDL. The biologic reaction necessitates a minimum amount of stress that is created by the forces and moments acting on the bracket bonded to the crown of the tooth. For a certain interval the speed of tooth movement will increase with increasing stress in the PDL. A further increase in the stress will only lead to very small additional increases in the velocity of tooth movement and at a certain upper limit, tooth movement will actually come to a halt because the pressure in the PDL will exceed the capillary blood pressure and therefore the supply of nutrients will be compromised. Such excessive forces can also lead to sterile necrosis of tissue in the PDL which will hinder tooth movement further [15].

Orthodontists have used thinner wires, loops, multistranded wires, and NiTi wires without superelastic behavior (Family-A type) to reduce the forces and moments and avoid the above mentioned problems. However in presence of non-superelastic wires, the forces and moments increase steeply with decreasing length of the wire segment between the brackets and with increasing irregularity of the teeth. In addition, it is extremely difficult for the clinician to estimate the forces that are developed by the appliance.

With superelastic wires of either family B or C type the force is not as closely related to the inter-bracket distance and to the degree of irregularity. Because of the plateau in the force-deflection-diagram of a good superelastic wire the force will stay on the plateau level or increase only very slowly no matter how much the wire is activated. Only if the material is deformed so much that it has completed the stress induced martensitic transformation or if there is binding in the bracket slots due to the narrow curvature of the wire, can there be a substantial increase in the acting force. This effect represents a self-limitation of the maximum force and can be used to ensure that no excessive forces act on the teeth. As opposed to the "fuse"-effect that can be achieved by using multistranded stainless steel wires that will permanently deform on excessive forces and thus lose part of their springback, these superelastic wires will remain effective until almost complete deactivation. Figure 6 shows the effect of superelastic wire materials on varying inter-bracket distances. Here the force on the deactivation curve at 1 mm is given for different inter-bracket distances and for two different wire materials. Sentalloy light is a superelastic material of *Type B* while Nitinol SE is a material that basically belongs to *Type A*. It will only react in a superelastic fashion when the deformation is extreme. Figure 6 shows that the truly superelastic material gives

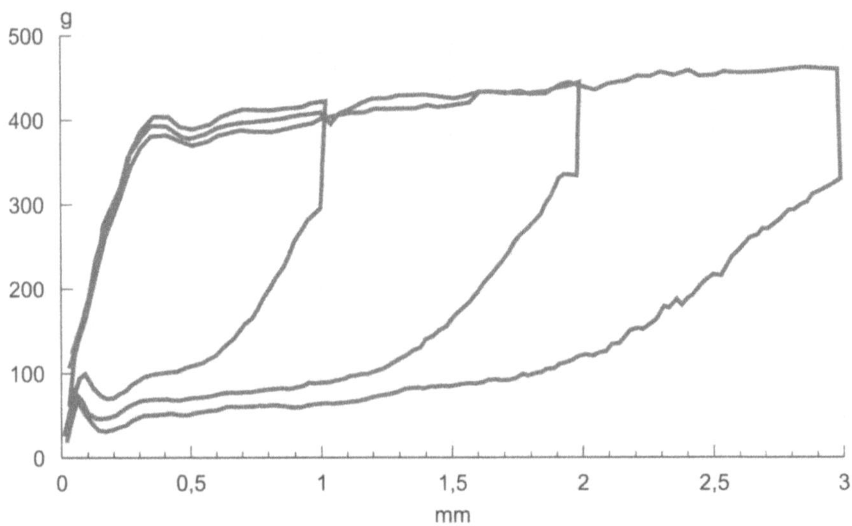

Fig. 6. Forces at an activation of 1 mm for different inter-bracket distances and two wire materials. Sentalloy light is a superelastic Type B material while Nitinol SE is a Type A material which displays superelastic characteristics only with extreme deformations that will not be present in orthodontic applications. Please note how the force is not related to the inter-bracket distance for the super-elastic material. The Type A material reacts more like a conventional material where the force increases with reduced inter-bracket distances. Wire specimens were activated to 2 mm, deactivated to 1 mm, and then measured

almost the same force even if the inter-bracket distance is reduced from 5.2 mm to 2.6 mm. The non-superelastic material shows an increase of the force by a factor of almost 2.5.

As the effective force of these wires is almost completely dependent on the material and not on how it is used and in what situation it is used the selection of a suitable material becomes all important. As can be seen from Figure 7, two materials of the same nominal cross-sectional dimension but from different manufacturers might develop dramatically different forces although both claim to be superelastic. Precise information on the available materials and quality control by the manufacturers would be extremely important to enable the clinician to select an appropriate wire. However, due to the complexity of the biomechanical behavior, due to a lack of understanding, and due to marketing claims this information has not been available to most clinicians.

It has to be kept in mind that typical force limiting measures used with linear elastic materials will not work with superelastic materials. An incomplete ligation, where the archwire is not fully engaged in the bracket slot, would reduce the force level in conventional materials. In superelastic materials this will usually not be the case as the force level on the plateau is almost constant. The elasticity of the periodontal ligament and the corresponding force diminution (tooth "gives way" to excessive forces) will result in a decrease in linear materials but not in superelastic materials. Lastly the effect of the force, the movement of the tooth, will lead to the lowering of the force in conventional materials, but again, not with the superelastic materials. Only the choice of the wire material is cap-

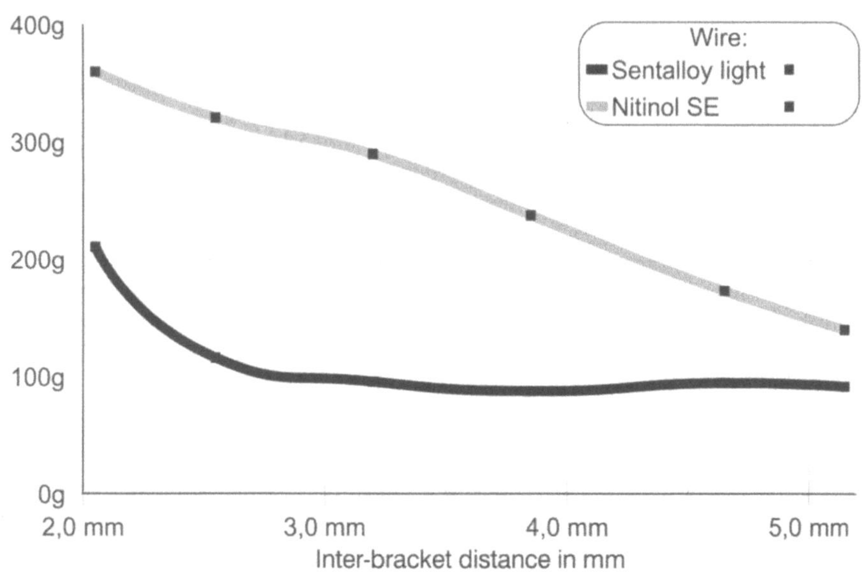

Fig. 7. Force deflection curves of the same wire dimension supplied by two different manufacturers. Note that one of the wires has a very short plateau with a force level of around 400 g, which is generally considered much too high for incisors. The other wire has a much longer plateau with a force level of about one third that of the other wire

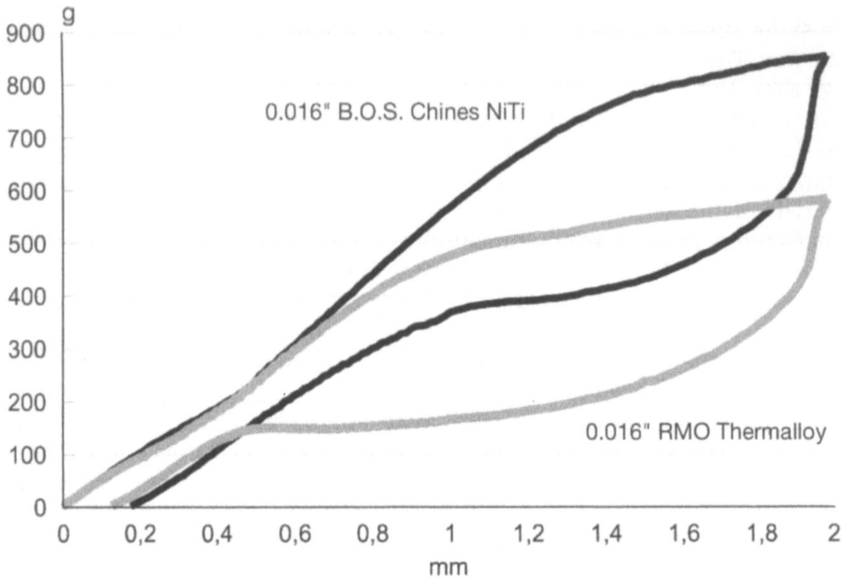

Fig. 8. Force deflection curves of a typical superelastic archwire (Sentalloy F80; 0.016"x0.022") tested in three-point bending mode with 12 mm between outer posts at 37°C. Activations were carried out to 1, 2 and 3 mm deflection. For the 3-mm activation the force on the deactivation plateau is actually close to 80 g between 2 mm and 0.3 mm

able to effectively influence the force level and therefore becomes a much higher importance than the choice of different brands of stainless steel wires.

A small problem with superelastic wire materials is the fact that in the first part of the deactivation process the forces are still rather high because the superelastic plateau on the activation curve is on a much higher level (Fig. 8). If for example, an activation of 2 mm is required by the dental situation the forces with the wire in Figure 8 will initially be in the range of 250–330 g. The plateau and its force level will only be reached after a deactivation due to tooth movement of 0.7 mm has taken place. To avoid this initial force peak the wire should be *over-activated* before it is ligated to the relevant teeth. If it is over-activated to 3 mm and then deactivated to 2 mm before being ligated, the force level would be close to the plateau level of 80 g from the beginning on. In many cases, such over-activation can be achieved clinically, but in some in cannot.

6 Different Force Requirements for Different Teeth

A wire with good superelastic properties has a characteristic plateau on the deactivation curve of the load-deflection diagram. For the clinical application of superelastic orthodontic archwires the force level on this plateau should be compatible to the physiological requirements. This has to be seen with regard to the different teeth the orthodontic wire is transferring the force to via brackets and

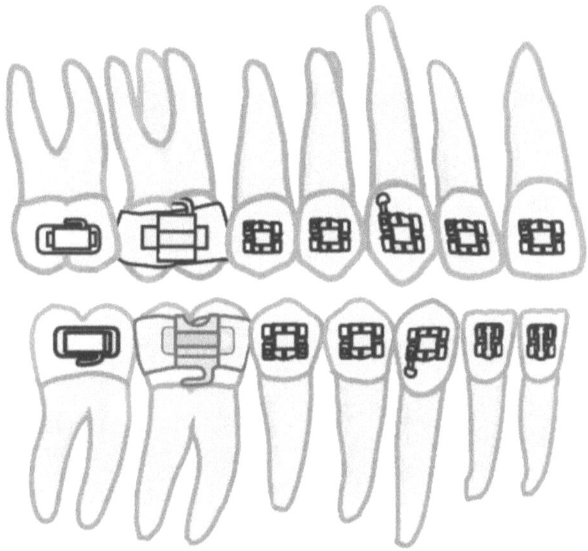

Fig. 9. Root surface areas of different teeth

ligatures. The human permanent dentition (Fig. 9) is characterized by teeth with varying numbers of roots and differing root surfaces as published by Bench et al. [16].

Responsible for the biologic reaction in the periodontal ligament is not the force acting on the tooth, at least not directly, but rather the pressure the osteocytes, osteoclasts and osteoblasts are subjected to. Pressure is defined, as force divided by area the application of the same amount of force on different teeth will create different loads per unit of root area (Table 1). The application of a conventional orthodontic archwire might therefore produce forces that are too high for selected teeth and as a result create side effects such as pain or root resorptions while on the other hand forces might be too low for effecting movement in other teeth. Using arch wires that are able to transfer different force levels to different teeth may be more physiological and safe for the teeth involved. It would therefore make sense to try to reduce the force for those teeth with the least root surface. Examples of superelastic wires that address this problem are *Bioforce*, *Multiforce* or *Triple force*. All are able to deliver different amounts of force to different teeth within the arch due to additional work hardening and/or heat treatment processing in the individual segments (Fig. 10) of the arch wire.

Table 1. Relationship between applied force and pressure for different teeth [16]

	Root surface (cm^2)	Force (g at 100 g/cm^2)	Force (g at 200 g/cm^2)
Molar	1.10	110	220
Premolar	0.60	60	120
Canine	0.75	75	150
Incisor	0.25	25	50

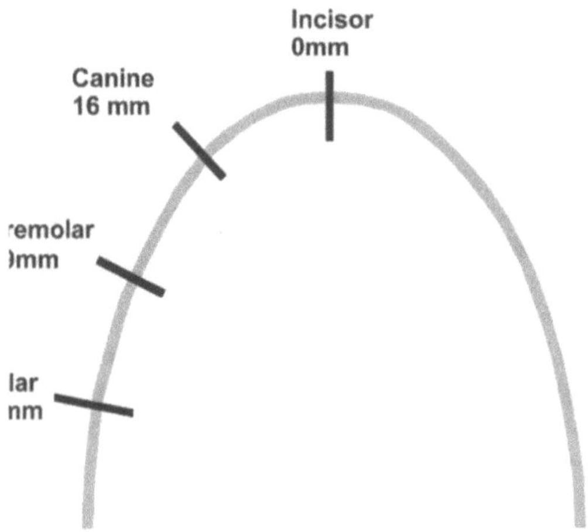

Fig. 10. Different segments on the archwires tested

The results of a study comparing orthodontic arch wire materials with different force levels within one arch by Ibe and Segner 1997 [17] showed significant differences between the intra-arch force levels as well as between the overall force levels of the different manufacturers. Figure 11 gives a schematic overview of the test device used. Table 2 summarizes the forces found in the anterior, canine, and premolar segments as a percentage relative to the forces in the molar segment, at body temperature.

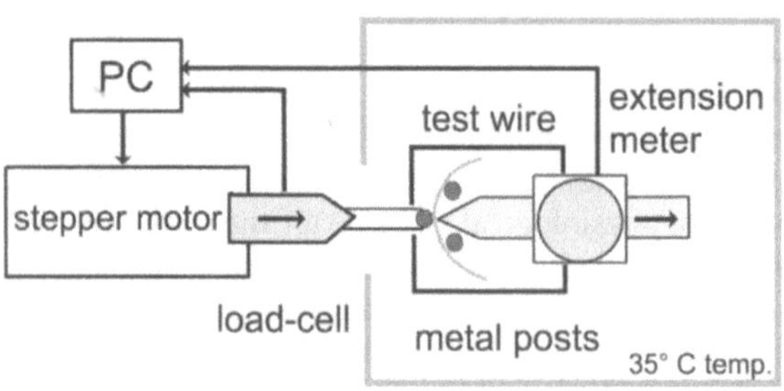

Fig. 11. Illustration of test device for measuring curved archwires in the three point bending test. *PC, Personal computer*

Table 2. Relative force reduction for the different segments in the tested wires

	Incisor (%)	Canine (%)	Premolar (%)	Molar (%)
NEO Sentalloy F 80	96.1	97.7	97.7	100
BIO force Sentalloy	14.2	40.9	96.9	100
Multi-Force archwire Titanol	89.1	90.4	96.7	100
Triple-Force archwire Titanol	50.0	56.4	72.4	100
Tri-force Arch	77.2	78.4	85.9	100
Multi-Force Arch	71.7	72.5	78.3	100

The figures give the force in the segment as a percentage of the force in the molar segment of the same wire.

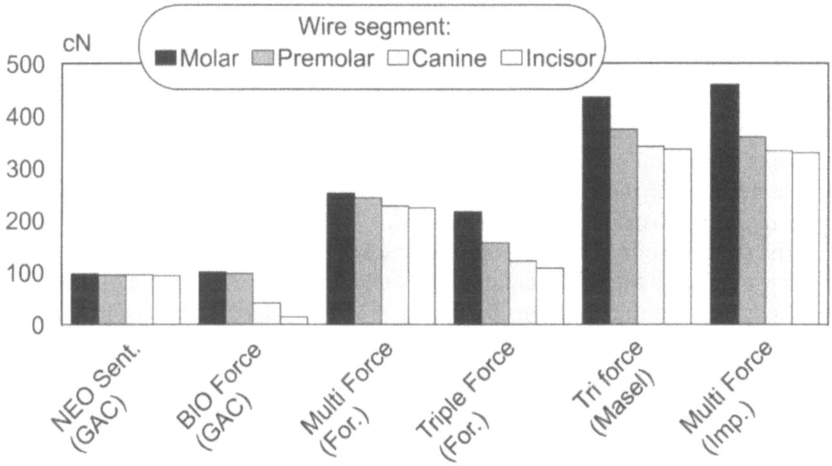

Fig. 12. Mean force of the different segments in the archwires tested. 1 cN[a]1.02 g

Superelastic properties of several different orthodontic wires are evaluated by the load-deflection curves of the wires: after 1, 2 and 3 mm activation and consecutive deactivation a clear superelastic plateau should be visible. This was the case for all wires studied. Regarding the level of this plateau, the results show differences in the overall amount of force delivered to the teeth and the precise graduation within the wires varying among different manufacturers (Fig. 12).

Burstone [18] claims that different types of teeth require the delivery of different magnitudes of force. This statement is congruent with the suggestions of Bench et al. [16] with reference to the correlation between orthodontic force and root area. As long as the most probable reason for root resorption is assumed to be the application of excessive forces, archwires with an extremely low deactivation curve are the most likely remedy against such unwanted side effects. A force level well within the physiological limits of each tooth opens up the possibility of initiated orthodontic treatment with a rectangular wire from the beginning leveling stage of fixed appliance treatment, which may lead to a shorter total treatment time. Whether the segmentation into three or more segments with different force levels is of clinical importance can be the subject of extensive discussion. In asymmetric extraction cases for example, the segments might not agree precisely with the tooth position. Because the occurrence of root resorptions is more often observed in the anterior regions of the maxilla and mandible, it may be safe to hypothesize that a force reduction in these areas may help in the prevention of these undesired side effects. The lower incisors with their small root surfaces are a perfect example, and the question of applicable force level deserves consideration. Not only is the inter-bracket distance much lower than elsewhere in the mouth, but if a wire of large rectangular cross section is used early in treatment, alignment irregularities in the rotational alignment and/or axial inclination will lead to additional deformation of the archwire and additional stress. As the clearance in the bracket slot is also smaller [11] all of this would tend to increase the force were it not for a good superelastic wire material of a sufficiently low force level. Of clinical importance would be the fabrication of orthodontic wires with at least two force levels: a lower force level for the incisor region and a higher level for the canine to molar areas. A finer distinction is probably not necessary due to non-average tooth sizes or extractions. It also appears that the need for archwires with graduated force levels is greater in the mandible than in the maxilla because of the greater difference between incisor root surface and premolar/molar root surface in the former jaw.

In the study by our group [9], the lowest force level and the highest differentiation is provided by the GAC Sentalloy BIOFORCE arch wire, which closely resembles the force recommendations by Bench et al. [16]. Here the forces in the anterior segment were reduced to such a degree, that there might even be too little pressure to bring about any vascular changes in the PDL (periodontal ligament) to instigate movement. Similar graduations, with slightly higher force levels were provided by the Forestadent TripleForce wire.

Clinical advantages of such wire systems could be the application of rectangular superelastic wires with a lower force level in the anterior segment as the first archwire in the leveling stage of fixed appliance treatment. This seems to be beneficial with respect to root resorption and patients' level of discomfort at the beginning of treatment. As the cross section of these rectangular archwires almost completely fills the bracket slots, they are able to achieve a more complete leveling in all three dimensions and therefore can eliminate one or two archwire changes with the time, cost, and discomfort involved.

7 Other Superelastic Elements in Orthodontics

During the last decade, several other applications for superelastic materials in the daily orthodontic practice have been found, in addition to archwires. Some of the most commonly used auxiliaries are open and closed coil springs for moving teeth into the desired position (Fig. 13). These auxiliaries replace similar spring mechanics made out of stainless steel, chromium-cobalt, or elastic elements fabricated from latex or other polymers. As opposed to these elements the superelastic coil springs have the advantages that they do not show any force decrease over time and that hardly any tedious length adaptation is necessary. The former is relevant because elastomeric elements loose more than 50% of their original force within days and coil springs from conventional metals loose their force in relation to the effected movement of the teeth. The superelastic coil springs will maintain an almost constant force, even if the teeth move by 7 mm. No need for reactivation will arise for at least 3 months. Because the force will be the same iregardless of the activities, standard lengths can be manufactured and used, which saves the orthodontist time, and allows the manufacturer to produce universal clasps on the ends of the springs, which are well appreciated by patients and orthodontists.

Superelastic springs are even used to enhance inter-maxillary anchorage control in the same way as, for example, class-II intermaxillary elastics from upper canine to lower molars. A problem in these applications is the risk of breakage

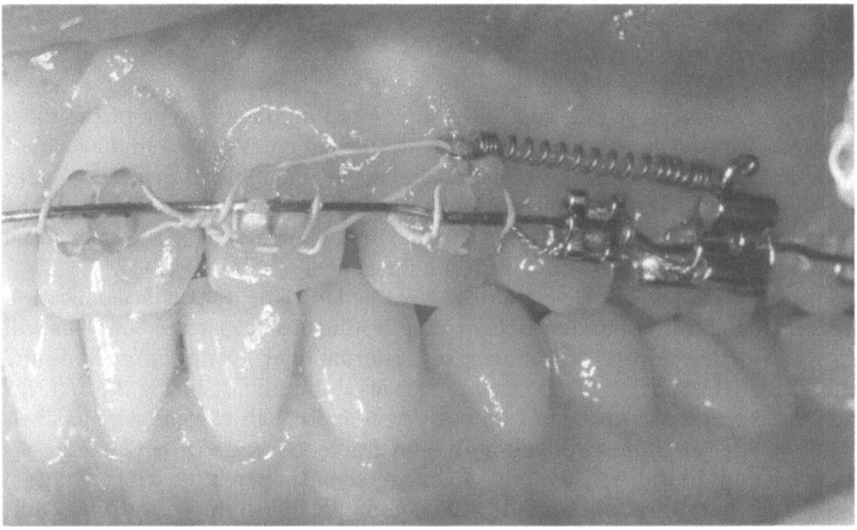

Fig. 13. Superelastic spring to close remaining spaces between teeth. Note the eyes at the ends of the spring, which is of standard length

due to fatigue or impact of cusps of teeth. These problems tend to be more pronounced in cases with unreliable co-operation, which are exactly the ones that would benefit most from such non-compliance devices.

One of the most frequently used orthodontic tools for anchorage control is the headgear, where forces are directed on the teeth via elastic elements, which attach an external cap to an intra-oral face bow, with the retention extra-orally on the neck or occipital part of the cranium. The device can be set up with different directions of pull in order to obtain a specific linear tooth movement or to transfer a rotational moment to the teeth. Usually the headgear consists of a metal inner bow soldered to an outer bow with hooks at the ends of both sides. Here the force generating modules are hooked up to deliver a distally directed force on the teeth or maxillary jawbone. Attempts had been made to use the superelastic properties in form of modules with superelastic coil spring for a constant force application in the use of headgears. As movements of the head have a substantial influence on the activation and force of these modules [19], superelastic shape memory alloys are sometimes used. Here the heat sensitivity of these alloys creates problems because the temperature of the spring varies between body temperature when the patient has his neck on a pillow during sleeping and almost room temperature if there is no pillow above the spring. The precise amount of force delivered by the headgear depends on factors other than the superelastic properties of the spring.

Another application of superelastic materials in orthodontics is in expansion screws for removable orthodontic appliances, for example Forestadent Memory expansion screw. These forces generate a more constant force between activations of the screw. As result patient comfort is increased and the speed of movement somewhat increased.

8
Conclusions and Outlook

The precise delivery of well controlled forces and moments is one of the most important aspects in orthodontic treatment. Shape memory alloys in the form of superelastic wires and/or superelastic auxiliary elements lend themselves very well to this task. Using superelastic wires that exhibit a physiologic force on the plateau can ensure a limitation of the maximum force and at the same time remain highly effective in achieving tooth movement of optimal speed.

Problems arise in the selection of wires suitable for a specific application because up to now no standardized labeling/testing has been agreed on. This will change with the advent of national and international norms on orthodontic wires (i.e., ADA Spec. 32). Further problems arise in individualizing the arch for the task at hand. As opposed to stainless steel and cobalt-chrome wires nickel-titanium wires are rather difficult if not impossible to form and bend using conventional tools like pliers. Methods to form and bend superelastic wires of type B and C exist, but differ substantially from common clinical practice and are cumbersome and expensive.

Difficulties in the handling of this material require a thorough understanding of its mechanical principles and behavior on behalf of the orthodontist. Modifications of treatment techniques are required, not only for basic forming and bending, but also for seemingly simple problems like dealing with the ends of the archwire that stick out of the bracket tubes on the last teeth in the dental arch. To avoid irritation of the buccal mucosa it is necessary to bend these ends towards the gingiva, which becomes a problem due to the shape memory and the difficulties in achieving sharp bends.

The fabrication of superelastic auxiliary elements is still in an early phase with the exception of open and closed coil springs. These coil springs represent highly effective instruments in moving teeth with constant and controlled forces. It is to be expected that most active elements in orthodontics will be available with superelastic force generating elements in the near future.

The question of unfavorable side effects of treatments with superelastic wires cannot be answered easily. Clinical observations show no evident differences in the amount of root resorptions between superelastic (type-B and -C) wires and classic steel or type-A nickel titanium wires. Few clinical studies have been conducted [20–22] but until now no long-term controlled study is available in which superelastic wires of moderate force were used in a suitable way. Unless such hard data is available the working hypothesis is that superelastic materials induce the same amount or less root resorption if the force levels are low enough. If, on the other hand, the force levels are too high, an increase in root resorption may be caused.

The many advantages that can be realized through the use of superelastic wires in the early stages of treatment (Fig. 14) far outweigh such problems as cost and modifications of the treatment technique. The latest generations of wires that deliver very delicate forces that are adapted to the requirements of individual teeth increase the safety and comfort of orthodontic treatments. In the later stages of treatment, however, these advantages decrease as factors such as friction

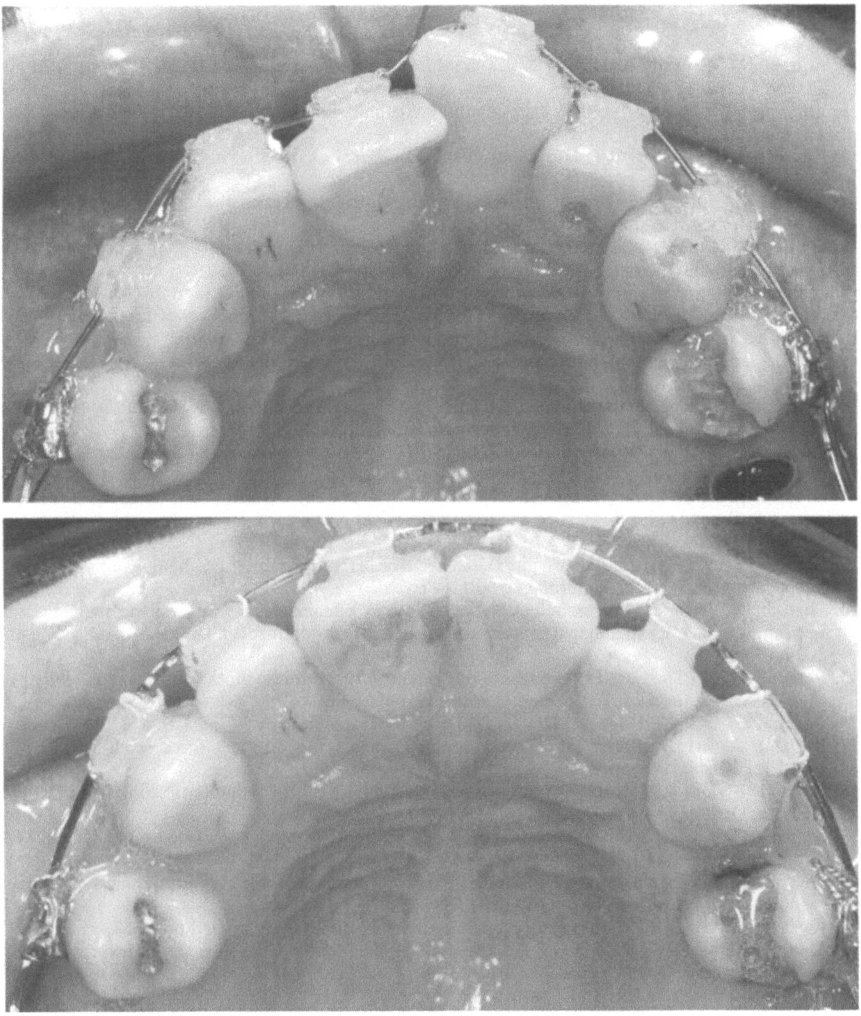

Fig. 14. Example of the leveling of an irregular upper arch using a superelastic archwire of Type B. *Top picture* taken directly after insertion of the fixed appliance, bottom picture 4 months later

and the ease of making precise finishing bends becomes of paramount importance. As of today, most patients will experience at least one archwire made of stainless steel during their treatment. In the future, new treatment procedures or new product developments might extend the indications of shape memory alloys in orthodontics.

References

1. Andreasen GF, Brady PR (1972) A use hypothesis for 55 Nitinol wire for orthodontics. Angle Orthod 42:172–177
2. Andreasen GF, Barret RD (1973) An evaluation of cobalt-substituted Nitinol wire in orthodontics. Am J Orthod 63:462–470
3. Andreasen GF, Morrow R (1978) Laboratory and clinical analyses of nitinol wire. Am J Orthod 73:142–151
4. Kusy RP, Greenberg AR (1982) Comparison of the elastic properties of nickel-titanium and beta titanium arch wires. Am J Orthod 82:199–205
5. O'Brien WJ (1989) Dental materials: properties and selection. Quintessence, Chicago, p 386
6. Burstone CJ (1981) Variable-modulus orthodontics. Am J Orthod 80:1–16
7. Burstone CJ (1985) Chinese Ni–Ti wire: a new orthodontic alloy. Am J Orthod 87:445–452
8. Miura F, Mogi M, Ohura Y, Hamanaka H (1986) The super-elastic property of the Japanese NiTi alloy wire for use in orthodontics. Am J Orthod Dentofac Orthod 90:1–10
9. Ibe D, Segner D (1998) Superelastic materials displaying different force levels within one archwire. Fortschr Kieferorthop 59:29–38
10. American National Standards Institute/American Dental Association ASC MD156 Specification 32 (1977) J Journal of the American dental association 95:1169
11. Ødegaard J, Meling T, Meling E, Holte K, Segner D (1995) An evaluation of the formulas for bending with respect to their use in estimating force levels in orthodontic appliances. A theoretical and in vitro study. Kieferorthop Mitt 9:73–88
12. Segner D, Ibe D (1995) Properties of superelastic wires and their relevance to orthodontic treatment. Eur J Orthod 17:395–402
13. Segner D (1995) Kraftniveau pseudoelastischer Nivellierungsdrähte in Abhängigkeit vom Interbracketabstand. Fortschr Kieferorthop 56:34–40
14. Plietsch R, Bourauel C, Drescher D, Nellen B (1994) Ein rechnergestützter Biegemeßplatz zur Bestimmung der Elastizitätsparameter hochflexibler orthodontischer Drähte. Fortschr Kieferorthop 55:84–95
15. Reitan K (1951) The initial tissue reaction incident to orthodontic tooth movement as related to the influence of function. Acta Odontol Scand (Suppl 6)
16. Bench RW, Gugin CF, Hilgers JJ (1977, 1978) Bioprogressive therapy. J Clin Orthod 11, 12
17. Ibe DM, Segner D (1998) Superelastic materials displaying different force levels within one archwire. J Orofac Orthop 59:29–38
18. Burstone CJ, Qin B, Morton JY (1985) Chinese NiTi wire – a new orthodontic alloy. Am J Orthod Dentofac Orthop 87:445–452
19. Segner D, Bonowski S (1999) Variationen in der Kraftentfaltung des Nackenzug-Headgears. Kieferorthop 13:135–144
20. Owman-Moll P, Kurol J, Lundgren D (1995) Effects of increased force magnitudes on tooth movement and root resorption. Eur J Orthod 17:347
21. Rygh P, Brudvik P (1993) Root resorption and new wire qualities. Eur J Orthod 15:343
22. Maltha JC, Dijkman GEHM (1996) Discontinuous forces cause less extensive root resorption than continuous forces. Eur J Orthod 18:420

Orthodontic Application of NiTi Shape-Memory Alloy in China

Chu Youyi, Zhu Ming, Yang Fengzhi

1 Introduction

Based on the finding of unique superelasticity of NiTi shape memory alloy [1], a Nitinol wire was introduced for orthodontic usage by Andreasen et al. in 1971 [2]. In the beginning of the 1980s, NiTi orthodontic archwires started being produced commercially for clinical use in China [3]. Due to NiTi's superior performance over traditional wires made of stainless steel, NiTi orthodontic devices have become the naterial of choice in the initial correction of dental malocclusion. In recent years, the domestic demand for the NiTi orthodontic products grew rapidly, increasing annuallly at a rate of 30–40%. A production line for NiTi wires has been set up in the Shape Memory Centre of Beijing's General Research Institute for Nonferrous Metals (BGRINM) with annual production of many million sets of orthodontic wires. This centre is proud to have already achieved ISO9001 certification. The production of this line accounts for an overwhelming share of the Chinese market of NiTi orthodontic devices, and a certain amounts of the products are being been exported abroad. Improvements of these NiTi orthodontic devices include : material quality, heat treatment, size, shape and performances [4, 5]. Now a variety of NiTi orthodontic products are available to meet different clinical requirements, and will be described below.

2 Superelastic Archwire (SE Type)

The superelastic NiTi orthodontic archwire has been widely used to correct the dental malocclusion, such as open bite, cross bite, torsion and crowding. The archwires are available in two possible cross sections: round shape and rectangular shape. The latter is more effective while aligning the teeth. The size specifications (in inch) of the archwires are as follows:

– Round archwire (in diameter).	0.012, 0.014, 0.016, 0.018, 0.020
– Rectangular archwire (widthxlength).	0.016 x 0.016, 0.016 x 0.022, 0.017 x 0.022, 0.017 x 0.025, 0.018 x 0.022, 0.018 x 0.025, 0.019 x 0.025, 0.021 x 0.025

The transformation temperatures of the wires measured via electrical resistance are presented in Table 1. For the SE type archwires, the A_f temperature is at 30°C

Table 1. Transformation temperatures (°C) of SE- and RTF-type NiTi orthodontic archwires made by BGRINM

Type	M_f	M_s	A_s	A_f
SE archwire	-150	-75	-5	30
RTF archwire	-89	-43	16	33

and A_s temperature is −5°C. Both in the mouth and room temperatures, the wire is mostly in the austentic parent phase, thus in the superelastic state.

The bending force–deflection curve of a 0.02" round NiTi archwire is shown in Figure 1. The span of the specimen is 0.5 inch. In comparison, the bending result of a stainless steel archwire at the same experimental condition is also shown in the Figure 1. It can be seen that for the stainless steel wire much higher force is needed for bending to 90° and abrupt drop of the force occurs upon unloading with large residual deformation. In contrast, the NiTi wire exhibited significant superelastic behavior. A plateau deformation appeared at a low force when bending to 90°. Upon unloading, a retentive force was kept in a wide range and little residual deformation was found with recovery rate of about 96%. Obviously, this superelastic performance is desirable for orthodontic therapy producing light, constant biologically optimal tooth movement.

When compared with stainless steel archwire, NiTi archwire offer the following advantages:

- Gentle and constant force for tooth movement
- Less pain and damage to tooth and soft tissues
- Reduction of appointments necessary for device adjustment

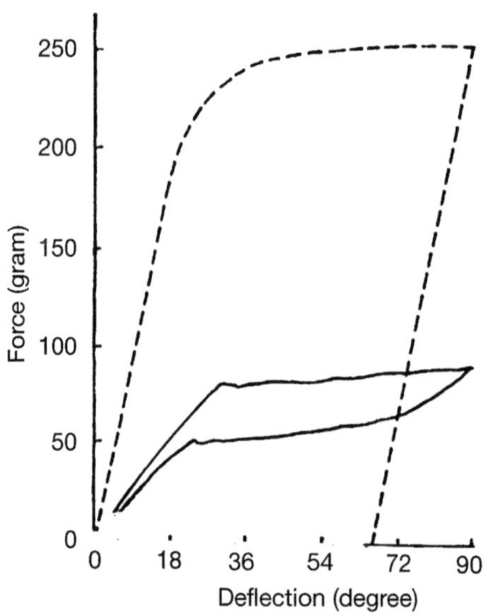

Fig. 1. Bending curve of NiTi superelastic archwire (*solid line*) in comparison with that of stainless steel archwire (*dashed line*). Specimen diameter of 0.02 in and gauge length of 0.5 in

- Reduction of therapeutic period by more than 50%
- Ease of use

  Thus, the application of NiTi archwire has been more and more popular in the orthodontic field.

3 Memory Archwire (RTF Type)

This type of NiTi archwire has the same composition, shape and size specifications when compared to SE type archwire mentioned above. The difference is that it is produced with a different thermomechanical treatment, which gives rise to higher transformation temperatures. As shown in Table 1, the A_s temperature at 16°C is 21°C higher than that of SE type, which is raised to above or nearby room temperature. Its A_f temperature at 33°C is still below mouth temperature of 37°C. Therefore the wire is kept in martensite state by the dentist during operation at room temperature. It is easily educated into complicated curvature and engaged in brackets. After deployment in the mouth, the wire is fully transformed to its parent phase and then exerts a recovery force on the teeth. Ease of deformation at room temperature (RT) is the main characteristic of this type of archwire, and gives rise to its name RTF (room temperature formation).

The structure change was examined with X-ray diffractometry at different temperatures from 0°C to 50°C. The results of the RTF archwire are given in

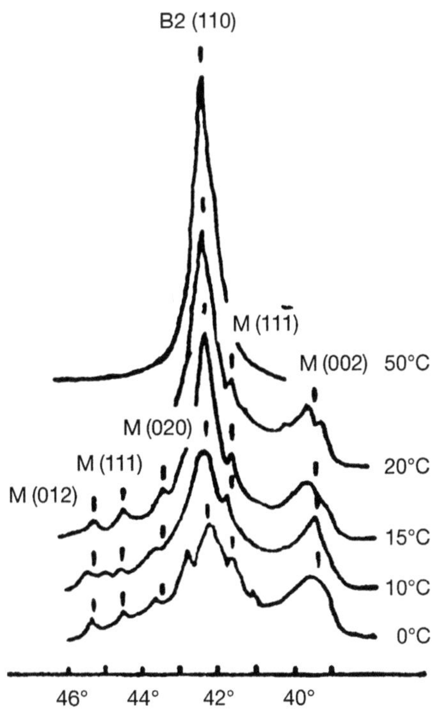

Fig. 2. X-ray diffraction patern at different temperatures for RTF type NiTi archwire

Figure 2. It can be seen that there are strong martensite peaks M(002) and M(11(−1)) at temperatures up to 20°C, which proves the existence of large amount of martensite.

The mechanical behavior was measured with bending and torsion tests at following temperatures: 4, 22, 37 and 50°C. Rectangular specimens with across section of 0.018x0.025 in^2 were used. In the bending test, the gauge had a 0.5-in length and the bending angle had a maximum of 90° at rate of 55°/min. In the torsion test, the gauge length was 1 in and the torsion angle had maximum of 720° at rate of 180°/min. As shown in Figures 3 and 4, the results of the bending and torsion tests had the same tendency. At lower temperatures of 4°C and 22°C, less moments were needed to bend and torque the wires and considerable permanent deformations were obtained upon unloading; when the wire was heated above the A_f temperature recover occurs by shape memory effect. When the temperature was raised to 37°C and 50°C, the wire exhibited full superelasticity with a return to original form shape.

4 Rocking-Chair Archwire

This archwire is shaped in the form of a rocking chair. It is specially designed for the levelling of deep curves of speed. It simplifies the dentist operation and eliminates over-correction and patient trauma. This kind of device is available in round and rectangular NiTi superelastic wires. The experimental results show

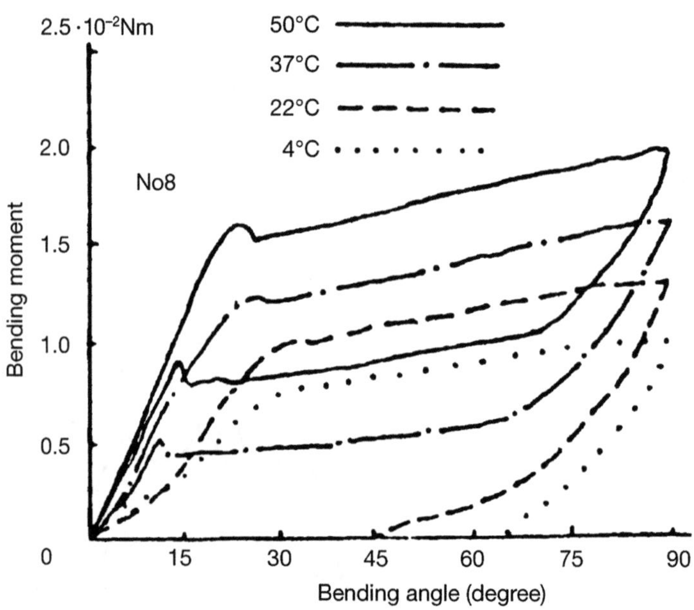

Fig. 3. Bending curves at different temperatures for RTF type NiTi archwire. Specimen size of 0.018x0.025 in^2 and gauge length of 0.5 in

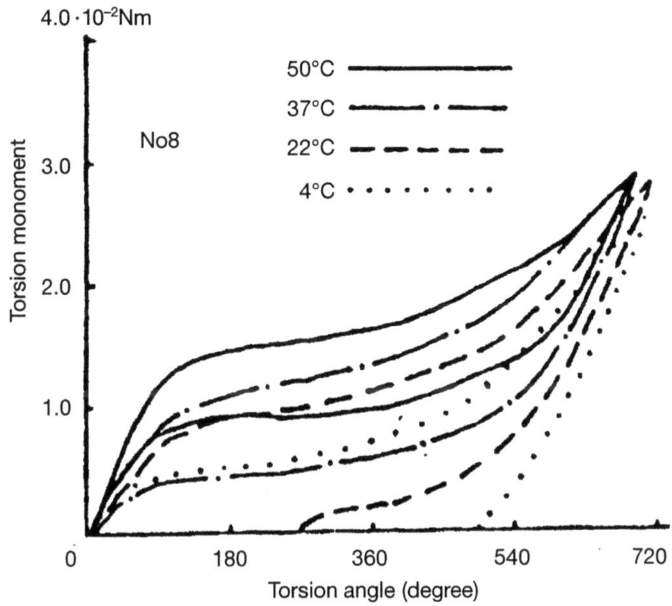

Fig. 4. Torsion curves at different temperatures for RFT type NiTi archwire. Specimen size of $0.018x0.025\ in^2$ and guage length of 1 in

that after compression flatening and maintaining for up to 4 months, the wire's recovery of its original form was an impressive 98%. This performance can satisfactorily meet the requirement of the clinical application.

5 Superelastic Orthodontic Springs

Two types of superelastic orthodontic springs exist: they are termed close coil springs (to close spacer) and open coil springs (to open spacer). Both springs (1.45 mm, outer diameter) are winded with NiTi superelastic wire (0.3 mm, diameter). The close coil spring is tightly wound, 10 mm in length with a hook on each end, made of the same material. The open spring has a screw pitch of 0.9 mm and a length of 180 mm, which can be cut and tailored at the discretion of the orthodontist. The application of NiTi superelastic orthdontic springs can overcome the shortcomings of high initial forces and large residual deformations of the stainless steel-type springs.

Tensile and compression tests were carried out for the close and open coil springs, respectively. For statistical reason of comparison, same sizes of NiTi and stainless steel (SS) springs were used. The close coil springs were 10 mm in length, the open coil springs measured 25 mm. The results are presented in Figures 5 and 6. It can be seen by Figure 5 that the force required to achieve an elongation of 25 mm was approximately 4.5 times higher for SS than that of the NiTi. Upon unloading, the tensile force of SS spring decreased rapidly and a large resi-

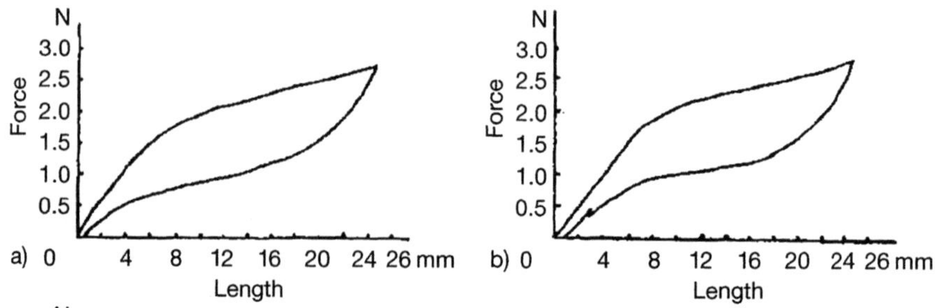

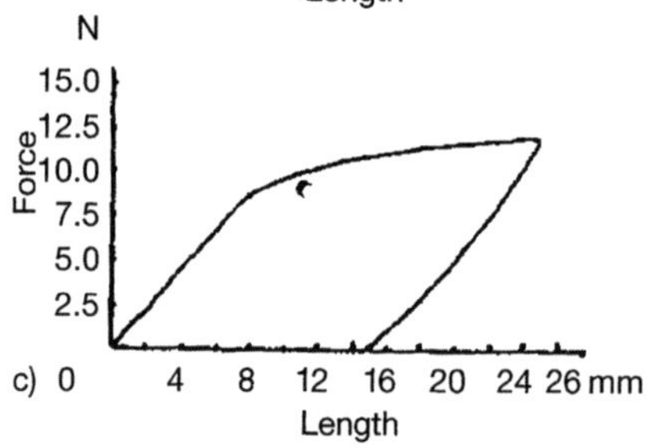

Fig. 5a–c. Tensile curves of NiTi close spring (**a**), NiTi close spring after used for 6 months (**b**) and stainless steel close spring (**c**). Specimen length of 10 mm

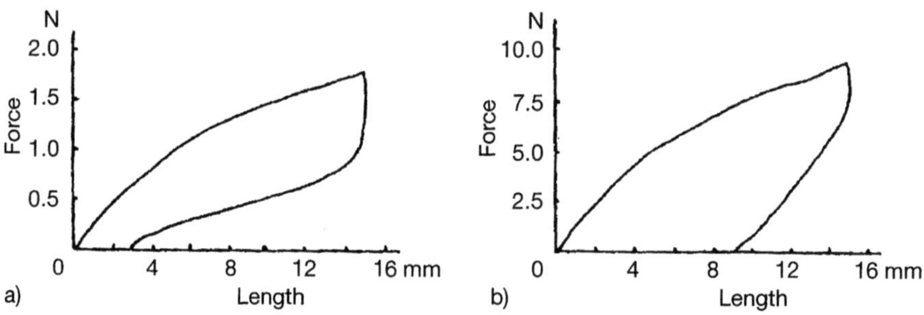

Fig. 6a,b. Compression curves of NiTi open spring (**a**) and stainless steel open spring (**b**). Specimen length of 25 mm

dual deformation remained in the wire. In contrast, the tensile force of the NiTi spring reduced gradually and little permanent deformation was found. The compression test results for NiTi and SS open springs have almost the same tendency and are shown in Figure 6.

The relationship of recovery rate versus tensile duration for these two kinds of close coil springs is given in Table 2. The results show that when duration of the tensile force are increased from 3 days to 30 days, the recovery rate of SS spring

Table 2. Recovery rate varied with tensile duration for NiTi and stainless steel (SS) close springs

Spring	3 days	5 days	20 days	30 days
NiTi	98%	98%	98%	98%
SS	57.2%	20%	10%	0%

was decreased from 57.2% to 0%. However, the recovery rate of the NiTi spring remained a constant 98%. These properties afford the luxury to reactivate the same spring without losing efficiency. As shown in Figure 5b, after a 6-month activation the NiTi close coil spring behaves like new.

It has been proved clinically that the NiTi close coil spring made from superelastic wire 0.3 mm in diameter is desirable for retracting cuspids with a force in the range of 1.74–1.65 N (177–168 g), and therefore can meet the physiological requirement of tooth movement and minimize adverse effects. Thus the whole course of the orthondontic treatment can be shortened and satisfactory curative effect obtained.

References

1. Buechler WJ, Gilfrick JV, Wiley RC (1963) Effects of low temperature phase changes on the mechanical properties of alloys near composition TiNi. J Appl Phys 34:1475–1484
2. Andreasen GF, Hilleman TB (1971) An evaluation of 55 cobalt substituted nitinol wire for use in orthodontics. J Am Dent Assoc 82:1373–1375
3. Tian HC, Yang FZ, Burstone CJ, et al. (1986) Chinese NiTi wire – a new orthodontic alloy. In: Chu YY, Hsu TY, Ko T (eds) Proceedings of the International Symposium on Shape-Memory Alloys, 6–9 Sept 1986. China Academic, Beijing, pp 421–426
4. Qin KQ, Yang FZ, Liu KF, Chen R (1994) Investigation on the properties of Chines NiTi wire RTF and three other products. In: Chu YY, Tu HL (eds) Proceedings of the International Symposium on Shape-Memory Materials, 25–28 September 1994. International Academic, Beijing, pp 681–685
5. Gao N, Luo B, Wang D, Tiang T, Cao Y, Qin K, Chen S (1994) Clinical application of NiTi wire spring to orthodontics. In: Chu YY, Tu HL (eds) Proceedings of the International Symposium on Shape-Memory Materials, 25–28 September 1994. International Academic, Beijing, pp 677–680

Progressive Damage Assessment of TiNi Endodontic Files

Yoshiki Oshida, Farrokh Farzin-Nia

1 Introduction

The occlusal surface of the tooth is formed with a coronal body made of hard enamel structure under which the dentine structure is found. Beneath the dentine structure there is a pulp chamber that extends to a root canal toward the root apex. Caries cause dental decay by attacking the enamel surface. Dental caries continue their attack until the decay progresses to the root chamber.

Endodontic treatment can simply be defined as the precaution taken to maintain the health of the vital pulp in a tooth or as the treatment of irreversibly damaged or necrotic pulp in a tooth which will allow the tooth to remain functional in the dental arch [1]. Debridement is the removal of existing or potential irritants from the root canal system. These irritants include bacteria, bacterial byproducts, necrotic tissue, organic debris, salivary byproducts, hemorrhage, and others [2]. The concept of treating the pulp of the tooth to preserve the tooth itself is a relatively new development in the history of dentistry. The object of this treatment is to clean the root canal system from infected and toxic debris and to shape the root canal to receive a filling material such as Gutta-percha, which will seal the entire canal system from the periodontal tissues. The procedure creates an environment, which is aimed at preserving normal periradicular tissues and restoring these tissues to health.

The pulp space or pulpal morphology is complex, and canals may branch out or divide and rejoin. Many roots have additional canals and a variety of canal configuration [3]. Pulpal morphology has been shown to have a limitless arrangement and distribution within the root canal space. The complex anatomy of the root canal system, especially the apical delta, makes a complete debridement virtually impossible. Eight separate pulp space configurations have been identified [3]. Although it is not always possible to reach all of these areas within the root canal space, proper instrumentation and irrigation techniques will remove sufficient amount of the offending irritants so that successful healing can occur.

Due to complexity and irregularity of root canal morphology, the complete endodontic therapy is complicated and challenging during both the reaming and filing actions. In order to accommodate cleaning of such complicated and irregularly shaped root canals, the endodontic files should be very elastic, particularly in the tip portion. Furthermore, variety of file sizes with varying tip diameter, stem diameter, and lengths are available. If the filing action (counter-clockwise rotation) is used, the flutes scrape against the walls, gouging a portion of dentin and removing it from the canal. If the reaming action (clockwise rotation) is used, the flutes contacting the walls scrape and shave the dentin to widen the preparation.

Files are the most efficient instruments in endodontics for the removal of hard tissue and canal enlargement. They are often manufactured by twisting a blank that has a square or diamond cross section. This produces a series of cutting flutes. Other files are manufactured by grinding the flutes in a tapered rod. Mechanical or engine-powered instrumentation has developed in the sonic and ultrasonic fields leading to a reduction in preparation time.

Recently, titanium-nickel (TiNi) files have been introduced which have more flexibility than stainless steel files. This flexibility (i.e., superelasticity) allows endodontic files to remain more centered in the canal reducing the chance for procedural accidents, such as zipping, leading, transportation and perforation [4]. While the superelastic properties of TiNi instruments are highly desirable, instrument breakage within the canal can still occur. Breakage can cause anxiety for the operator and compromise the success of the endodontic therapy [5]. It has been suggested that, if an endodontic file was broken and left in the root canal, it cannot be considered malpractice. However, if the endodontist fails to record this evidence on the patient's chart and fails to mention this incident to his or her patient, it is malpractice. If the broken instrument cannot be retrieved, surgical endodontic procedures may be necessary.

Non-surgical endodontic therapy has been shown to have a very good success rate [6, 7]. Some authors have reported that the success of this treatment is decreased when separation of endodontic files occurs within the root canal [8–10]. Periapical surgery has been recommended when separation of file soccurs in the canal [11]. Grossman [12] stated that the root canal instruments should be examined before their insertion into a canal to ensure that the spiral twists are regularly aligned. Cutting blades that are not spaced equally indicate that the instrument has been strained and that the torque forces has caused the blades to become irregularly spaced.

Instrument damage may include one or a combination of the following: bending, stretching or straightening of twist contour without bending, peeling or tearing of metal at the cutting edges of the instrument without bending or straightening of the twist contour, partial reverse twisting of the instrument, cracking along the file axis, and fracture of the instrument [13, 14]. The aforementioned types of file damage can be attributed to many factors, namely: stress concentrations [15], torsional forces [16–23], bending moments [16, 17, 24], wear [24, 25], inherent flaws [26], and metallurgical factors [21, 27].

Haikel et al. [28] analyzed dynamic fracture of some endodontic hand instruments. Traditional files (K and H files) were compared to Hybrid files (K-Flex, Flexofile, Unifile, Bois Colombes, and Helifile). The instruments were held in a lathe and placed in a tempered steel groove with a 60° angle at 1900 rpm. SEM photographs showed that the motion-induced stress forces are first generated at the peripheral edges or, possibly, at the sites of surface imperfections. Triangular cross-sections, particularly Flexofile, were found to be most resistant to fracture.

Dederich and Zakariasen [29] studied cyclic axial motion on endodontic instruments in motor-driven techniques. Pyrex capillary tubes were used after bending them to simulate moderately curved root canals. The authors concluded that cyclic axial motion may extend the life span of rotary motor-driven files. Also, small instruments may have a longer life span than larger ones due to different degrees of work hardening of the stainless steel files. The flexibility of TiNi endodontic instruments is far greater than that of stainless steel [30, 31]. Despite increased flexibility, unexpected fracture is still a concern. A recent article by Pruett et al. [32] studied the fatigue life and resulting separation of TiNi endodontic instruments under different operating rpm, shaft diameters and canal curvature. Artificial canals were constructed with 18-gauge needles at 30°, 45°, and 60° curves, at rotational speeds of 750, 1300 and 2000 rpm. File sizes 20 to 80 were used for each clinical use. This study supports the concept of cyclic fatigue and recommends that dynamic operation of files should be included in standardized testing of rotary motor-driven endodontic instruments. Operation at varying rotational speeds was not a factors in instrument fatigue.

If the file is evaluated as non-damaged, it will be sterilized and re-used (or recycled) for the next patient. The commonly practiced sterilization method is an autoclave-sterilization at 121°C for 30 min at 20 psi pressure. It was found that the stainless steel files have shown a decrease in fracture resistance following repeated autoclave sterilization [33]. On the contrary, it was observed that regardless of the size of TiNi rotary files, multiple (five) cycles of autoclave sterilization resulted in a significant increase in torsional fracture resistance [34]. The present authors speculate that differences in the effects of steam sterilization on the remaining life of stainless steel files as compared to TiNi files might be due to the following reasons:

- internal oxidation,
- crystalline/re-crystalization structure and the chemistry of surface oxides of the spinel type formed on stainless steel and tetragonal type of rutile on TiNi material,
- different mechanisms of dislocation annihilation between stainless steel and titanium materials, and
- transformation-related relaxation phenomenon.

As mentioned before, although the breakage and abandonment of a portion of endodontic file in the root canal may be considered an option, it may cause many problems to certain types of patients. If the patient is hypersensitive to nickel, an allergic reaction may occur. In a more serious case, the dissolved nickel element will act as a carcinogen. Because nickel content in the TiNi files (~50 weight %) is higher than that of stainless steel files (~8 weight %), the potential problems may be exasperated when the superelastic TiNi files are broken and left behind in the root canal. Although the biocompatibility of TiNi implants is well documented, the studies have been based on in vitro evaluation using artificial saliva, which has a chlorine concentration of about seven times less than that of dentinal liquid. Accordingly, it can be speculated that the dissolution of nickel out of a broken TiNi is likely to be much greater than in the intraoral environment.

Even with the materials containing potentially harmful elements, there are no scientific standards to assess the cumulative damage on TiNi or stainless steel files before mechanical failure occurs. The degree of bending of the tip portion of the files and the cutting efficiency of used files are the two major parameters that are considered by the endodontist for evaluating the progressive damage of the files and for judging whether or not it can be recycled or abandoned. These observations are normally performed by either the naked eye or under relatively low magnification lenses (x 20). These macroscopic inspections are limited only to the surface layers.

It appears that low-cycle or high-stress fatigue process causes the major damage on endodontic files under torsional, bending or combination of these stresses. The deformation that takes place during fatigue stressing of TiNi materials is a combination of strain due to elastic slip deformation of the parent phase and the plastic deformation due to slip formations. Apparent plastic deformation caused by stress-induced martensitic transformation is elastic slip deformation. Relative proportions of each deformation depend on test temperature, phase transformation temperature, stress amplitude, and strain amplitude.

For example, TiNi exhibited a fatigue endurance limit (F_T) at the stress ratio (F_{app}/F_{UTS}) of 0.2 starting around 5 x 10^5 cycles when the material was treated to have its transformation temperature (A_f) at 343°K (70°C). The F_T was improved to stress ratio of 0.4 at 5 x 10^4 cycles with transformation temperature at 283 K (10°C), and it was furthermore improved to stress ratio of 0.7 at 1 x 10^4 cycles when it was heat-treated having its transformation temperature of 243 K (–30°C) [35].

TiNi possesses several unique characteristics. They include shape memory effect (SME), superelasticity (SE), a good corrosion resistance, and high damping capacity. Different heat treatments can be designed to produce any of the properties mentioned above.

When the superelastic property of TiNi material is utilized, fatigue deformation occurs above A_f (austenitic transformation finish) temperature. The greater the difference between the test temperature and the M_s (martensitic transformation start) temperature, the greater the energy requirements for inducing the martensitic transformation. Under such a condition, stresses will be concentrated at the interface between the martensite phase and the parent phase. Stress concentration might also take place at grain boundaries because of anisotropy of deformation strain. Generally, fatigue stress is more severe for TiNi materials with superelastic condition when compared to the shape-memory effect (SME) condition.

There are two sub-processes involved with fatigue life in TiNi files: period consumed for the crack initiation (N_I) and the subsequent crack propagation period (N_P). The total fatigue life N_F is the sum of these two periods, $N_F=N_I+N_P$. Relative proportion of these sub-periods depends on whether low-cycle fatigue (LCF) regime or high-cycle fatigue (HCF) regimes were the dominant mode leading to failure. During the entire fatigue life, the internal microstructure as well as surface/subsurface microstructure will progressively be affected and altered. These progressive changes can be monitored by means of mechanical property evaluation or physical property measurements. Figure 1 shows the damage evolution curves measured by eight different methods [36]. Both damage fraction axis and life fraction axis are normalized to show unity as the 100% cumulative damage and onset of N_F, respectively.

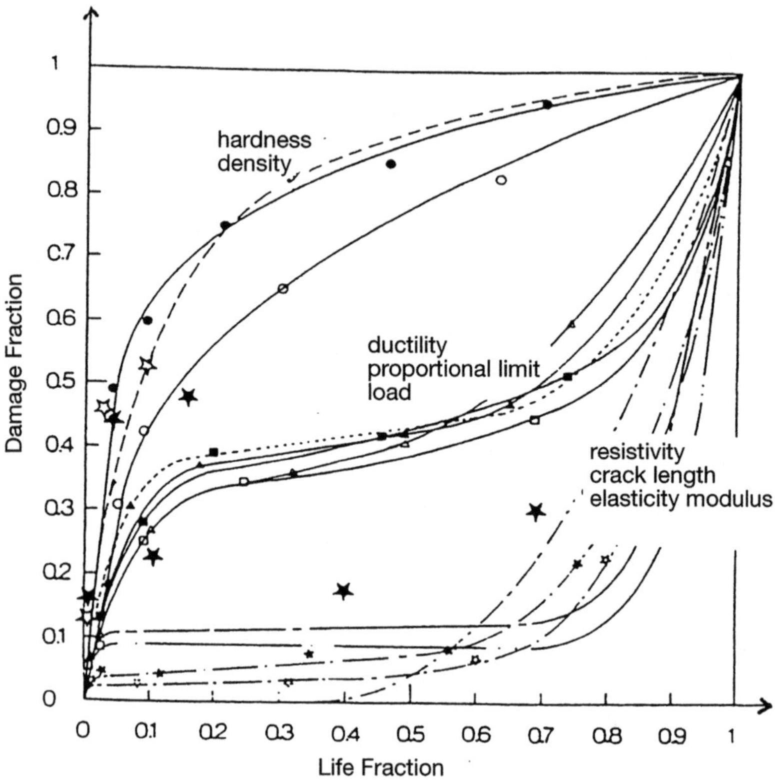

Fig. 1. Nonlinear relationship between damage fraction and fatigue life fraction of various fatigue damage parameters [36]

As seen in Figure 1, there are three distinct groups of fatigue damage indicators. The first group (hardness, density) exhibits a rapid accumulation of damage up to about 20% of the N_F followed by small increases. In the second group (resistivity, crack length, elasticity modulus), an opposite trend is shown when compared with the first group; while the third group (ductility, proportional limit, load) appears to be a mixed feature of the first and second groups.

The above indicators are non-linear. Ideally, the fatigue damage indicator should change linearly as a function of life fraction. Oshida et al. [37–39] employed the x-ray diffraction parameters as a fatigue damage indicator. Superficial layer and interior (un-damaged, serving as self-reference) were simultaneously subjected to x-ray diffraction with different wavelengths. From the analysis on progressive changes in line broadening, the dislocation densities at surface (D_s) and interior (D_i) were calculated. It was found that the dislocation density ratio, D_i/D_s, is linearly related to the life fraction.

In light of this background, it is obvious that there is a need for a reliable endodontic file inspection method that is usable by endodontist and safe for the endodontic patient. Current work may provide the opportunity to establish a method by which life fraction of endodontic files can be predicted. It is therefore the purpose of this study to establish a reliable non-destructive evaluation of cumulative fatigue damage that progressively develops in superelastic TiNi endodontic files. One option by which the fatigue damage assessment on TiNi files can be accomplished is x-ray diffraction. However, this method has its limitations that are generally caused by irregular surface morphology of the flutes. It has been demonstrated that the electric resistance of the metallic materials is linearly related to their surface/subsurface dislocation density. Hence, monitoring the variations in the electric resistance of the files may be an option that can be considered for determining the extent of the fatigue damage to the files and screening out the files that have the greatest potential for failure during endodontic debridement.

2 Materials and Methods

ProFile No. 4 (0.25-mm diameter) and No. 6 (0.35-mm diameter) TiNi files were used in this study. Extracted human mandibular molars with natural root apices and three different degrees of root canal curvatures were used. The teeth were disinfected by storing them in 10% formalin for at least two weeks to comply with OSHA standards. The clinical crowns of each molar were removed to the pulp chamber to facilitate file placement and measurements. The teeth were radiographically imaged using digital radiography. The degree of curvature for each specimen was determined from digital images of mesiodistal and buccolingual views using the Scheider method (Figs. 2, 3) [40]. Three different degrees of root canal curvatures (α=25°, 40° and 55°) were used.

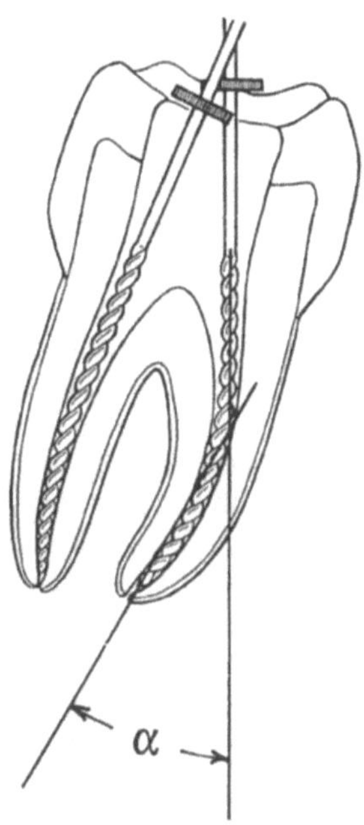

Fig. 2. Schematic definition of curvature of root canal, <α>

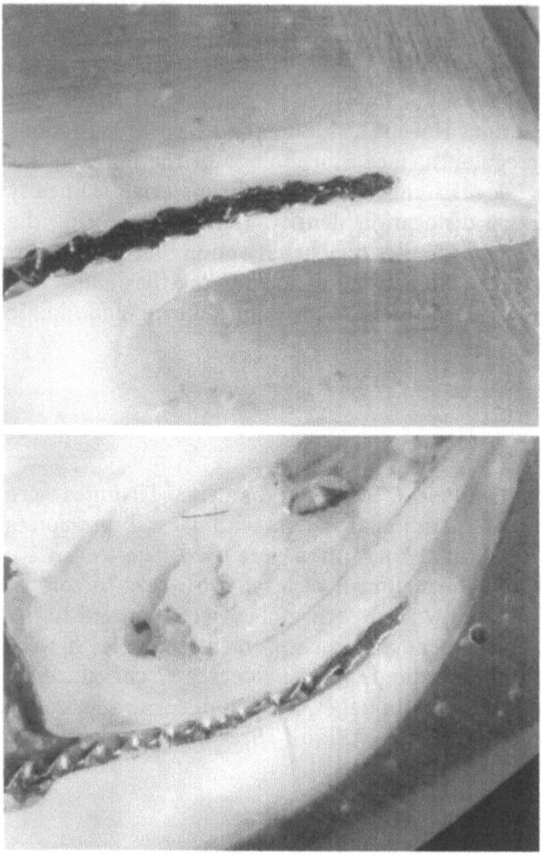

Fig. 3. Examples of files in root canal cavities

The following procedure was followed for the insertion of the files into the root canal. First, the clinical crown was flattened. Then, the mesiobuccal canal was located and a size-15 K-type file inserted until its tip could just be seen at the apex. A stopper was placed on each file at the flattened surface on the molars coronal end. The file was measured and 1 mm was subtracted from the total length that was inserted in canal to establish the working length.

After the working length was determined, the coronal aspects of the acceptable canals were treated with the ProFile orifice shapers in the following manner. The No. 2 orifice shaper was coated with RC prep, while rotating at approximately 300 rpm in a MicroMega air-driven handpiece, and inserted into the canal to approximately 6–8 mm or until resistance was felt through the handpiece. Next, the No. 3 orifice shaper was used in the same manner and inserted to about 4–6 mm or until resistance was felt through the handpiece. Then the No. 4 orifice shaper was used in the same manner and inserted to about 2–4 mm or until resistance was felt through the handpiece. Finally, in the same manner the No. 5 orifice shaper was inserted to about 1–2 mm or until resistance was felt through the handpiece.

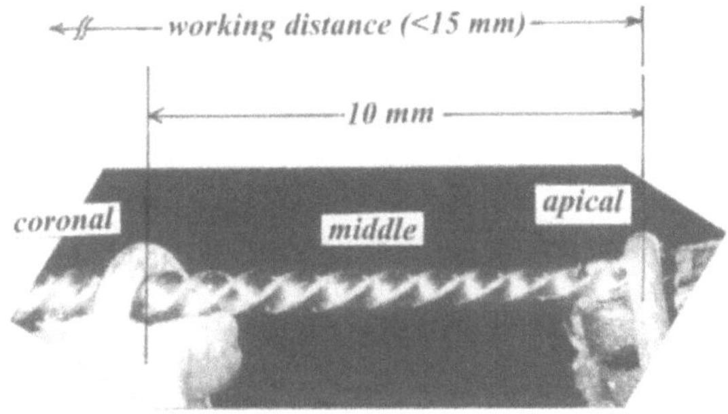

Fig. 4. Resistance measuring distance on endodontic file

Once the orifice shapers had opened the coronal one-half to three-fourths of the canal length, the size-four ProFile was inserted with RC prep up to the working length or until resistance was felt through the handpiece. The R (resistance) was measured using Wheastone's bridge type galvanometer (4.5 DCV) to an accuracy of 0.001 ohm at room temperature (Fig. 4).

Instrumentation of each No. 4 and No. 6 ProFiles were conducted at every, predetermined, 50 rotational cycles increments. Between each instrumentation, the autoclave sterilization was performed at 121°C in distilled water.

Resistance measurement, instrumentation and sterilization sequences were as follows:

1. Initial resistance (R_0) measurement
2. Instrumentation for 50 rotational cycles
3. R_i measurement
4. Autoclave-sterilization
5. $R_{i=2,3,4,5....}$ measurement

Steps 2 through 5 above were repeated for a total 500 cycles or until fatigue failure occurred.

When the test sample was fractured into two pieces, it was considered that the sample has reached its fatigue life (N_F) by being fractured. However, as described previously, there are several other modes in TiNi files that can demonstrate a permanent fatigue damage. Hence, it is necessary to define the fatigue failure in this study. After every 50 rotational cycles, the extent of bending and flute elongation were examined under optical microscope with x20 magnification. If a failure or permanent deformation was noticed, the last cumulative cycles (=nx50 cycles) was determined as N_F in this study.

3 Results and Discussion

Figure 5 shows the progressive changes in electric resistance for No. 4 and No. 6 files being instrumented at α=25°. One No. 4 file and all of No. 6 files survived up to a total of 500 cycles; while two No. 4 files were determined to reach their N_F at 450 and 500 cycles.

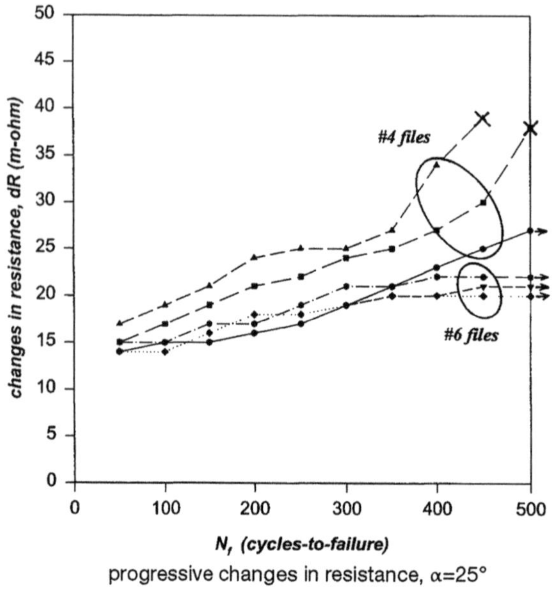

Fig. 5. Changes in resistance versus number of fatigue cycles of files 4 and 6 for <©2>=25°

If the root canal curvature becomes more severe, the cumulative damage appears to manifest itself quickly, as seen in Figures 6 and 7 for cases of α of 40° and 55°, respectively. All of No. 4 and No. 6 files were broken before the pre-set of 500 total cycles. Manufacturers recommend that the files must be discarded after a maximum of 500 cycles. It was found that if the resistance increases to a range between 35 and 40 (m-ohm), the file is fatigue-fractured regardless of the initial reading of resistance, file size and degree of curvature of root canal (Figs. 5–7).

Figure 8 shows typical failure patterns for files, (a) fracture, and (b) flute elongation. The initial resistance (R_o) of the files can be different depending on (a) detailed chemical compositions of TiNi materials, (b) heat-treatments, (c) manufacturing process, (d) conditions of previous use(s), etc. However, the progressive changes in the electric resistance of the files were found to be the only meaningful fatigue damage indicator. The fatigue damage, D_F, can be defined as

$$D_F=(R_i-R_o)/R_o \times 100\%.$$

The term D_F includes the variations in the size of TiNi file as well as the root canal curvature (α).

For the same reason, the number of cycles-to-failure (N_F), in Figures 5–7 should be normalized to a cycle ratio, N_i/N_F. Accordingly, every file which was broken at or prior to 500 pre-determined rotational cycles should have $N_i/N_F=1$.

When the results presented in Figures 5–7 were re-plotted as D_F versus N_i/N_F, a linear relationship was obtained, as demonstrated in Figure 9. Therefore, according to Figure 9, if the initial resistance (R_o) is known, a 10% change in the resistance (D_F) would clearly indicate that approximately 50% of the damage that leads to failure has occurred.

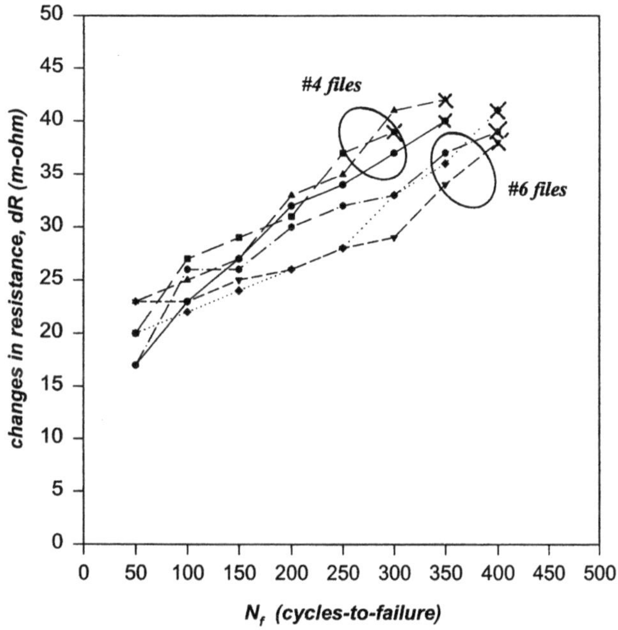

Fig. 6. Changes in resistance versus number of fatigue cycles of files 4 and 6 for <©2>=40°

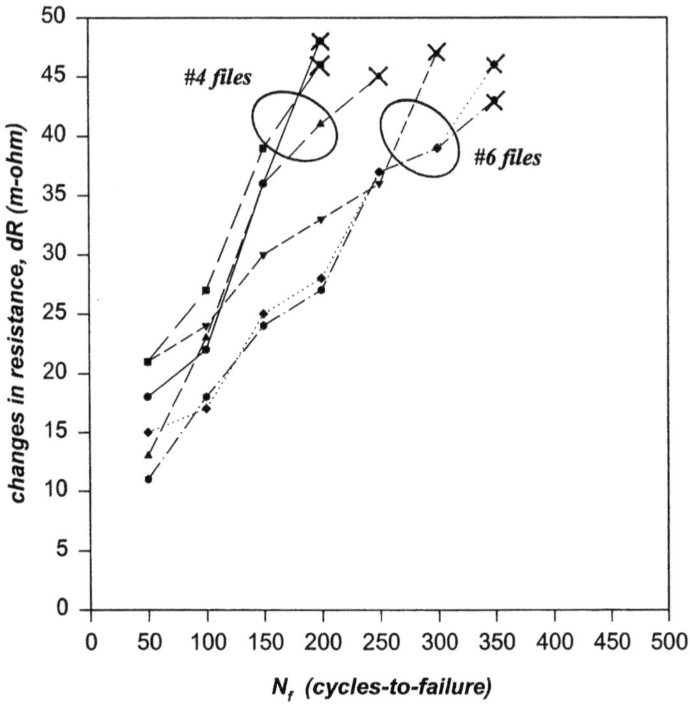

Fig. 7. Changes in resistance versus number of fatigue cycles of files 4 and 6 for <©2>=55°

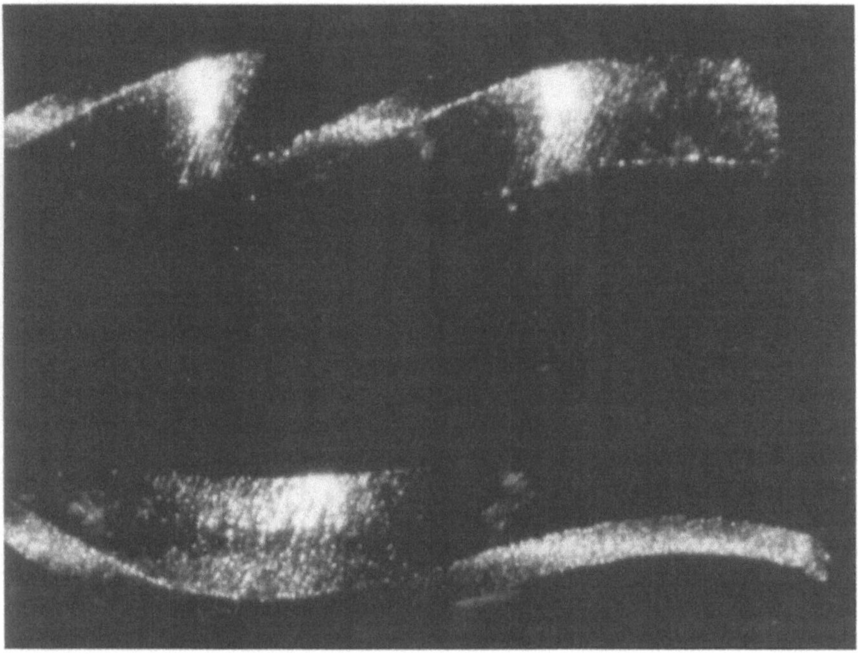

Fig. 8. Examples of fatigued-fractured files; top is a case of failure along with a slight stretching and bottom is a case of failure with unwinding flutes

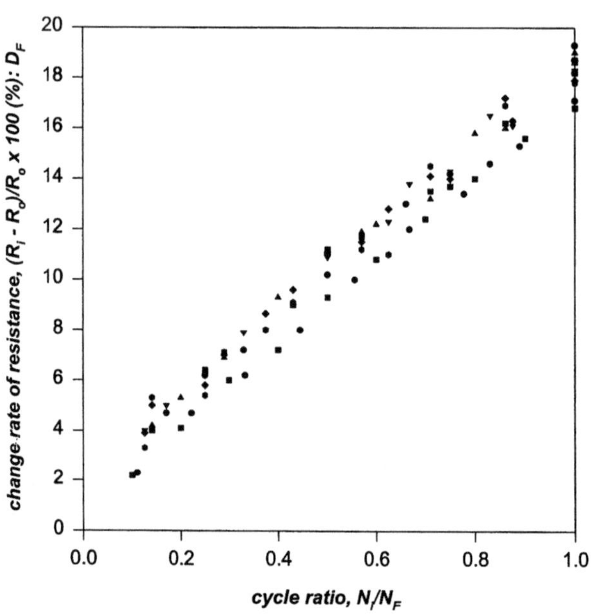

Fig. 9. Linear relationship between normalized damage parameter (changes in resistance) and normalized cycle ratio for all fatigue-fractured files and all angles of curvature of root canal

It is important to note that the above observation is limited to a case when the same stress level is applied on the same file for the entire fatigue life. However, in a clinical situation, a single file might experience different stress levels by being used on different root canal configurations. Hence, to complete the cumulative fatigue damage assessment, the validity of the cumulative fatigue rule should be examined,

$$3 \text{ x } (d_i/D_i)=1.$$

This relationship is similar to the Miner's cumulative fatigue damage rule,

$$3 \text{ x } (n_i/N_i)=1.$$

Figures 5–7 demonstrate that autoclave-sterilization did not appear to have any influence on changes in electric resistance for each instrumentation. This result does not agree with previous observations made for stainless steel files [33] and TiNi files [34]. This discrepancy may be explained by the fact that Ti based alloys are extremely reactive and sensitive to surface contamination which in turn may effect their resistivity. It was reported that titanium-based alloy was discolored under certain conditions [41]. This phenomena was first observed on a titanium box used for storage of titanium implants during autoclaving and surgical procedures. After autoclaving, the box was found to be severely discolored indicating some kind of contamination or chemical attack. It was found that an oxide thickness of up to 650 Å was formed on the box, i.e., more than ten times thicker than on normal implants. Microanalytical investigations by ESCA showed that these oxide films contained considerable amounts of fluorine, alkali metals and silicon. In cases where discoloration was observed in clinical situations, the source of fluorine was the textile cloths in which the titanium implant storage box had been wrapped during the autoclaving procedure. The cloths contained residual Na_2SiF_6, which was an additive to the rinsing water used in the last step of the cloth laundry procedure [41].

4
Conclusions

With the limited results obtained from this study, the following conclusions were made:

1. Electric resistance measurement can serve as a reliable and non-destructive cumulative fatigue damage indicator.
2. All the fatigue-fractured files, that were tested, demonstrated an electric resistance ranging between 35m-ohm and 40 m-ohm.
3. A linear relationship was found between the damage fraction, D_F, and the cycle ratio, N_i/N_F, for all the fatigue-fractured files.
4. Autoclave sterilization did not seem to have any effect on any of the samples investigated in this study.

References

1. Ford TRP (1997) Introduction, history and scope. In: Ford TRP (ed) Endodontics in clinical practice, 4th edn, Wright, Oxford, pp 1–7
2. Walton R, Torbinejad M (1989) Principles and practice of endodontics. Saunders, Philadelphia, p 196
3. Vertucci FJ (1984) Root anatomy of the human permanent teeth. Oral Surg Oral Med Oral Pathol Oral Radiol Endod 58:589–599
4. Esposito P, Cunningham W (1995) A comparison of canal preparation with nickel-titanium and stainless steel instruments. J Endod 21:173–176
5. Stirndberg L (1956) The dependency of the results of pulp therapy on certain factors. Acta Odont Scand 14:21
6. Stringer L (1956) The dependence of the results of pulp therapy on certain factors. Acta Odont Scand 14:175–176
7. Adenubi J, Rule D (1976) Success rate for root fillings in young patients. Br Dent J 141:327–329
8. Nicholls N (1967) Endodontics. Wright, London, p 227
9. Sommer R, Ostrander F, Crowley M (1966) Clinical endodontics, 3rd edn, p 227
10. Siskin M (1967) Surgical techniques applicable to endodontics. Dent Clin North Am 1967:747–749
11. Stringer L (1956) Dependence of the results of pulp therapy on certain factors. Acta Odontol Scand 14:1
12. Grossman L (1969) Guidelines for the prevention of fracture of root canal instruments. Oral Surg Oral Med Oral Pathol Oral Radiol Endod 28:746–752
13. Sotokawa T (1988) An analysis of clinical breakage of root canal instruments. J Endod 14:75–82
14. Zuolo M, Waloton R, Murgel C (1992) Canal mater files: scanning electron microscopic evaluation of new instruments and their wear with clinical usage. J Endod 18:336–339
15. Lilley J, Smith D (1966) An investigation of the fracture of root canal reamers. Br Dent J 19:364–372
16. Lautenschlager E, Jacobs J, Marshall G, Heuer M (1977) Brittle and ductile torsional failures of endodontic instruments. J Endod 3:175–178
17. Craig R, McIiwain E, Peyton F (1968) Bending and tension properties of endodontic instruments. Oral Surg Oral Med Oral Pathol Oral Radiol Endod 25:239–254
18. Dolan D, Craig R (1982) Bending and torsion of endodontic files with rhombus cross-sections. J Endod 8:260–264
19. Luebke N, Brantley W (1990) Physical dimensions and torsional properties of rotary endodontic instruments. Part 1. Gates Glidden drills. J Endod 16:438–441
20. Lausten L, Luebke N, Brantley W (1993) Bending and metallurgical properties of rotary endodntic instruments. Part 4. Gates Glidden and Peeso drills. J Endod 19:440–447
21. Brantley W, Luebke N, Luebke F, Mittchell J (1994) Performance of engine-driven rotary endodntic instruments with a superimposed bending deflection. Part 5. Gates Glidden and Peeso drills. J Endod 20:241–245
22. Dieter G (1986) Mechanical metallurgy, 3rd edn. McGraw-Hill, New York, pp 262–345
23. Wolcott J, Himel V (1997) Torsional properties of nickel-titanium versus stainless steel endodontic files. J Endod 23:217–220
24. Campus J, Pertot W (1995) Machining efficiency of Ni–Ti K-type files in a linear motion. Int Endod J 28:239–243
25. Scott G, Walton R (1986) Ultrasonic endodontics: the wear of instruments with usage. J Endod 12:279–283
26. Chernick L, Jacobs J, Lautenschlager E, Heuer M (1976) Torsional failure of endodontic files. J Endod 2:94–97
27. Brick R, Pense A, Gordon R (1977) Structure and properties of engineering materials, 4th edn. McGraw-Hill, New York, pp 337–339
28. Haikel Y, Gasser P, Allemann C (1991) Dynamic fracture of hybrid endodontic hand instruments compared with traditional files. J Endod 17:217–220
29. Dederich D, Zakariasen K (1986) The effects of cyclic axial motion on rotary endodntic instrument fatigue. Oral Surg Oral Med Oral Pathol Oral Radiol Endod 61:192–196
30. Walia H, Brantley W, Gerstein H (1988) An initial investigation of the bending and torsional properties of Nitinol root canal files. J Endod 14:346–351
31. Serene T, Adams J, Saxena A (1995) Nickel-titanium instruments: Applications in endodontics, 1st edn. Ishiyaku EuroAmerica, St. Louis, p 16
32. Pruett JP, Clement DJ, Carnes DL (1997) Cyclic fatigue testing of nickel-titanium endodontic instruments. J Endod 23:77–85

33. Mitchell BF, James GA, Nelson RC (1983) The effect of autoclave sterilization on endodontic fils. Oral Surg Oral Med Oral Pathol Oral Radiol Endod 55:204–207
34. Kiss EP, Murchison DF, Davis RD (1997) Effect of sterilization on torsional fracture resistance of nickel-titanium rotary files. J Dent Res 76:82
35. Melton KN, Mercier O (1979) Fatigue of NiTi thermoelastic martensites. Acta Metallica 27:137–144
36. Pluvinage GC, Raquet MN (1983) Physical and mechanical measurements of damage in low-cycle fatigue: application for two-level tests. ASTM STP811 21st journées des Aciers speciaux. Colloque Intern sur les aciers inoxydables, 1982 S'Etienne Soc Française de Metallurgie Paris, pp 139–150
37. Oshida Y, Daly J (1990) Fatigue damage evaluation of shot-peened high strength aluminum alloy. In: Meguid SA (ed) Surface engineering. Elsevier Applied Science, London, pp 404–416
38. Oshida Y, Chen PC (1990) Non-destructive low-cycle fatigue characterization of multi-layer thin film structures. J Non-Destructive Eval 8:235–245
39. Oshida Y, Chen PC (1991) High and low-cycle fatigue damage evaluation of multi-layer thin film structure. Trans Am Soc Mechanical Eng J Electro Packaging 113:58–62
40. Schneider SW (1971) A comparison of canal preparation in straight and curved canals. Oral Surg Oral Med Oral Pathol Oral Radiol Endod 32:271–275
41. Lausmaa J, Kasemo B, Hanson S (1985) Accelerated oxide growth on titanium implants during autoclaving caused by fluorine contamination. Biomaterials 6:23–27

Endovascular Applications

Effects of Surface Modification Induced by Sterilization Processes on the Thrombogenicity of Nickel-Titanium Stents

B. Thierry, M. Tabrizian, Y. Merhi, L. Bilodeau, O. Savadogo, L'H. Yahia

1 Introduction

Based on promising reports from international randomized trials such as BENESTENT, vascular stent implantation has been increasingly used – with more than 500,000 implantations per year worldwide – and has become a procedure of choice in the treatment of atherosclerotic vascular disease [1, 2]. Since the first implantation in 1986, many devices, mostly balloon expandable stainless steel stents, have been approved by regulatory agencies, for both peripheral and coronary revascularization. In the early 90's, many studies have reported experimental temporary or permanent use of NiTi stents [3, 4]. Due to the thermoelastic properties of the alloy, optimal deployment of NiTi self expanding stents can be easily achieved with high expansion ratio and less longitudinal shortening than most common stainless steel devices. In addition, their flexibility and radiopacity along with their good biocompatibility and resistance to corrosion have contributed to their increasing use in cardiovascular applications [5–7]. Despite its good biocompatibility, like all metallic materials NiTi is prone to adsorption of plasma protein such as fibrinogen, which is a potent promoter of platelet adhesion [8]. In a recent study, Makkar et al. have shown that surface topography significantly affects the thrombogenicity of NiTi stents, which may in turn have an effect on complications observed much after implantation [9]. Indeed, platelets adhesion is the first step of thrombus formation which may lead to acute and sub-acute thrombosis. Stent thrombosis still represents a major complication after stent implantation, especially for small vessels, thrombus containing lesions and bifurcation stenoses [10]. Moreover, through the release of growth factor by activated platelets, thrombus formation is expected to participate to the restenosis process initiation [11]. Many strategies have been developed to reduce stents thrombogenicity such as improvement of surface characteristics or coating with pharmacological drugs and polymers [9, 12, 13].

The measurement of surface-adsorbed plasma proteins mostly fibrinogen and of surface-adhering cells such as platelets in *ex vivo* arterioveneous shunt models have been widely used to study the thrombogenicity of different materials and designs [14]. These studies suggest that many parameters are involved in the thrombogenicity of a stent. Among all, the design seems to be very important [15]. However, the chemical composition of the alloy and the surface properties of the material have a prominent effect on the thrombogenicity of a stent [6, 16]. The cellular interactions between any biomaterial and blood are modulated by several parameters such as the chemical composition, the surface energy, topography, and potential. These factors are strongly related together, and it is very difficult to find out which one has the strongest effect on the host plasma protein layer adhering on implants surface in contact with blood and therefore on the cellular response. Before or during the use of biomaterials, many deeds can induce surface modification, leading to different biological reaction. Sterilization process is one of them, which has been demonstrated to modify the topography and the chemical properties of the surface as well as the biocompatibility of metallic biomaterials such as NiTi [17, 18]. Consequently, the effects of sterilization must be well characterized before using biomaterials, which will be in contact with blood.

Conversely, depending on the flow condition, contradictory results have been reported on the relative thrombogenicity of NiTi and stainless steel [6, 19, 20]. This is not surprising since thrombus formation in vivo is a dynamic phenomenon highly dependent on various parameters and among these the wall shear stress is of primary importance. On one hand, increasing blood flow increases the number of circulating platelets and thus enhances thrombus formation, and on the other hand, it intensifies the shearing force capable of fragmenting thrombogenic material [21]. The wall shear stresses tested in this study are representative of those encountered in vivo in medium size arteries through reproduction of coronary blood flow in properly sized AV shunt tubing.

There is an obvious lack of data regarding such interactions between flow conditions and surface properties as well as the surface composition, especially for new biomaterials such as NiTi. The aim of this work was therefore to evaluate these points by using an ex vivo AV shunt porcine model, which was constituted with a multichannel perfusion chamber, under physiological conditions. First, the relative thrombogenicity of NiTi stents following various sterilization processes, namely steam autoclave, ethylene oxide, Sterrad and liquid peracetic acid sterilization were compared. Second, a comparison between thrombogenicity of NiTi and stainless steel stents of identical design was achieved under physiological flow conditions (wall shear rate of 228 s^{-1} and 456 s^{-1}). The chemical composition and the topography of NiTi and stainless steel stents were characterized using AES and SEM to achieve a better comprehension of the effect of surface properties on biological response to NiTi cardiovascular implants.

2 Materials

The stents were 3 mm diameter and 30 mm-long slotted-tube geometry devices (Fig. 1) that were laser cut from NiTi tubing (Cordis Corporation – Nitinol Devices and Components, USA) to duplicate commercially available Palmaz stents design (P294M, J and J, Cordis Corporation, USA). They were processed, electropolished and cleaned according to standard procedure for metallic biomaterials (Table 1). The NiTi stents were divided in five groups (n=7). Four have been steri-

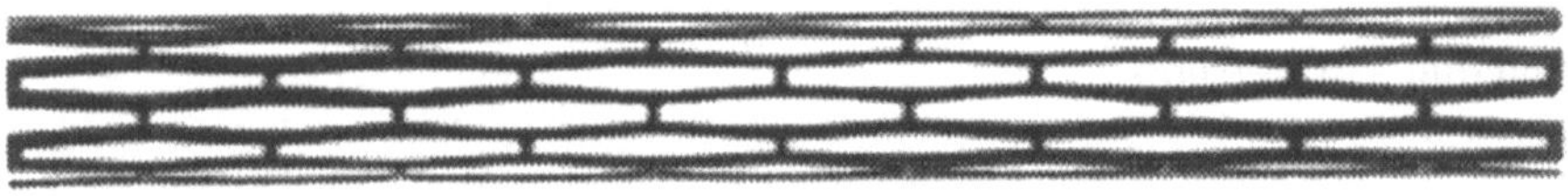

Fig. 1. NiTi stent measuring 30 mm in length and 3 mm in diameter reproducing the design of the Palmaz stents (J and J, Cordis Corporation, USA)

Table 1. Characteristics of stainless steel and nitinol stents used in the study

	Geometry	Metal-to surface area	Materials	Treatments	Surface composition
Stainless steel	Laser-cut tube	~8%	Stainless steel (316 L)	Electropolishing and cleaning	Ti oxide with few amount of nickel
Nitinol	Laser-cut tube	~8%	Nitinol (nickel: 55.8 wt%)	Electropolishing and cleaning	Cr and Fe oxide with few amount of nickel

lized one cycle by steam autoclave (SA; Amsco 3043), ethylene oxide (EO; Steri Vac 5XL), hydrogen peroxide plasma (PS; Sterrad-100S) and peracetic acid decontaminator (PA; Steris I) under clinical conditions while one group has been used as control (EP). However, non-sterilized NiTi stents and commercially available balloon expandable Palmaz stents were only cleaned in soapy water and methanol for the second part of the study regarding the thrombogenicity of NiTi in comparison to stainless steel (n=10).

3 Methods

3.1 Animal Preparation

All procedures followed the American Heart Association Guidelines for Animal Research and were approved by the animal ethic committee of the Montreal Heart Institute. Experiments were performed using ten pigs weighing 25±5 kg. Animals were sedated by an intramuscular injection of 20 mg/kg ketamine (Rogarsetic, Rogar/STB Inc, Montreal, Canada) and 2 mg/kg azaperone (Stresnil, Janssen Pharmaceuticals, Mississauga, Ontario, Canada). Anaesthesia was maintained with 0.5–0.75% halothane (Fluothane, Ayerst, Montreal, Quebec, Canada) following endotracheal intubation. The left femoral artery and the right femoral vein were canulated to establish an extracorporeal circuit. Arterial pressure and ECG were monitored continuously during the experiment. Animals were administrated 50 U/Kg of heparin initially and additional 25 U/kg prior to each perfusion to reach an Activated Clotting Time (ACT) of 200±30 s (Hemochron, ITC, USA).

3.2
Isolation and Labeling

Using the method described by Mehri et al., platelets were isolated and radiolabeled with chromium-51 (^{51}Cr, Merck Frosst Canada Inc., Montreal, Canada) [22]. An autologous blood sample of 120 ml was collected 3 h before the beginning of the experiment and anticoagulated with acid-citrate-dextrose (ACD). The sample was then centrifuged at low speed to obtain a platelet-rich plasma. This platelet suspension was washed, eliminated from contaminated red blood cells by a low-speed centrifugation. The suspension was then incubated with ^{51}Cr for 30 min before being resuspended and reinjected into the animal 1 hour prior to the experimentation. The quantification of fibrinogen deposition was achieved with ^{125}I-human fibrinogen (Amersham International plc, Buckinghamshire, England) injected one hour before the experiment (~10 μCi).

3.3
Stent Insertion

Prior to each perfusion, stents were inserted manually in Silastic tubes (Medical grade, Dow Corning, USA) using talc free surgical gloves and a sterile filament as a positioning device. The tubes were 1/8 inch in internal diameter (ID) and 9 cm long. A deflated 3.0 mm x 20 mm semi-compliant coronary angioplasty balloon catheter was then inserted in stented tubes and inflated at 12 atm for 18 s to optimize stent deployment. Even though nitinol stents are self-expanding devices and do not require balloon expansion for deployment, they were implanted using the same method as for stainless steel stents to avoid differences during manipulation of each device.

3.4
Extracorporeal AV Shunt

The extracorporeal AV shunt was composed of a silicon tubing circuit connecting the left femoral artery to the perfusion channels and returning to the right femoral vein (Fig. 2). The main extracorporeal AV shunt was separated in four (effects of sterilization processes on NiTi) or two (effects of shear rate on NiTi versus stainless steel) parallel channels systems in the perfusion chamber and linked to four (or two) Silastic tubes (3.1 mm in diameter, 9 cm length) in which the tested stents were inserted. The blood flow was maintained by a roller pump (Easy-load, Cole Palmer Inst. Co, USA) at a stable rate of either 40 ml or 80 ml per minute in the tubing containing the stents. The perfusion chamber was placed in a water bath maintained at 37 ± 1°C.

One hour after reinjection of the labeled platelets, the stents were mounted in each channel of the circuit perfusion chambers and the whole was rinsed for 60 s using saline solution. Subsequently, blood was then allowed to circulate in the circuit for 15 min at a wall shear rate of either 228 sec^{-1} or 456 sec^{-1}, corresponding to 40 ml/min and 80 ml/min in 3.1 mm ID tubes, respectively. At the end of the perfusion, saline solution was used to wash off unattached cells and blood

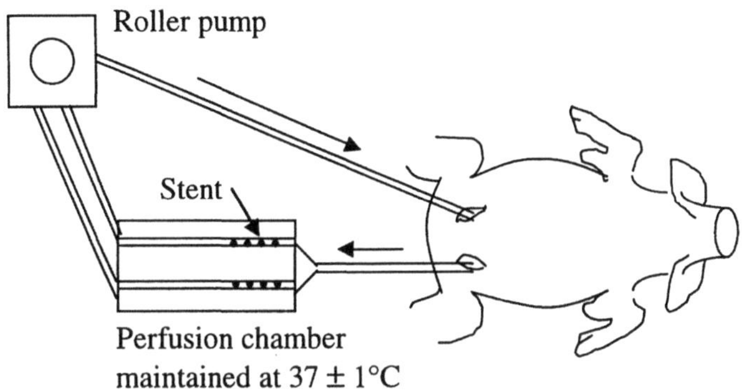

Fig. 2. Schematic representation of the ex vivo AV shunt model: an extracorporeal circulation is carried out between the left femoral artery and the right femoral vein

from the stents and the perfusion set-up. Tubing segments containing the stents were then removed and cut at both distal and proximal ends. A segment of 1.5 cm in each proximal tube was then cut and used as a negative control. These tubing samples were fixed in 1.5% glutaraldehyde solution and kept at room temperature until quantification. The extracorporeal circuit was washed again with saline and a new series of stents was tested using the same procedure. At the end of the daily experiment, the amount of ^{51}Cr -platelet deposition and ^{125}I-fibrinogen adsorption was measured for each tubing segment containing or not a stent using a Minaxi 5000 γ counter (Packard Instruments Co.). Student's *t* tests were used to determine the statistical significance of results.

3.5 Auger-Electron Spectroscopy

AES analyses of the stents (n=2) prior to *ex vivo* testing were performed by a JAMP-30 Auger system (JEOL, Japan) in derivative mode at 10 keV with an electron beam current of 0.5 μA. AES survey spectra (50–1500 eV) were recorded from at least three different locations on the samples. Depth profiles (0–500 nm) were measured by combining AES spectra analyses and ion sputter etching.

3.6 Scanning Electron Microscopy

After quantification of platelet and fibrinogen, two stents (one nitinol and one stainless steel), chosen on the basis of platelet and fibrinogen quantification as representative of the mean level. The stents were post-fixed in a solution of buffered-cacodylate and washed twice in distilled water before being dehydrated in increasing concentrations of absolute ethanol. They were then transferred from absolute ethanol to absolute acetone before final drying with liquid CO_2 as the transfer medium. They were then observed after metallization with a JEOL SEM at 15 kV accelarating voltage.

4 Results

4.1 Surface Analyses of Electropolished NiTi Stents

Using AES, we investigated the surface chemistry of electropolished NiTi and stainless steel stents. Spectra analyses, presented in Figure 3 illustrate the average composition of the NiTi stents surface, i.e. titanium (~418 eV), nickel (~848 eV), and oxygen (~510 eV). The main surface contamination was carbon (~272 eV). The amount of titanium and nickel on the surface of NiTi stents was 15% and less than 1% respectively (Table 1). AES depth profile indicated that the oxide layer thickness averaged 4 nm for electropolished NiTi. Conversely, stainless steel stents were covered by a chromium and iron oxide layer with an average thickness of 2 nm (Fig. 3b). SEM of the stents prior to ex vivo testing did not reveal any differences in surface topography.

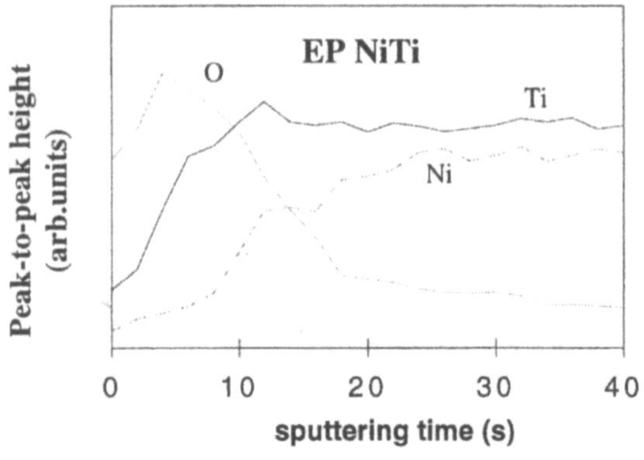

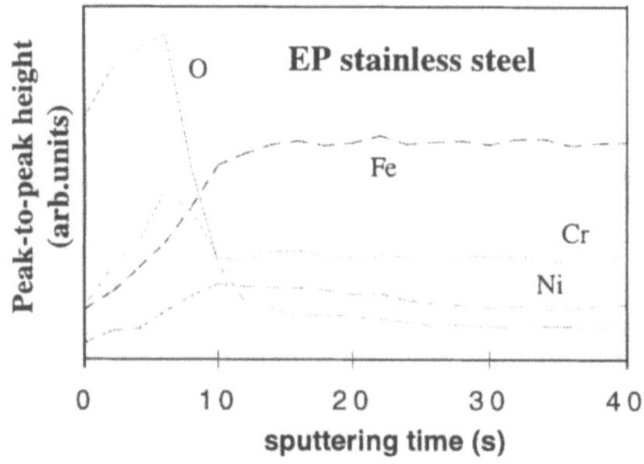

Fig. 3a,b. Depth profile for electropolished NiTi (a) and stainless-steel (b) stents

4.2
Effect of Sterilization on Thrombogenicity of Electropolished NiTi Stents

The effect of sterilization processes on fibrinogen adsorption and platelet deposition at a wall shear rate of 228 s^{-1} are presented in Figure 4. The fibrinogen deposition averaged 71.6±56 cpm/stent for electropolished NiTi stents (n=10). Steam autoclave- and Sterrad- sterilized stents (n=7) presented fibrinogen deposition values similar to that of electropolished. There was a statistically non-significant trend to lower fibrinogen adsorption for the ethylene oxide (23.1±13.9 cpm/stent; n=7) and peracetic acid (18.9±13.5 cpm/stent; n=6) processed stents in comparison to electropolished stents (P>0.05). The platelet adhesion pattern was similar to that of fibrin(ogen) deposition. The mean of platelet adhesion was 300±226 x 10^6 platelets/stent for electropolished group, and no statistically significant difference was observed with sterilized NiTi stents (P>0.05) despite trend for EO and PA groups toward decreasing platelets adhesion.

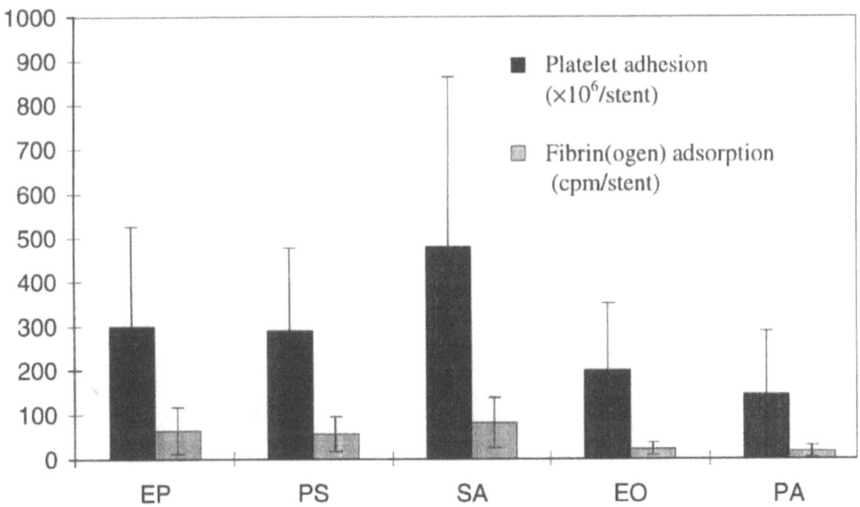

Fig. 4. Effect of sterilization processes on the ^{125}I-fibrinogen adsorption and ^{51}Cr-platelet adhesion on electropolished NiTi stents after 15 min of perfusion

4.3
Effect of Blood Flow on Platelet Adhesion of Electropolished NiTi Stents in Comparison to Stainless Steel

The platelet adhesion was dramatically affected by blood flow for stainless-steel stents. At a wall shear rate of 228 s^{-1}, no significant difference on the amount of ^{51}Cr-labelled platelets adhesion on stainless steel (159±86 x 10^6 platelets/stent; n=5) and NiTi stents (103±90 x 10^6 platelets/stent; n=5) was measured. As shown

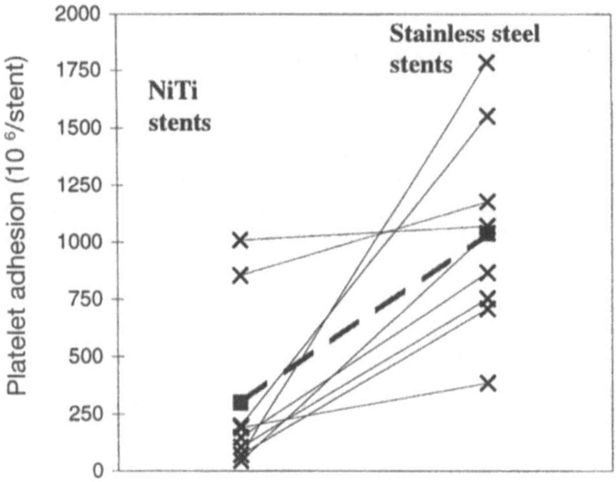

Fig. 5. Comparison between the amount of platelets adhesion for each perfusion (n=9) at a wall shear rate of 456 s^{-1} for NiTi stents and stainless-steel stents (the *broken line* indicates the mean value)

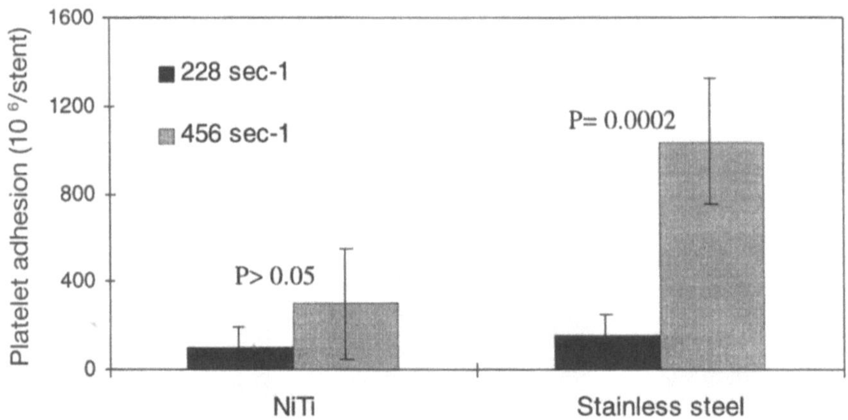

Fig. 6. Effect of the shear rate on the adhesion of ^{111}I-platelet on NiTi stents, stainless steel stents and control Silastic tubes after 15 min of perfusion

in Figures 5 and 6, at a wall shear rate of 456 s^{-1}, NiTi stents presented significantly less adhered platelets than stainless steel ($298 \pm 252 \times 10^6$ platelets/stent, n=9 for NiTi versus $1037 \pm 280 \times 10^6$ platelets/stent, n=9 for stainless steel, P=0.0035). Despite an increase in the amount of adhered platelets on NiTi stents, there were no significant differences between 228 s^{-1} and 456 s^{-1} (P>0.05). However, platelets adhesion on stainless steel stents was dramatically blood flow-dependent since an increase in the shear rate significantly increases the adhesion during the perfusion (P=0.0002; Fig. 7).

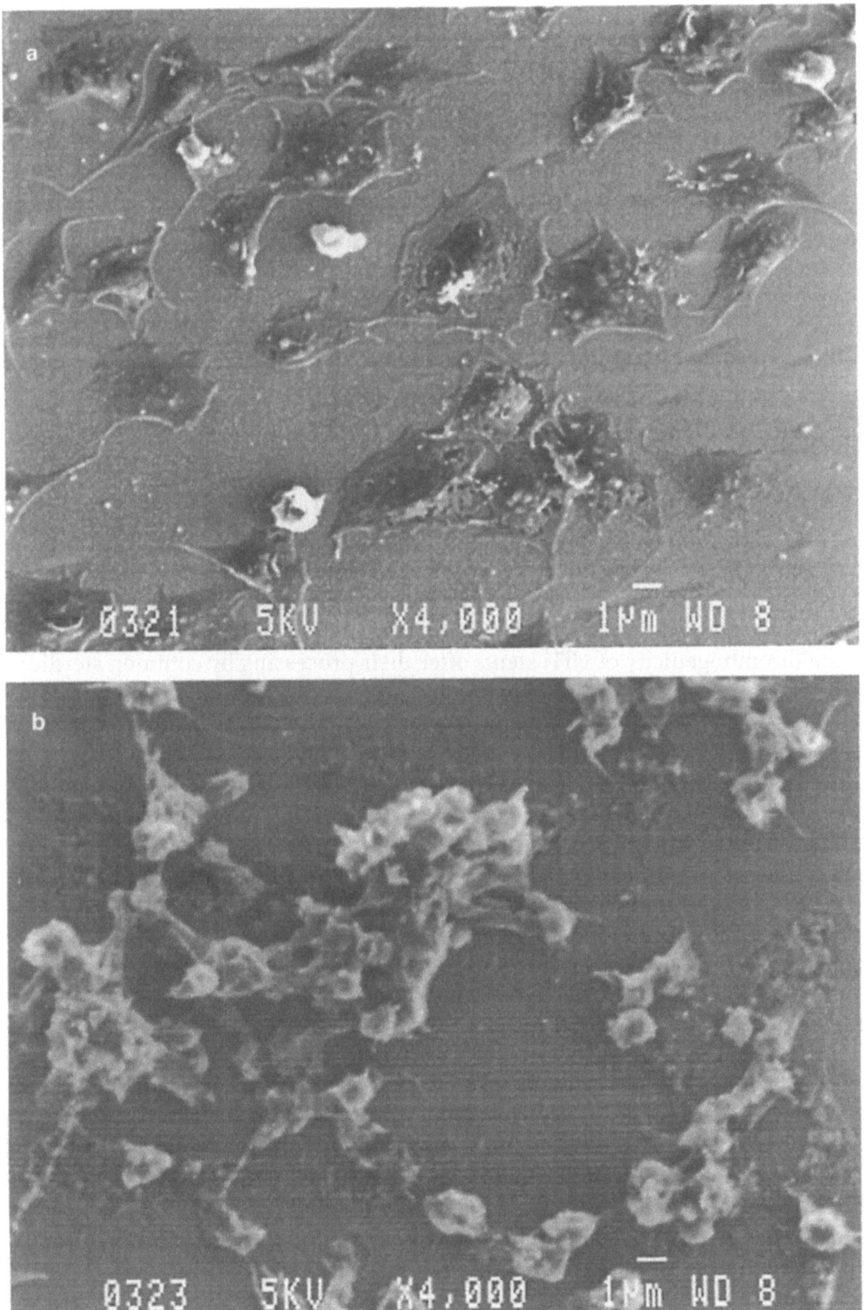

Fig. 7a, b. Scanning electron micrographs of thrombus adherent to NiTi stents (**a**) and stainless-steel stents (**b**)

4.4
Morphological Analyses of the Stents Post-Perfusion

Macroscopic observation of the stents confirmed these differences. Where NiTi stents presented only few amount of white and/or red thrombus principally located at the strut intersections, stainless steel stents showed clearly more thrombus. SEM surface observation of NiTi stents after perfusion are presented in the Figure 7. Morphological analysis of microphotograph of stents suggested different thrombus formation patterns for nitinol and stainless-steel stents. Adhered platelets on nitinol stents presented spreading morphologies uniformly distributed on the surface. Meanwhile, stainless-steel stents seemed to induce clustering of pseudopodial platelets, which tended to be located on the strut intersections of the stents.

5
Discussion

The aim of this study was to determine the effect of surface modifications induced by sterilization techniques on the thrombogenicity of NiTi alloy. In addition, it aimed to compare the relative thrombogenicity of NiTi and stainless steel stents of same design under wall shear rates in the range of those encountered in vivo [23]. In general, our ex vivo studies have shown no significant modifications in the thrombogenicity of NiTi stents after their processing by common sterilization technologies. Among investigated techniques, Sterrad and steam autoclave did not affect the thrombogenicity of electropolished NiTi stents. Techniques using Peracetic acid and ethylene oxide tend to slightly decrease the NiTi stents thrombogenicity as indicated by the lower mean fibrinogen adsorption and platelet adhesion on the EO and PA processed stents. Although these modifications remained under statistical significant value ($p>0.05$), it can not be excluded that a more powerful protocol with more samples would conclude on significant differences among processed samples. Still, it can be expected that sterilization and the resulting surface modifications would not dramatically modify the thrombogenicity of NiTi stents. Our previous works have shown that both EO and PA techniques have induced relatively high surface energy and smoother surfaces that might explain the trend to lower thrombogenicity [17]. Indeed, thrombus formation is a dynamic process, which involves many parameters. It is initiated by adsorption of plasma-protein such as fibrinogen leading to the deposition of platelets. The formation of an irreversible thrombus is related to the denaturation of adsorbed fibrinogen into fibrin monomer and fibrinopeptides and activation of platelets [24, 25]. Many studies have investigated the effects of surface properties of metallic devices in contact with blood [26, 27]. It is well known that the surface energy have a strong influence on the surface-adsorbed plasma proteins and therefore, on the surface-adhering cells such as platelets. Moreover, Hehrlein et al. suggested that the topography was the most important factor for coated stainless-steel stents [28].

In addition, stents design was reported to influence thrombus formation [29]. Some studies have reported significant differences in the thrombogenicity of commercially available stents such as balloon expandable Palmaz stents and self expandable Wallstents which have different design (slotted tube versus mesh) and composition (stainless steel versus cobalt alloy with platinium core) [16].

Besides design and surface chemistry, the optimal deployment of the device to minimize disturbances in local rheology, is also of primary importance in reducing mural thrombus formation and hence, decreasing the rate of acute and subacute thrombogenic occlusion [30]. However, high-pressure used to obtain optimal deployment may increase vessel media and intima trauma, and as a consequence, increases the risk of restenosis. Due to their superelastic structure, self-expanding NiTi stents can theoretically be achieved without balloon inflation. However, specific atherosclerotic vessels may require adjunctive balloon post-deployment to assure efficient dilatation of calcified plaques. When considering the numerous specificities of endoluminal stents, it is very difficult to determine which one of their characteristic influence more the thrombogenicity of the device after implantation.

Based on the above, non-sterilized NiTi and stainless steel stents of same design were used to characterize the relative thrombogenicity of both materials. The statistically significant difference between NiTi and stainless steel in terms of platelet adhesion combined to our macroscopic and SEM observations of the stents after perfusion indicate that stainless steel stents induced more thrombus formation, especially at a wall shear rate of 456 s^{-1}. Moreover, stainless steel stents were highly blood flow dependent. They presented a significant increase of adhering platelets of almost seven folds at 456 sec^{-1} ($P=0.0002$), while NiTi stents showed a non statistically significant increase ($P>0.05$). This difference cannot be explained by surface topography because no differences were observed with SEM on the stents topography prior to the experiment. Furthermore, both types of stents were electropolished, which has been demonstrated to decrease stents thrombogenicity by improving their surface characteristics. Since the stents design was similar, these results may particularly be related to their respective surface chemistry. AES analyses have shown that stainless steel stents were covered by chromium and iron oxide with limited amount of nickel, while the NiTi stents surface was composed of titanium oxide (mainly TiO_2). Therefore, differences in surface compositions may explain the less thrombogenic properties and also the difference behavior of platelet of NiTi stents adhesion as observed on SEM micrographs. Nygren et al. have been demonstrated the dependency of TiO_2 characteristics on the surface fibrin(ogen) adsorption and platelets adhesion [27]. Moreover, it has been recently suggested that TiO_{2-x} oxide films may prevent the denaturation of fibrinogen by inhibiting the transfer of charges from fibrinogen to the surface of the material [24]. Accordingly, the TiO_2 rich surface of NiTi stents may have prevented the formation of an irreversible platelet-rich thrombus and, as a consequence, promoted its fragmentation by blood flow. An increase in the flow increases the number of circulating blood components, and thus enhances the thrombus formation. It also increases the shearing force able to fragment this thrombus [23]. In agreement with results of Sheth *et al*, obtained with NiTi and stainless steel stents of different designs in a rabbit carotid artery model [6], our results demonstrate that independently of the design and extent of vessel wall injury, the surface materials itself has a obvious influence on the thrombogenicity of metallic stents.

Many investigations have been devoted to reduce the metallic stents thrombogenicity [12] to avoid acute and subacute clinical complications. Moreover, thrombogenic material accumulation within the stented vessel are expected to contribute to neointimal proliferation processes through platelets-derived growth factors release and expression of the Gp IIb/IIIa receptor on the platelet surface [11]. The EPIC, EPILOG and EPISTENT trials have shown that prevention of thrombus formation by IIb/IIIa antagonist have a beneficial effect on both short term and long term complications in PTCA procedures [31, 32]. Reduction of restenosis through prevention of thrombus formation is also suggested by the success of anti-platelet therapy after coronary stenting in the ISAR and STARS trials [33, 34]. In addition, the decrease in neointimal thickening reported in a recent study following thrombus formation reduction by surface characteristics improvements, suggests clinical relevance of this issue. Still, further studies is required to conclude whether or not the lower thrombogenicity observed for NiTi stents may have an effect on acute or subacute complication rates and/or on long-term need for re-vascularization.

6 Conclusions

By using an ex vivo extracorporeal model, we determined that sterilization did not significantly modify the thrombogenicty of NiTi stents. Stainless steel stents were more blood flow dependent than NiTi stents, and significantly more thrombogenic at a wall shear rate of 456 s^{-1}. The latter was related to the surface chemistry of NiTi, namely titanium oxide, which may prevent thrombus growth within the stents. Along with the favorable thermoelastic properties of NiTi alloy, the good surface heamocompatibility of NiTi alloy, as shown in our study provide more information in justifying the increasing use NiTi as a biomaterial for peripheral and coronary stents. However, when considering the clinical relevance, our results should be carefully analyzed in regard to the methodology used for our study. Some limitations such as the use of inert Silastic tube instead of vessel wall and the lack of anti-platelets therapy may decrease the clinical relevance of our results.

Acknowledgements. This work was partly supported by NSERC. We thank Cordis Corporation Nitinol Devices and Components (Fremont, Calif.) for their technical and financial support. The authors also wish to thank Dr. T. Ahn, P. Thai and J. F. Théorêt from the Montreal Heart Institute for their excellent technical assistance and St. Eustache and Charles Lemoyne hospitals for sterilization process.

References

1. Serruys PW, De Jaegere P, Kiemeneiji F, et al. (1994) A comparison of balloon expendable-stent implantation with balloon angioplasty in patients with coronary artery disease. N Engl J Med 331:489–495
2. Holmes DR, Bell MR, Holmes DR, et al. (1997) Interventional cardiology and intracoronary stents. A changing practice: approved vs. nonapproved indications. Cathet Cardiovasc Diagn 40:133–138

3. Sutton CS, Tominaga R, Harasaki H, Emoto H, Oku T, et al. (1990) Vascular stenting in normal and atherosclerotic rabbit: studies of the intravascular endoprothesis of titanium-nickel alloy. Circulation 81:667–683
4. Carter AJ, Scott D, Laird JR, Bailey L, Kovach JA, Hoopes TG, et al. (1998) Progressive vascular remodeling and reduced neointimal formation after placement of a thermoelastic self-expanding Nitinol stent in an experimental model. Cathet Cardiovasc Diagn 44:193–201
5. Henry M, Amor M, Beyar R, Henry I, Porte JM, Mentre B, et al. (1996) Clinical experience with a new nitinol self-expanding stent in peripheral arteries. J Endovasc Surg 3:369–379
6. Sheth S, Litvack F, Dev V, Fishbein MC, Forrester JS, Eigler N (1996) Subacute thrombosis and vascular injury resulting from slotted-tube Nitinol and stainless steel stents in a rabbit carotid artery model. Circulation 94:1733–1740
7. Rechavia E, Fishbien MC, DeFrance T, Nakamura M, Parikh A, Litvack F, Eigler N (1997) Temporary arterial stenting: comparison to permanent stenting and conventional ballon injury in a rabbit carotid artery model. Cathet Cardiovasc Diagn 41:85–92
8. Horbett TA (1994) The role of adsorbed proteins in animal cell adhesion, colloids and surfaces. Biointerfaces 2:225–236
9. Makkar RR, Kaul S, Nakamura M, Dev V, Litvack FI, (1995) Park Modulation of acute stent thrombosis by metal surface characteristics and shear rate. Circulation 92:I-86
10. Keane D, Azar AJ, Serruys PW, Macaya C, Rutsch W, Sigwart U, Comlombo A, Marco J, Klugmann S, Crean P (1995) On behalf of the BENESTENT investigators. Outcome following elective stent implantation in small coronary arteries. Eur Heart J 16:335
11. Komatsu R, Ueda M, Naruko T, Kojima A, Becker AE (1998) Neointimal tissue responses at sites of coronary stenting in humans – Macroscopic, histological and immunohistochemical analyses. Circulation 98:224–233
12. De Scheerder I, Verbeken E, Van Humbeeck J (1998) Metallic surface modification. Semin Interv Cardiol 3:139–144
13. Schurmann K, Vorwerk D, Bucker A, Neuerburg J, Klosterhalfen B, Gunther RW, et al. (1997) Perigraft inflammation due to Dacron-covered stents grafts in sheep iliac arteries: Correlation of MIR imaging and histopathologic findings. Radiology 204:757–763
14. Makar RR, Eigler NL, Kaul S, Frimerman A, Nakamura M, Shah PK, Forrester JS, Hebert J-M, Litvack F (1998) Effects of clopidogrel, aspirin and combined therapy in a porcine ex vivo model of high-shear induced stent thromobosis. Eur Heart J 10:1538–1546
15. Rogers C, Edelman ER (1995) Endovascular stent design dictates experimental restenosis and thrombosis. Circulation 91:2995–3001
16. Siegerstetter V, Krause T, Haag K, Ochs A, Hauenstein K-H, Moser HE (1997) Transjugular intrahepatic portosystemic shunt (TIPS) thrombogenicity in stents and its effects on shunt patency. Acta Radiol 38:558–564
17. Thierry B, Tabrizian M, Savadogo O, Yahia L'H (2000) Effects of sterilisation processes on NiTi alloy: surface characterizations. J Biomed Mater Res 49:88–98
18. Shabalovskaya S, Andereeg J (1995) Surface spectroscopic characterization of NiTi equiatomic shape memory alloys for implants. J Vacuum Sci Technol A 13:5
19. Armitage DA, Grant DM, Parker TL, Parker KG (1997) Haemocompatibility of surface modified NiTi. In: Pelton AR, Hodgson D, Russell SM, Duerig TW (eds) Proceedings of SMST 1997. Shape Memory and Superelastic Technologies, Pacific Grove, pp:411–416
20. Sutton CS, Consigny PM, Thakur M (1994) Thrombogenicity of intravascular stent wires. Circulation 90(Suppl I):I-9
21. Hanson SR, Sakariassen KS (1998) Blood flow and anti-thrombotic drug effects. Am Heart J 135:S132–S145
22. Merhi Y, King M, Guidoin R (1997) Acute thrombogenicity of intact and injured natural blood conduits versus synthetic conduits: neutrophil, platelet, and fibrin(ogen) adsorption under various shear-rate conditions. J Biomed Mater Res 34:477–485
23. Hanson SR, Sakariassen KS (1998) Blood flow and antithrombotic drug effects. Am Heart J 135:S132–S145
24. Baurschmidt P, Schaldach M (1977) The electrochemical aspects of the thrombogenicity of a material. J Bioeng 1:261–278
25. Nan H, Ping Y, Xuan C, Yongxuan L, Xiaolan Z, Guangjun C, et al. (1998) Blood compatibility of amorphous titanium oxide films synthetized by ion beam enhanced deposition. Biomaterials 19:771–776
26. Palma VEDE, Baier RE (1972) Investigation of three-surface properties of several metals and their relation to blood biocompatibility. J Biomed Mat Res 3:37–75
27. Nygren H, Erikson C, Lausma J (1997) Adhesion and activation of platelets and polymorphonuclear granulocyte cells at TiO_2 surfaces. J Lab Clin Med 129:35–46

28. Hehrlein C, Zimmermann M, Metz J, Ensinger W, Kübler W (1995) Influence of surface texture and charge on the biocompatibility of endovascular stents. Coron Artery Dis 6:581–586
29. Rogers CR, Edelman ER (1995) Endovascular stent design dictates experimental restenosis and thrombosis. Circulation 91:2995–3001
30. Goldberg SL, Di Mario C, Hall P, Colombo A (1998) Comparison of aggressive versus nonaggressive balloon dilatation for stent deployment on late loss and restenosis in native coronary arteries. Am J Cardiol 81:708–712
31. The EPIC investigators (1994) Use of a monoclonal antibody directed against glycoprotein IIb/IIIa receptor in high-risk coronary angioplasty. New Engl J Med 3330:956–961
32. The EPILOG investigators (1997) Platelet glycoprotein IIb/IIIa receptor blockade and low-dose heparin during percutaneous coronary revascularization. New Engl J Med 336:1689–1696
33. Schomig A, Neumann FJ, Kastrati A, et al. (1996) A randomized comparison of antiplatelet and anticoagulant therapy after the placement of coronary-artery stents. New Engl J Med 334: 1084–1089
34. Leon M, Baim D, Popma J, et al. (1998) A clinical trial comparing three anti-thrombotic drug regimens after coronary-artery stenting. Stent Anticoagulation Restenosis Study Investigators. New Engl J Med 339:1665–1671

X-Ray Endostenting Surgery of Vessels and Hollow Organs

I.Y. Khmelevskaya, I.K. Rabkin, E.P. Ryklina, S.D. Prokoshkin

1
X-Ray Endovascular Stent Surgery

For a long time, the most radical method of vessel pathology treatment (narrow bright interval or its complete closing) was surgical intervention. The surgical reconstructed operations most used were autovenous plastic and plastic by the synthetic stents with the intention of creating a new, temporal or permanent, way of blood flow.

Next step of treatment for occlusive or stenotic illnesses of different located arteries was balloon expanded dilatation–an effective, little bit traumatising and relatively simple method, which does not demand some difficult surgical accesses or a general anaesthesia. However, a well-founded opinion appeared in the literature. The effect of the X-ray endovascular dilatation of vessels is less resistant than during the reconstructed surgical interventions.

A general law for all body tissues, also for all hollow organs, is the reconstitution of their initial form in case of their deformation thanks to the elasticity and to the plastic resistance of tissues. Particularly, in case of the balloon expanded dilatation of a vessel, also like in widening of an hollow and narrow organ (oesophagus, trachea, bronchi, bile and urinary ducts, vagina), because of the excess of resistance and plastic deformation of normal and especially pathological (healed, fibrous, swelling) tissues, without counting the recurrence of the pathomorphologic process itself, it arrives a reconstitution of their initial form after the removal of a widening and balloon expanded load which bring a restenosis and reocclusion. As a rule, the pathologic process brings an irreversible deformation of the vessel and hollow organ. The recurrence frequency of stenosis after the dilatation had served as a stimulus for a continued research of more perfect transcatheter methods to avoid the repetitive vessel's strictures.

The discovery and study of memory shape alloys' properties, their unique capacity of taking a compact form and to change it with a temperature modification, and the appearance in the literature of data about a possible use of NiTi alloys in medicine [1] brought the idea of endovascular stenting surgery. The NiTi alloys, which have a high degree of corrosion resistance and a high degree of biological compatibility, proved to be goods enough to medicine use.

Two independent American radiologist groups [2, 3] had simultaneously announced an experimental development of an into-vessel stent made of NiTi. Their studies showed important difficulties during the transfer and installation of stent; because of that, the studies were not continued. In spring 1983, the doctors from Scientific Surgery Centre (Rabkin IE and his colleagues) proposed to the Moscow Steel and Alloys Institute the idea of using memory shape alloys to perform stenting surgery of vessels. The jointly conducted studies found a new tendency in the X-ray endovascular stenting, which is X-ray endostenting surgery of vessels and hollow organs.

First stents were made of thread and had a spiral shape with an ear at the end. The stent's surface was covered by an anti-thrombosis coat, which is a special silicon coat. The developed delivering and installing device had a lock to fix the stent at one end and a fixing device to separate the stent at the other end. In a special prototype of human being artery's system, the physico–mechanical characteristics of stents were studied and the methodology of its installation into different parts of vessel's canal was worked.

The initial positive opinion of biological compatibility and hemocompatibility of stents and successful studies permeated to begin the X-ray endostenting use in experiments on the animals [5–7]. The goal of these experiments was to improve the methodology of transport and precise implantation of stents into different parts of vessel's canal and to define the optimal size ratio between the stent and vessel's diameter.

The experiments were conducted on 53 dogs. Between them, 32 dogs received an acute experiment and 21 dogs were dynamically observed and studied several times after the surgery (12 h to 15 months). In total, 85 NiTi stents were implanted into chest and stomach aortas, kidney's, iliac and femoral arteries (51 acute and 34 chronic experiments).

The essential of study includes the following: The dogs were conducted into an anaesthesia state, were intubated, and had undergone artificial ventilation of lungs. Later, in sterilised conditions, the femoral artery was separate: the vessel was puncted by Seldinger's needle and, in accordance with conductor, the Edmann-Ledin's catheter, which is 3-mm diameter, was introduced. Under the fluoroscopic control, the catheter was transported until the supposed place of stent installation and an electronography was realised. The catheter's diameter was measured on picture. After, an X-ray electroarteriography was realised by a manual introduction of contrast substance. The vessel's diameter was measured in the supposed part of implantation. Knowing real interior diameter of catheter, the coefficient of projected increase was calculated, the real interior diameter of vessel was defined and the stents of corresponding size were chosen.

The stent was fixed on conductor, was cooled by chlorine-ethyl until the metal become "soft", was stretched into a right thread and was introduced through catheter in the supposed place of stent installation into the vessel. Under the fluoroscopic control, the thread's end with fixed stent in the lock was installed in a necessary arterial part, the catheter was removed slowly and the conductor with the stent were fixed on a defined level thanks to orientation by the X-ray contrast signs on the animal's leather and on the anatomic bone structure.

After the exit of rear end of stretched NiTi spiral from catheter, the stent under the effect of human blood returns to its initial given form of spiral in 5–10 s. Furthermore, the stent is fixed in the vessel thanks to the uniform pressure of turns on the vessel's wall. After placing the stent in a necessary place of vessel and its separation from the fixing lock, the conductor was removed. To define the going past of stents, an angiographic study was realised. After, the catheter was removed and some stitches were applied on the punction hole. The operation wound was sewed up by several coats.

Under the radiological control studies, a long period after the operation, it was found that any of 34 stents installed did not moved from its initial implantation place, the spiral form corresponded to the initial form. The angiographic studies of vessel's parts after the surgery did not show any characteristics of thrombosis formation or any stricture of bright interval in places of spiral installation among 19 of 20 dogs. The following observations of dogs during 15 months had shown that if during the stent installation processes the endothelium's integrity of vessel's neointima is not perturbed and all the rules of choosing a spiral are respected, the thrombosis formation on a surgery part does not arrive.

During the operation and after, the anti-coagulators and anti-agregants were not used. This factor is important because dogs have more active coagulation blood system than human beings. The main substratum of blood coagulation, which is fibrinogen and fibrinstabilising factor remain stable during the whole operation. By the method of electrophorus of plasma's albumen (II and III blood type), it was defined that albumen adsorbs on the stent's surface. This albumen, which inhibits the blood coagulation process and the thrombosis adhesion, avoids the development of thrombosis formation and of intimate growths.

To study the regeneration steps of vessel's wall around the stent's turns, the capacity of stent's material to provoke body's tissue responsive reaction and to define the criteria of this reaction–biological inertia of implant, the morphological studies were realised [5]. 23 stenting parts of arterial vessels obtained from 15 dogs after 1, 3, 5, 7, 10, 14 days and 2, 3, 4 (two dogs) 1, 5, 6, 8, 9, 15 months after the surgery were studied under the optic and electronic microscopes.

It was found that on the endostents' turns the thin albumen coat accumulates in falling from the blood flow. In the contact places of spiral's turns with interior arterial coat, the pressure marks were noticed. 3–5 days later, the fibrocoats growth up trough albumen coat and the granular young unification tissue forms around the endostents' turns. Later, this tissue become fibrous unification tissue, which has a fibrous structure tendency to the formation of fibroses in form of muff, which is 0.03-to 0.1-mm thickness around each particular stent's turn. At the end of the second week, the formed capsule above the endostents' turns covers from the blood flow side by a continued coat of real vessel's endothelium, which comes from the arterial intima and which is not seen in the vessel's bright interval. The unification muff creates a gradual going past from the arterial wall to the endostents' turns in bringing better hemodynamic conditions. During a longer observation period, the interior capsule above the stent's turns become more similar to the real intimate artery according to morphofunctional link. In this case, the neointima's thickness corresponds to the normal sizes of an interior coat of vessel's operation part (4–21 μm). By this way, at the end of the second week yet, the implanted stent does not represent a real endostent. It is rather an auto-vessel. The conducted experiments had found the basis to introduce into clinical practice a new direction–X-ray endovascular stenting surgery.

In March 1984, first in the world clinic practice, an implantation of X-ray endovascular stent, which has shape memory effect, was realised on a human being [6]. The patient T, 56 years old, suffers of inferior limbs sclerosis and is lame. He has been sick for two years. When the pain appeared in the legs, he was obligated to stop each 100–150 metres because of pain. He followed a traditional treatment (he took the pills to widen the vessels) but without positive results. When he came to the hospital, the pulse on the femoral artery was weak, at the right, the pulse was relatively good, on the popliteal and feet arteries, it was impossible to determine it. The graphical perturbations were absent. The 21.03, an aortoarteriography, was realised; an asymmetric stenosis of left exterior iliac artery and of right surface femoral artery was confirmed. The 27.03.84, an X-ray endovascular dilatation of left exterior femoral artery trough leather access by the balloon-expanded catheter, was realised. After, as a widening device, a scaffold made of memory shape alloy was introduced trough the same access. This scaffold was installed at the ex-stenosis place. As a result, the pulse appears on the foot's artery of the rear tibia's artery at the left. Plus, the pulsation was very good on the femoral artery.

The second step consisted of an X-ray endovascular dilatation of the femoral surface artery at the right with a reconstitution of vessel's bright interval and an appearance of pulsation on a popliteal artery. During the reography, all the indicators increased at all levels of both inferior limbs. According to the functional studies, the perepheric blood circulation of inferior limbs (after the X-ray endovascular dilatation and operation) was completely compensated at the left and the permpheric blood circulation of right inferior limb became more intensive at all levels. Two months later, the patient does not have any complaint, the pain in legs is absent. He walks 2 km without stopping. The foot's pulse at the left and the pulse on the popliteal artery at the right were conserved [7].

From the first publication, the construction of endostent and delivering device had sustained several modifications [8–10]. Actually, an endostent presents a NiTi spiral, which is 3–12 mm diameter and has two ears at the ends. The stent is transported in a compact form to a necessary place by means of original delivering device, which provides quick and precise implantation.

The operation is performed trough a punction hole in femoral artery under fluoroscopic guidance without blood, anaesthesia or scalpel. The operation is realised in accordance to the following method. First, a balloon-expanded dilatation of a stenosis vessel's part is practised. After, the stent in its compact form (Fig. 1a) is transported to an insertion site with the help of a delivering device. If we are sure about the correct position, the stent's distal node is separated (in this case, the possibility of correction of stent position with the help of fixed proximal node remains). Under the body temperature, the stent recovers ("reminds") its initial given shape of cylinder scaffold (Fig. 1b) and reconstitutes the bright interval of affected vessel. Stent is firmly fixed in the insertion site due to the precisely chosen diameter and additional fixing elements. The proximal node is separated from the stent and the delivering device is removed (Fig. 2).

The implanted stent arms firmly the vessel and represents a supporting scaffold, which prevents the following fall of vessel's walls (stenosis recurrence). The delivering device may be used repeatedly, but the constructive elements contacting with the patient's blood must be substituted.

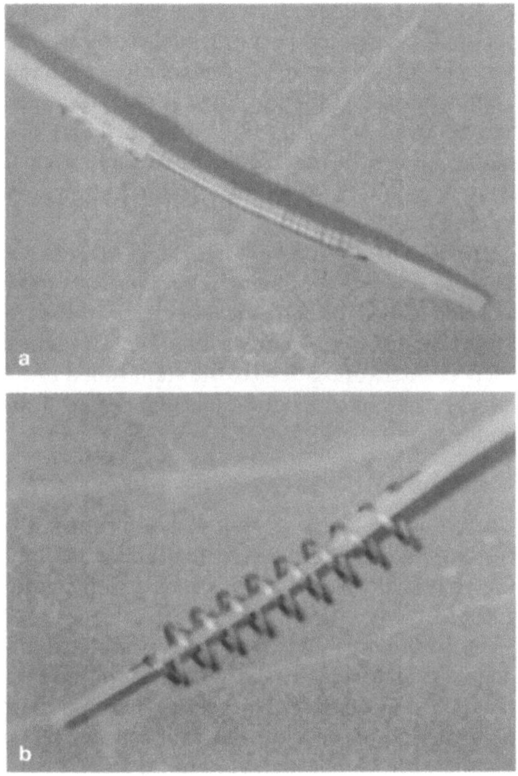

Fig. 1a, b. X-ray endovascular stent for femoral artery in its compact form before the implantation (**a**) and in its recuperated initial form (**b**) on the distal end of delivery device.

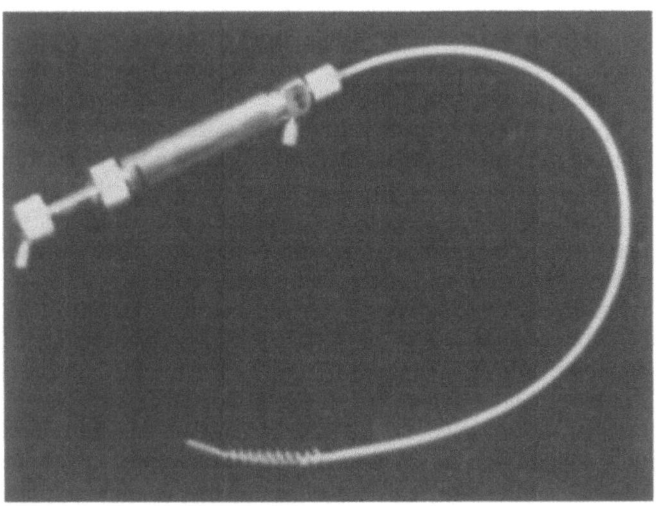

Fig. 2. Delivery device

An X-ray endostenting operation usually takes 45 min and is ten times cheaper than traditional surgical interventions. Furthermore, the followed up recovery period of a patient decreases more than ten times.

Long term therapeutic effect (5 years) was observed among 102 patients, which had received 118 stents into different parts of arterial system (general iliac: 27 stents; exterior iliac: 32; general femoral: 3; surface femoral: 36; popliteal: 9; subclavian: 4; brachio-cephalic stone: 1; kidneys: 6). Eight patients had received two stents each.

In spite of first positive results of vessel's stents, the field of X-ray endostenting use was limited during a long period of time by several constructive imperfections of delivering device; because of that, implantation of stents bigger than 8-mm diameter was impossible. The problem consisted in initial constructions used as a delivering device: the stent rolled into a smaller diameter "in place", e.g. in case of its fixed length. The stent deformation degree is calculated by this formula:

$$\Delta\varepsilon = d(R_1 - R_2)/R_1 R_2 \quad (1)$$

where d is thickness of stent's ribbon (mm), R_1 is initial radius of stent (mm), R_2 is stent's radius after its deformation by the rolling (mm) and is 7.5% (as it known the maximal size of reconstituted deformation of NiTi used in medicine is 7–8%).

From this formula, the increase of stent's initial diameter D_1 brings a higher degree of deformation $\Delta\varepsilon$ in case of rolling the stent in a fixed diameter D_2, e.g. the deformation of stent bigger than 8-mm diameter brings a significant residual deformation and a stent like that can not reconstitute its initial diameter at the temperature higher than at point Ak (Af).

To wide the X-ray endostenting use for bigger diameter vessels, particularly in case of aorta, which is 12-mm interior diameter, a new universal-delivering device, which permits to roll the stent into smaller diameter stent simultaneously with its stretching until a desired length, was developed. The diameter of stretched stent beside the cylindrical stent's axis significantly decrease in comparison to the its initial diameter. By this way, it became possible (if the objective is to implant a bigger diameter stent in a patient) to always choose a certain stretch degree of a stent before the rolling in order that its deformation does not exceed the degree of elastic reconstituted deformation, e.g. 7–8%.

In 1993, at the Invalids of Second World War Hospital, at the X-ray Endovascular Surgery Department, first in the world practice, the patient K., 68 years old, who had following diagnosis: panetherosclerosis, stenosis of a terminal aorta's part 75%, occlusion of iliac arteries at the right, multiple stenosis of iliac arteries at the left, occlusion of surface femoral arteries at both sides, had sustained an X-ray endovascular dilatation of terminal aorta's part with a following implantation of Rabkin's NiTi endostent, which is 12-mm diameter [11]. This was the biggest diameter endovascular stent (Fig. 3).

More than 300 stents produced in our laboratory have been implanted in a number of hospitals in CIS (Table 1). Patients who had not any complications in the early post-operation period (96.2%) recovered in 15 h. The results of implantation just after stenting were evaluated by the angiographic control on the table

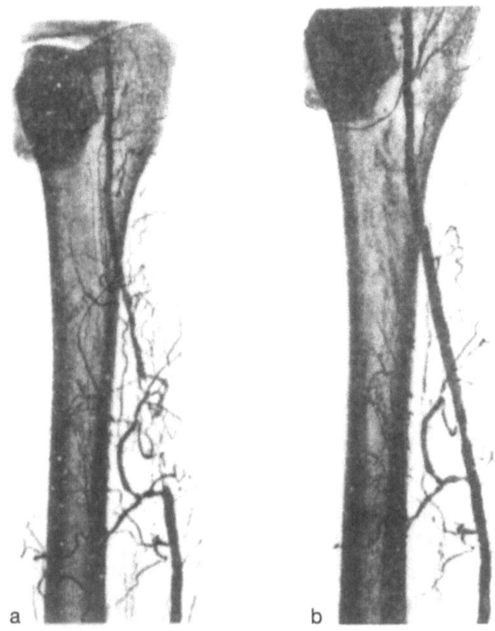

Fig. 3. Aorto-artery picture before (a) and after (b) implantation

Table 1. The dimensions of stents applied in X-ray endosurgery

Field of use	Endostent diameter (mm)	Endostent length (mm)	Delivering device[a] length (mm)
Brachio-cephalic and sub-clavian artery	5 6 7 8	25–30	1200
Iliac artery	7 8 9 10	25–30	350
Femoral artery	4 5 6 7	25–30	600
Popliteal artery	4 5 6	25–30	500
Bile ducts	4 5 6 7 8	40–80	350

[a]The handle of the delivering device is manufactured in two variants: for endovascular and for endobiliar stents. The total length of delivering device shown in the table varies during assemblage because of replacement elements

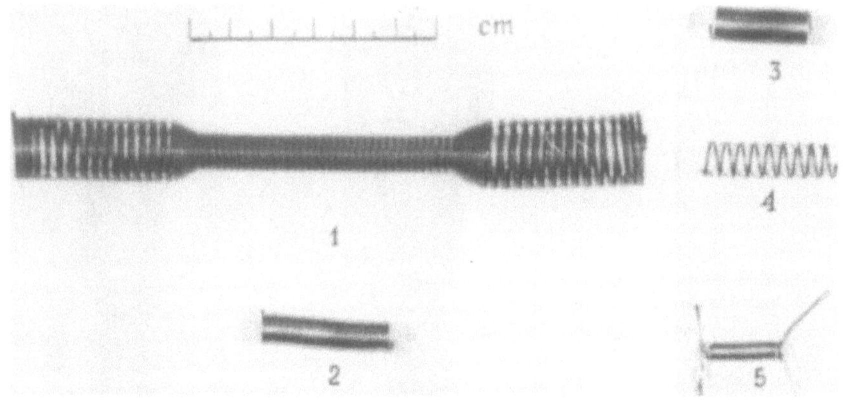

Fig. 4. NiTi endostents. *1,* oesophagus; *2,* biliar ducts; *3, 4,* blood vessels; *5,* cervical canal of uterus

and then every year by the angiographic control and by not invasive methods of evaluation of the blood flow.

Unsatisfactory, such as restenosis and reocclusion have been observed in 1.5% cases of general and exterior iliac, general femoral, subclavian and kidney's arteries. Analysing the operation results of mentioned arteries, all the patients did not have a satisfying blood flow in the distal canal, which is defined in 50% cases by grave diabetes.

By this way, X-ray endostenting surgery permits to reconstitute adequate blood flow and to create a significant hemodynamic canal among the large number of patients who suffer of arterial stenosis and occlusion. If the distal blood flow is satisfying, a positive result is obtained in 98.5% of cases.

The great economical effect of the application of the new method to medical practice is composed of the decrease of patient's hospital days before operation and of the shortening of the followed up recovery period (7–10 days). In addition to the endovascular field, other X-ray endostenting directions have been developed: endobiliar, endobronchi, endooesophageal and endocervical. The correspondent modifications of stent (Fig. 4) and of delivering device have also been developed. The experimental studies preceded the clinic applications of mentioned types of stents.

2 Stenting Surgery on Bile Ducts

In the field of bile pathology, the strictures of bile ducts are the most important treatment difficulty. Dead in cases of bile ducts operation, according to B.V. Petrovski, is 13.3% of cases and the stricture recurrence is 10.9% of cases [12]. The causes of stricture appearance in 60% of cases are iatrogen and swelling affectations of bile ducts, or the pressure exercised on bile ducts by metastases' nodes.

The operative treatment most popular in case of this pathology is characterised by traumas and by frequent recurrences. The balloon-expanded dilatation of a stricture is less traumatic, but has a larger number of recurrences.

From 1986, under I.K. Rabkin's guidance, the study of possible X-ray endobiliar surgery using NiTi stent had began. Forty-nine acute and chronicle experiments were realised in the X-ray endobiliar stenting field on five rabbits and 35 dogs. Fifty-two NiTi stents were installed in the different part of bile tree; ten of these stents were installed in a before created part of a bile stricture of bile flow. The NiTi stents, which are 0.5–3 cm length and 3–6 mm diameter, were used.

The stent was installed by an intra-operational method. After the secretion of hepaticoholedoh from the ligament's elements of lever, a punction by a needle was performed and the transportation catheter was introduced into the bright interval of bile flow by Seldinger. NiTi stent was cooled by chlorine-ethyl, was deformed into a smaller diameter spiral, was fixed on the conductor and, under the X-ray control, was introduced in the desired place of stent installation. The stent was separated from the conductor. Under the bile temperature effect, the stent reconstituted its given form and was fixed by scaffold's turns on the bile flow's walls. To determine the going past of surgery zone, several holangio-charts were realised.

The histological studies were effectuated some days later and 2, 3, 6, 10, 12 months later. In the uncomplicated cases, during the 6-month period, the stent's spiral enters under the mucous membrane of bile flow. Furthermore, any significant modification in the flow's wall-spiral presence place-was not noticed. In the stroma, a minimal lymphohistotionary infiltration with the presence of macrophages, which finishes to convert into a medium sclerosis, was detected.

All NiTi stents were removed from the bile flows of animals that had sustained an acute experiment. The surface state, size and force characteristics of stents remained the same (Fig. 5).

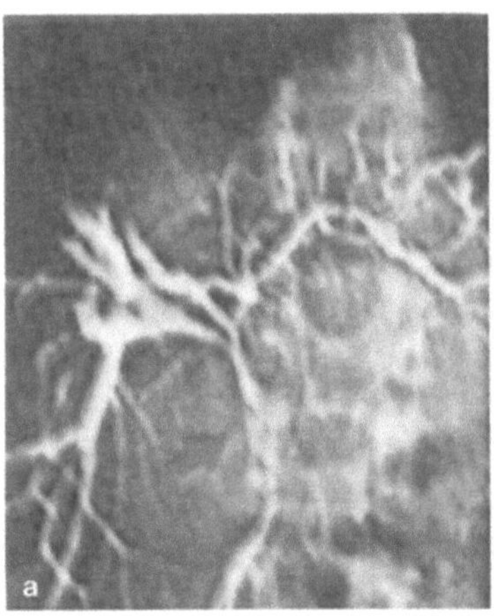

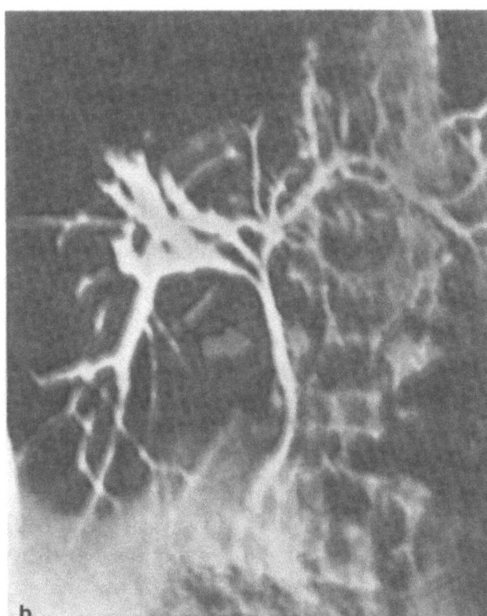

Fig. 5a, b. Biliar ducts before (a) and after (b) implantation

The clinical application of an endostent in case of X-ray endobiliar surgery has been begun in the Clinic of Surgery Spasokoukotski's Faculty of Pirogov's Second Medical Institute [13]. The surgery was performed through leather and lever among 38 patients who had swelling and healed strictures of bile ducts in different parts of bile tree.

Sometime, a patient had received two or three stents. The following observations had shown that stent is functional during 2 years or 3 years in case of strictures and during 7–12 months in case of swellings (the cause of lethal result among 21 patients was progressive growth of main illness, cancer intoxication growth and kakhesy growth. It is important to know that recurrence of jaundice was not noticed among these patients).

If it is necessary, the stent can be removed with the help of a simple manipulator for removing pieces. During the whole period, the implanted stent had not been dislocated and thereby maintained the going past of the reconstructed place. Particular results were followed during 1.5 year.

It was experimentally proved that NiTi alloys do not provoke absorption of bile oxides and do not provoke biliar stone formation. In case of a correct implantation and a corresponding stent's diameter to the bile ducts' diameter, the whole epitheliasation is attempted.

It is important to mention that an endostenting surgery of lever's door because that group of sick persons is not allowed to a surgical treatment [14]. These operations are successfully performed at the Moscow Diagnostic and Surgical Institute. The results of patients' observations had shown that endobiliar stenting surgery could be viewed as a new method of palliative treatment in case of cancer of lever's door, pancreatoduodenal zone. X-ray endobiliar stenting surgery of bile flows permits to improve the patients' state that can not follow an operative treatment or it is too dangerous for their lives.

3 Stenting Surgery on Oesophagus

The endostenting surgery on oesophagus in case of cancer is known for a long period of time. A lot of stents applied before have just an historical interest because they did not show expected positive results. Using all constructions, different complications were observed: erosion, decubituses, haemorrhage, obturation of bright interval by food masses, etc. The efforts of new construction researches are justified because it does not exist a perfect or universal construction (Fig. 6).

The X-ray oesophageal surgery experiments by a spiral NiTi stent were performed at a special experimental X-ray operation room [15]. For the experiments, 32 rabbits of chinchilla breed (weight: 1.7–4.0 kg) and 37 white linear rats (weight: 180–280 g) were used. X-ray endooesophageal surgery was practised on five rabbits, which have a burning stricture of oesophagus, which was created by surgical way (in the period of 23–119 days after modelling the burning stricture), on seven rabbits with oesophagus's stricture, which was created by surgical way (in the period of 14–35 days after modelling the stricture) and on 20 rabbits with an intact oesophagus. The rats used in experiments were intact. The NiTi spirals,

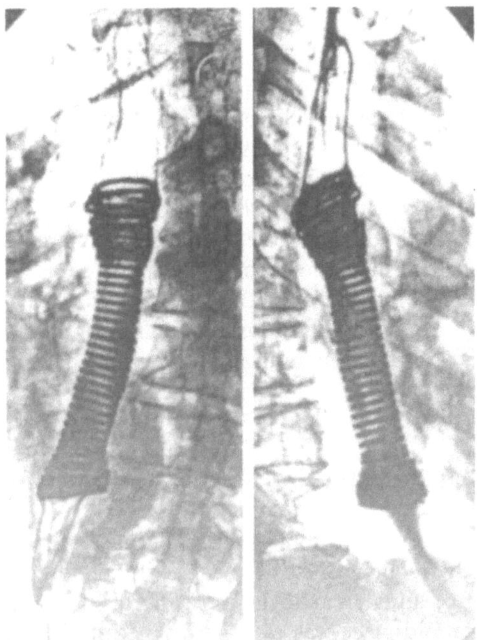

Fig. 6. Stenting of oesophagus

which are 4.2–11. 5 mm diameter, 7–35 mm length and 2–4.5 mm step of spiral's turns, were used. By the electronic charts obtained during the contrast study of oesophagus, the spirals were chosen according to the size. To install the spirals, the X-ray contrast angiographic catheters and the conductors with a special lock and a special oesophagus's moving device were used.

The histological investigations were effectuated 1, 2, 5, 14, 21 days and 1, 1.5, 2, 3 months after the intervention. Macroscopic modifications of tissue around the oesophagus were not noticed. One day after the stent installation, the spiral's turns are 1–1.5 mm into the oesophagus's wall. From 2 months to 3 months, the oesophagus's wall becomes thicker in the turns' place, the mucous membrane on the cut become whiter, its longitudinal pleats are not noticed. On the oesophagus's longitudinal cut, the spiral's turns are placed into the wall.

In case of histological investigations, the spiral's turns enter to the oesophagus's wall a day after the operation. At the third week, in these parts, some crater form depths, which are covered by multiple-coat flat epithelium and by granules, which are relatively infiltrated by round cells where some neutrophil leukocytes can be find, are created. Two months later, a canal, which is covered by multiple-coat flat epithelium, is created around the spiral's turns. The canal's wall is presented by soft fibrous tissue and by granules, which are ripening and which contain lymphoid cells, histocytes and thin-wall vessels. The obtained results allow us to introduce the X-ray endooesophageal stenting surgery in the clinical practice.

First in the world practice, in February 1988, the X-ray endooesophageal surgery using a NiTi stent in case of oesophagus's cancer was performed among three patients, 62–76 years old [16]. In case of oesophagus's contrast investigation after the operation, an expansion of bright interval was observed; the patients noticed a better going past of food.

The stent installation in the oesophagus does not require anaesthesia or big operation, as it is necessary in case of an intra-operational installation of plastic stents, constant X-ray control makes practically impossible the oesophagus's perforation, which is possible in case of oesophagus's stents installation by endoscopic method. The whole operation can be realised using just sedative components. The transportation process of endostent is less traumatic for tissue because NiTi stent is introduced in the oesophagus in rolled state (the diameter of rolled stent is 7–10 mm). The endostent application in case of oesophagus's cancer is advised in case when the radical interventions have not been applied because of patient's grave state.

4 The Endostenting Surgery on Trachea by NiTi Spiral

The loss of scaffold function because of trachea's cartilage half-rings is a relatively frequent cause of unsuccessful endoscopic and surgical treatments. The idea of creating an artificial intra-mural support scaffold of trachea on the destroyed scaffold place was realised by the introduction into the trachea's wall of a NiTi endostent [17].

To study the possibilities of endostenting surgery on the intact stenosis trachea, 82 experiments on rabbits were performed. In the first series of experiments, 23 animals had received various size implants in the intact trachea. The stents, which are 15–20 mm length and 5–8 mm diameter, had an ear at one end; this ear served to fix the stent on the conductor. The conductor with cooled stent was introduced in the trachea's bright interval trough X-ray contrast catheter, which was installed in the tracheotomic hole or trough an intubation cable. The stent in the trachea's bright interval was heated until body's temperature, and then it took a spiral form and was fixed by itself thanks to the pressure exercised by turns on the trachea's wall. All manipulations, which are linked to the stent implantation in the trachea, were controlled by electronic charts realised in side projection. The animals were under observation within the period from 2 days to 1.5 year (mean=3.3 months). After, the animals were removed from the experiment for a pathomorphological study.

A created not rigid stenosis of trachea was formatted on 40 rabbits (second series of experiments). This was attempted with the help of local perturbation of cartilage scaffold of trachea's neck part with an additional burn of mucous and sub-mucous membrane in the affectation part by 5% iodine. 1.5–2 months later, a part of not rigid stenosis of trachea was created, in which 19 rabbits had received NiTi stent (third series of experiments). After the stent installation in the stenosis trachea, the animals were observed within the period from 7 days to 150 days (mean=1.4 months). Finally, the animals had suffered euthanasia and had sustained a pathomorphological study.

The experimental investigation had shown that spiral NiTi endostent during its implantation, as in intact as in stenosis trachea, is biologically inert corresponding to the tracheal wall and does not exercise negative effect on lungs. During the first days, an inflammation reaction of tracheal wall was detected among all animals. Three to four weeks later and when the stent's turns were more implanted in the tracheal wall with a reconstitution of epithelial coat, the inflammation was not noticed according to all macroscopic characteristics, and a relatively weak inflammation beside the turns does not have any negative effect on the animals according to histological methods. Furthermore, the tracheal bright interval on the endostent level kept the same volume; endostent was placed into tracheal wall. To insure an adequate bright interval of narrow tracheal part, in case of which this endostent will be completely covered by tracheal mucous membrane and will effectively accomplish the artificial support tracheal scaffold function, the stent's diameter must be 2–3 mm bigger than bright interval diameter of trachea.

In clinical practice, the X-ray endotracheal stenting surgery is performed in cases of post-tracheostomic stenosis and tracheobronchiomaly among 15 sick persons. The implantation of cooled stent is realised with the help of a bronchoscope under the fluoroscopic guidance or manually through tracheal system. In the bright interval, the stent was heated up and took its given form. The results were observed during two years after operation.

As a result of a stent implantation in a soft tracheal part, the affected tracheal segment receives loosed scaffold, which serves as support for tracheal wall, which has thin elements (cartilage's fragments, fibrous tissue, etc.), and which prevents the trachea's fall. The new scaffold posses a satisfying density and elasticity to counterbalance the pressure of structures, which exist beside the trachea. An adequate choice of stent's size permits to avoid its additional fixation to the tracheal wall. Using stent, which is 2–3 mm bigger than bright interval diameter of trachea, the turns' pressure on mucous membrane is significant. The turns cut the mucous membrane and appear into the tracheal wall in remaining a linear fault, which cures by itself.

In case of an incomplete growth of stent into the tracheal wall and of a remained contact of its turns with tracheal bright interval, the continued inflammation process in the tracheal wall brings a growth of granulated tissue in the tracheal bright interval, a narrowness of air circulation way, which demands repeated endoscopic interventions. In this situation, an introduction in the tracheal bright interval of a cable-protector made of polyvinyl-chloride trough tracheal stoma for 4–6 months until the whole enter of all endostent's turns in the tracheal wall and until epithalisation of tracheal mucous membrane above the endostent is necessary.

5 The Stenting Surgery on Cervical Canal of Uterus

The accumulated experience of X-ray endostenting surgery by NiTi spiral stent of arteries, veins, bile ducts, oesophagus's trachea with a positive effect in 84.6% of cases allows us to pass to the X-ray cervical stenting surgery. The first clinical experiment of an X-ray endocervical stenting surgery was performed at Gertzen's Moscow Scientific Investigation Institute of Oncology in 1988 [18].

The treatment was applied among the patients who suffer of atresion and stricture of cervical canal, which were developed after the "soft" operations on uterus, which were performed in cases of malignant and benignant processes. The essential of treatment is the introduction of spiral endostent, which has special fixing elements (Fig. 4), to create an interior support scaffold for walls of cervical canal.

The length of stents' staying in cervical canal was defined by the time of formation of a new bright interval from the healed tissue. The stent was removed within the period of 63–193 days after the operation. From dynamic observation, we can confirm that for 3–4 months of spiral staying in the endocervixes, a "dense" canal is formatting with its complete epithalisation. Later observations had proved that stricture and atresion recurrence is absent until a year after operation. By this way, a temporal installation of stent permits to create a 100% functional cervical canal of uterus (Fig. 7).

The stenting surgery of cervical canal is determined [18] by reproductive age of patients, pathology of cervical canal (strictures, atresions, and synechia of canal cavity), that is accompanied by clinical symptoms and also by an impossibility of execution of aesthetic operations, and by ineffectiveness of moving device. The endostenting surgery is not recommended in case of swelling narrowness of uterus's canal.

It is important to mention that in addition to the works of Russian specialists, the development of X-ray endostents of very various metallic constructions in

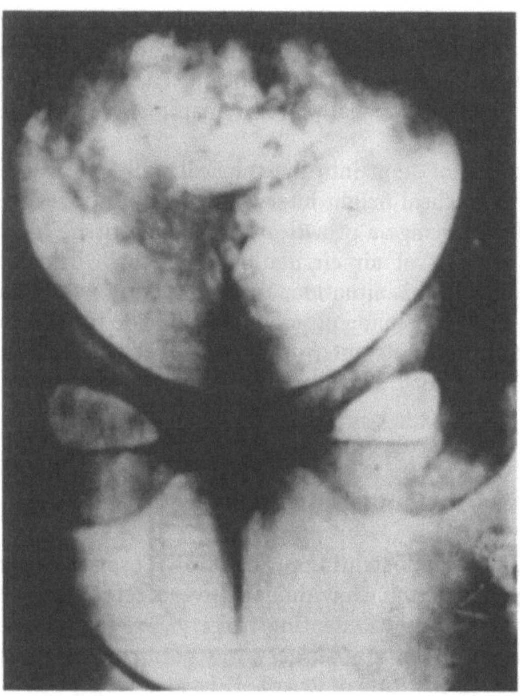

Fig. 7. Stenting of cervical canal of uterus

other countries was realised: stainless steel 316L, 316 LVM, tantalum, alloys based on Co [19]. Actually, these stents, which can become wider with the help of balloon expanded catheter, are produced and applied in clinical practice. In spite of that, as D. Lierman [20] mentioned during its speech at an international symposium on heart's vessel and intervention radiology, the future are the alloys which possess shape memory effect.

The advantage of these stents is the insurance of permanent and uniform compression on the vessel's wall. The stent made of stainless steel and which become wider with the help of balloon, in case of its removal, brings a "springs back" under the pressure of vessel's walls and become a smaller diameter stent, e.g. a stent like that can not fulfill perfectly its function of scaffold to support the bright interval of vessel canal. Because of that, the risk of restenosis is always present. In practice, the restenosis is usually observed. Always functional compression of shape memory stent, which become wider by itself, insure maintenance of hemodynamic significant canal thanks to its programmed diameter; that practically avoids all the stenosis risk.

References

1. Schmerling MA, Wilkov MA, Sanders AE, Wysley JE, et al. (1975) A proposed medical application of the shape memory effect: a NiTi Harrington rod the treatment of scoliosis. In: Perkins J (ed) Proceedings of the International Conference on Shape-Memory Effect in Alloys (Toronto). Plenum, New York, pp 563–568
2. Dotter CT, Buschmann RW, Mangomery K, Rosch J (1983) Non-surgical placement of arterial endoprostheses. A new technique using nitinol wire. Radiology 147:259–260
3. Cragg A, Lund G, Rysavy J (1983) Non-surgical placement of arterial endoprostheses. A new technique using nitinol wire. Radiology 147:259–266
4. Author certificate n0 1237201 from 15.02.1986.Intra-vessel scaffold
5. Rabkin DI, Minkina SM, Kadnikov AA, Khasenov BP (1986) The experimental morphologic justification of X-ray endovascular stenting surgery of vessels. Med Radiol (Mosk) 31:55–63
6. Rabkin IK, Zaimovskiv VA, Khmelevskaya IY, et al. (1984) Experimental justification and first clinical experiment of X-ray endovascular stenting surgery of vessels. Radiol News 4:59–64
7. Rabkin DI (1987) Experimental justification of X-ray endovascular stenting surgery. Ph.D. thesis, University of Moscow, Moscow
8. Author certificate 1768154 from 15.10.92. The stent of hollow organ. VNIIPI, Moscow: Petrovsky BV, Poluhin PI, Rabkin IKH et al.
9. Author certificate 1768068 from 15.10.92; BI n0 38. Delivering device for transportation of intravessel scaffold. VNIIPI, Moscow: Rabkin IKH, Bernstein ML, Lapkin KS et al.
10. Khmelevskaya IY, Ryklina EP, Morozova TV, Prokoshkin SD, Rabkin IK (1994) Application of NiTi shape memory effect alloys to X-ray endostenting and other medical fields. In: Pelton AR, Hodgson D, Duerig TW (eds) Proceedings of SMST 1994. Shape Memory and Superelastic Technologies, Pacific Grove, pp 495–498
11. Ryklina EP, Maximovitch IV (1993) First experiment of clinical application of shape memory effect NiTi stent in case of aorta surgery. In: Likhacher VA (ed) Materials of 29th inter-republican seminar on actual problems of solidity. Pskov, pp 145–147
12. Nelubin SP, Mednik GI (1989) Experimental justification of clinical application of X-ray endobiliar NiTi stent in case of bile ducts strictures treatment. In: New technologies in radiosurgery: theses of 9th USSR Symposium. Moscow, pp 37–39
13. Kapranov SA (1989) The immediate and later results of trough lever endostenting surgery of bile ducts. In: New technologies in radiosurgery: theses of 9th USSR Symposium. Moscow, pp 32–34
14. Krivenko EV, Momjan BK, Khartchenko VP (1995) A possibility of percutaneous transhepatic endobiliar stenting with Rabkin's NiTi Stent in malignant biliar strictures. In: International Symposium of Cardiovascular and Interventional Radiology. Moscow, p 49
15. Gigauri IS, Sheremeteva GF, Prozorov SA, Mednik GI (1989) Experimental justification of X-ray endoesophageal stenting surgery. In: New technologies in radiosurgery: theses of 9th USSR Symposium. Moscow, pp 41–42

16. Rabkin IK, Mamontov AS, Prozorov SA et al. (1989) First experiment of clinical use of X-ray endoesophageal stenting surgery. In: New technologies in radiosurgery: theses of 9th USSR Symposium. Moscow, pp 39–40
17. Kumraev SM, Rabkin DI (1989) Experimental justification of endostenting surgery of trachea by NiTi spiral. In: New technologies in radiosurgery: theses of 9th USSR Symposium. Moscow, pp 44–46
18. Rabkin IK, Novikova EG, Pronin AG, Prozorov SA, Kadnikov AA (1991) NiTi endostenting implantation in treatment of strictures and artesias of cervical canal of uterus. Med Radiol (Mosk) 36:35–38
19. Serruys PW (1997) Handbook of coronary stents. Rotterdam Thoraxcentre Interventional Cardiology Group, Rotterdam
20. Liermann D (1995) Future of the Stents. In: International Symposium of Cardiovascular and Interventional Radiology. Moscow, pp 39–40

Device for Extravasal Correction of the Function of Vein Valves Based on Nitinol Shape Memory and Its Clinical Application

S.D. Prokoshkin, A.P. Chadaev, E.P. Ryklina, I.Y. Khmelevskaya, A.C. Butckevich

1
Introduction

Lower extremities varicosity is widely spread. Statistical data presented by various authors show that in Japan, 8–9% of adult population suffer from varicosity, in Africa 7–10%, in the USA 20–30%, in England 10–17%, in Europe 20–30% [1–8]. In accordance with Goldmann (1990) [9], only in the USA 80 million of adults have varicosity.

Side effects, namely, trophical ulcer, chronic dermatitis, thrombophlibitis, bleeding from expanded veins, bring to the temporary disability that sometimes can result in total disability, where the patients become invalid. 0.2–3.9% of population [10–12] have the lower extremities trophical ulcer, 75–90% of these cases are caused by varicosity and post-thrombophlibitis [13, 14]. In accordance with Smith et al. (1990) [15] 500,000 in the United Kingdom and 800,000 in the USA suffer from chronic venous insufficiency complicated by trophic ulcers. Trophic ulcers resulted in significant loss of working days: 500,000 working days and 2 million working days were lost in England and Wales, and USA, accordingly [12]. Until recently, the chronic venous insufficiency progress was connected to the pathological expansion of hypodermic veins or thrombic occlusion of deep main veins. But researches conducted during the pass twenty years have revealed that the venous blood flow along deep veins requires correction or reconstruction.

As we can see from the materials of the Ninth World Congress on phlebology (Kyoto, Japan, 1986); 15th World Congress on angenology (Rome, Italy, 1989) and 12th World Congress on phlebology (London, England, 1995) the main idea was underlined that only high level of diagnostics of disturbance of venous blood flow in the system of hypodermic and deep veins can allow to discontinue standard operations and to begin to make operations based on pathogenic reasoning specific for every patient. All the above-mentioned conditions are of high social significance in research and would improve the treatment of the veins diseases.

Problem of elimination of pathological retrograde blood flow in the deep main veins constitutes the most significant moment of operative treatment of varicosity. This allows to reduce venous hypertension and to cut down the probability of further pathological changes in the system of veins [16]. In case the deep veins valves are certified to be in the normal state, it is reasonable to make a varictomy of hypodermic veins with perforant ligation [16]. But it has been revealed that 80–94% of the patients have ectasia of deep veins and relative insufficiency of their valve [17–20], which requires a correction of their function.

In case of the hypodermic vein removal and perforants ligation without further reconstruction of abnormal valve, severe misfuctions of microcirculation and tissues trophism [21, 22] in distal sections of lower extremities stay in force. Trophic ulcers treatment does not bring any reliable results [23, 24]. In accordance with Hopkins et al. (1992) [25] the stable venous pressure in the veins of the foot above 60 mmHg increases the risk of ulcers formation higher than 50%.

A variety of methods of venous valvular correction apparatus are known and still under permanent development. Today the methods of extravasal correction of valves are widely used because they are lowing traumatic and technically available. This method lies in the fact of vein narrowing in the valve area, which allows the closing of the cusps and the correction of the valve function. Zelenin and Kurakov (1979) [26], Kuo-Hau Zhang et al. (1993) [27] used muffs from autovein to correct valve functions through vein narrowing. Askerkhanov (1984) [28] used muffs from broad fascia; Raju et al. (1991) [29] used artificial muffs. But artificial muffs can cause the vein cicatrization and narrowing [30]. Automuffs do not cause such cicatrization but the veins become narrower [30]. Moreover, these materials are not flexible and they are skeleton-free, which makes it difficult to monitor the extent of vein narrowing.

Tsukanov [31] proposes to narrow the vein in the valve area by using fascial paravasual structures, while Zuev with co-authors (1986) [32] propose to perform peryvenous muscular plastic. But most patients do not have pronounced fascial structures and the muscles suturing leads to cicatrization. It is difficult to monitor the vein narrowing and the skeleton function is not reliable.

Lavsan skeleton spirals do not present limitations [19]. Lavsan is inactive and is not subjected to destruction. Correction techniques are simple. Spaces between loops provide quick revascularization of the vein wall. But the valve cusp anatomic defectiveness in 20–25% of the cases does not allow to perform complete correction of their function.

Moreover, the spiral positioning is time and labor consuming process. The vein should be mobilized along a significant length, which complicates cicatrization and requires to apply turnstiles. At the same time, the vein can be traumatized by the sharp ends of the spiral in the reeling process in the area of abnormal valve, which further prolongs the post-operative period. The most negative factor of any spiral (and any muff) circular in section is that it significantly reduces the vein lumen, which has negative impact on its passage capacity.

In 1975, Kistner [33] performed the first intravasal direct reconstruction of the valve cusps (valvuloplastic) in clinic. Positive results of valvuloplastic (VP) were certified by other authors [34–39]. Raju and Sottiurai [40] modified and simplified these operations techniques. They performed VP through transverse incision and valve cusp suturing through the vein wall. But this method of VP has not gotten a wide application in common surgical practice because of high-risk of post-operational thrombosis and technical complication. In accordance with Perrin [41], 10–13.8% of thrombosis were observed within 36 h after operation.

Bergan [42] considers that scientific analyses of signs for valvual reconstruction have not yet been worked out until now. Distant results of VP are quite different. In accordance with Eriksson et al. (1985) [38] in 18 cases of VP, five cases demonstrated reflux recidivation and repeated clinical symptoms in some distant time upon operation. It incited Kistner (1990) [43] to work out a method of closed VP through the vein wall gathering together in the area of commissurial eminence. Though the valvual sinus becomes shorter the cusp excess is not eliminated which is the main anatomic defect bringing to the valvual insufficiency. Analysis of all the existing methods and devices used in the lower extremities varicosity treatment allows us to conclude that no efficient method of correction of the deep main vein abnormal valves has been designed until now.

A new effective device [44] for extravasual correction of the lower extremities main vein valve function under chronic venous insufficiency is based on the use of the unique functional properties of titanium nickelide ("nitinol") alloy, displaying shape memory effect. These alloys have a remarkable reputation as an effective functional material for various constructions of medical equipment for surgical interventions. As it was mentioned above, these alloys are compatible with human tissues. They are highly resistant to corrosion in human biological liquids. Their special properties, such as superelasticity and shape memory, allow to realize certain possibilities which can not be achieved through the use of other materials. For substantiated selection of constructive characteristics necessary for extravasal corrector used in main vein valves, an anatomic examination of valves took place in the Fourth Municipal Hospital of Moscow.

2 Anatomic Examination of Main Vein Valves and Grounds for Corrector Shape Selection

Until now, the following problem is still unresolved – what is the leading factor in varicosity pathogenesis: progressing ectasia of veins bringing secondary abnormalities of the valves or the initial insufficiency of the valvual apparatus leading to ectasia under hypertension. Both mechanisms can act on the same level to induce pathogenesis.

As we can see from Lord (1978) and Borchberg (1967) [45], inherent insufficiency of venous valves is a contributory factor of varicosity. Bernshtein EF [46] (1986) examined healthy populations and noted that 16% have inherent abnormalities of the femoral vein valves. It is opinion of some authors [19, 47–49] that insufficiency of the valves represents one of the most significant aspects of varicosity pathogenesis, because the rising retrograde blood flow plays an indubitable role in the disease progress. Other researchers [50–52] consider that the venous valves do not play a significant role in the pathogenesis of varicosity and that their insufficiency is secondary.

The anatomic sizes of the valves cusps, the lengths of free edges, and the diameter of the vein in the valve area are specific for every given patient. But in normally functioning valves, these anatomic index correlations are constant. For example, if we take a normal valve, we can see that the relation of valve cusp free-edge length to vein diameter is quite constant.

Taking into consideration all the above-mentioned, an anatomic examination of normal valve constructions of cadaver veins segments (taken from patients who died of various diseases excluding the venous system pathology) and the veins of cadavers with pronounced signs of chronic venous insufficiency was conducted. In our opinion, examination of only two parameters, even though they are basic (length of the cusp free-edge and vein diameter) will not provide necessary understanding of a united and functioning complicated anatomic system, which has volume dimensionality. All parameters of this anatomic system are mutually connected, mutually caused and providing normal functioning.

Using the extravasal correction (EVC) of valve method requires a vein narrowing by one third to one fourth of initial diameter in the area of the valve. Empirically determined degree of vein narrowing has not found its explanation. In this case, the problem of maximum admissible reduction of the vein opening space maintaining efficient EVC was not considered. Anatomic examinations were conducted on cadaveric speciments in the Department of Anatomy of the Fourth Municipal Hospital, Moscow.

2.1 Methods of Anatomic Examination

Anatomic examinations were performed on cadaver vein segments removed within 24 h from death because of various somatic diseases. The patients selected were of age 56–73 years old. Their histories did not contain any remarks about vein diseases. At the same time, the veins were subjected to a visual examination, which allowed to estimate the lack of any signs of chronic venous insufficiency: hypodermic veins varicosity, edema, hyperpigmentation and lypodermatosclerosis of trophic ulcers.

Below the inguinal fold, a longitudinal incision of 15–20 cm was made along the Ken line. Layer by layer, the neurovascular fascicle proximal section was open. The hypodermic femoral vein (HFV) was mobilized by acute way. A HFV segment, 15 cm in length, was excised for examination. It was put in a glass bulb with isotonic NaCl solution. The vein was washed of any residual blood in physiologic solution until its clearness. Then the vein was cleaned from paravasual tissues with the help of microsurgical forceps and scissors. The small tributaries were ligated by caproon treads 4/0 at 2–3 cm from the wall.

The vein diameter was measured with the help of slide gage out of the valve zone (D_V) and on the level of valvual sinus (D_K) approximately at the middle of the valve between the two most distant points. Moreover, the last measurement was made in two directions: parallel and perpendicular to the free edge of the valve cusps. All the obtained measurements were grouped in two groups for normal and abnormal valves.

Then the vein segment was placed on a special table. It was subjected to longitudinal incision through the valve commissure with the help of microsurgical scissors and forceps. The quarterangular layer was fixed on the table with the help of nails.

The cusp free-edge length (L_C) was measured through accurate and maximum tension of cusps with the help of an elevator providing a uniform tension of the cusps under a 60–80° angle in relation to the table surface. Ellipsoid form of the cusp free edge became triangular under tension, which made the method of measurement simpler. Two sides of this triangle were measured with the help of ruler and divider. The figures obtained were put together. This index was determined for every valve cusp.

Then the cusp height (H_C) was measured for every cusp – distance between the middle of the cusp free edge and the most distal point of the connected edge. The valve height (H_K) measurement was performed as an averaged value of two indexes – distance from the proximal connection of two cusps near the commisures to the level of the most distal points of connected edges:

$$A_K=(A_{K1}+A_{K2})/2 \quad (1)$$

2.2 Results of Anatomic Examinations and Discussion

Every vein segment with valve from the examined groups (normal and abnormal) was subjected to the examination of five specific parameters: diameter of the vein out of valve=D_V; diameter of the vein in the valve area=D_K; length of the free-edges of the valve cusps=L_C; height of valve cusps=H_C; valve height=H_K. Averaged values with mean square deviations were calculated for every parameter of the venous valves for both groups.

Below are the examination results of the space anatomy of the femoral vein segments with valves taken from the first and second groups (Table 1). In this line, the valve height increase is statistically reliable (risk probability $P<0.05$), and the lengthening of the valve cusp free edge is highly reliable ($P<0.01$).

All the parameters of the normal and abnormal valves brought to the normal vein diameter and to one diameter of valve, as their relation does not have any significant differences under normal and pathological state, were compared to get more exact and obvious understanding. Coefficient K was determined equal to the relation between the mean diameter of pathologically changed (abnormal) vein (10.1 mm) to the diameter of normal vein (9.4 mm):

$$K=10.1/9.4=1.07 \quad (2)$$

Table 1. Averaged numerical meanings of parameters of the superficial femoral vein under normal state and varicosity

Number	Valve parameter	Vein segment with valve Normal ($M_1 \pm m_1$) (mm)	Varicosity ($M_2 \pm m_2$) (mm)
1	D_V	9.4±0.8	10.1±1.0.8
2	D_K	12.3±1.2	13.1±1.0
3	L_C	20.2±2.1	28.6±2.0*
4	H_C	7.1±1.1	4.9±0.6
5	H_K	13.0±1.3	16.8±0.7*

*Statistically reliable numerical meanings

Table 2. Parameters of pathologic deformations in a valve

Number	Valve parameter	Abnormal valve index	Index divided by K	Normal valve index	Relation of indexes	%
1	D_V	10.1	9.4	9.4	1	0
2	D_K	13.1	12.3	12.3	1	0
3	L_C	28.6	26.8	20.2	1.33	+33
4	H_C	4.9	4.6	7.1	0.65	-35
5	H_K	16.8	15.8	13.0	1.22	+21

The coefficient was used to bring the valve characteristics to the normal vein diameter through division of all values by this coefficient. The values obtained displayed an obvious evidence of pathological changes in the valve (Table 2).

Absolute values of various parameters of the valve (diameter, length of cusp free edge, etc.) characterize the valve capacity to work for the given vein. In case of proportional increase or decrease of all the parameters of the capable valve, the mutual relation of these parameters remains unchanged. So, the normally operating valve has stable characteristics based on the relation between the valve separate element sizes which can be subjected to a group averaging; the first group (normal) and the second group (pathological). Contrary to absolute sizes, relative sizes of the valve elements are more sensitive to the valve pathology characteristics.

A calculated relation between parameters is required for comparison of these two groups and more precise definition of the quantitative and qualitative changes in the valve elements leading to its abnormality. Parameters of the vein segments with valve were subjected to comparative estimation of every valve separately taken and of the first (normal) and second (pathological) groups. Then the parameters were calculated in pairs, followed by calculation of the averaging (mean arithmetic) values with mean square deviation: D_K/D_V; D_K/H_C; D_V/H_C; L_C/D_V; L_C/D_K; L_C/H_C; L_C/H_K; H_K/D_V; H_K/D_K; H_K/H_C. The results are shown in the Table 3.

Statistical processing of the obtained values has shown that only the difference of four relations of valve parameters are statistically reliable (*): relation of the cusp free edge length to the vein diameter ($P<0.05$), relation of the cusp free edge length to the valve diameter ($P<0.05$), relation of the cusp free edge length to the

Table 3. Averaged values of the valves elements taken in pairs for normal veins and in case of varicosity

Number	Relation between valve elements values	Normal valve	Abnormal valve
1	D_K/D_B	1.31±0.1	1.35±0.1
2	D_K/H_C	1.8 ±0.2	2.46±0.3
3	D_B/H_C	1.38±0.2	1.88±0.2
4	L_C/D_B	1.4 ±0.2	2.88±0.2*
5	L_C/D_K	1.64±0.2	2.21±0.2*
6	L_C/H_C	2.95±0.4	5.5 ±0.5*
7	L_C/H_K	1.5 ±0.2	1.69±0.1
8	H_K/D_B	1.05±0.2	1.29±0.1
9	H_K/D_K	1.39±0.2	1.67±0.1
10	H_K/H_C	1.9 ±0.3	3.18±0.3*

*Statistically reliable numerical meanings

valve cusp height ($P<0.01$), relation of the valve height to the valve cusp height ($P<0.01$). So, we can see that, in case of varicosity, the valve elements' relative sizes suffer from pathologic changes in comparison with normal sizes. The valve cusps free edges length was reliably larger in proportional relation to other elements in comparison with the normal one. It can be seen from three statistically reliable relations (L_C/D_V, L_C/D_K and L_C/H_C). The cusp free-edge lengthening is accompanied by the valve lengthening with simultaneous reduction of the cusp height. Perhaps the relative reduction of the cusp height is related to the fact that, in case of the vein narrowing and cusp lengthening, its height remains constant.

Consequently, the vein ectasia progresses in case of varicosity, but the venous valve injury is more significant. It is certified by the empirical supposition made by Kistner (1975) that extension and lengthening of the cusp free-edges occupy the leading place in the venous valve abnormality.

Availability of some accurate numerical values of the separately taken parameters and their relations allowed to create the scale cartogram of the normal and abnormal valves in plane (Fig. 1). Thus, it allows their visual comparison. Creation of the model of normal and abnormal valves occupies one of the most significant aspects of a more profound understanding of the normal and pathological venous valves structure and operation, which is very important for an improvement of surgical correction methods.

Anatomic examination allows to make the following conclusions: use of muffs round in section for vein wringing out and closing up the valve cusps is not efficient. Really, the fact that the valve cusps relative length increases (L_C/D_V; L_C/D_K) is one of the main factors of the valve abnormality, and proportional reduction of the vein diameter makes an additional contribution to the length increase. An obvious solution we can propose is switching over from the round to elliptic section. In such a case, if the direction of the large axis of ellipse coincides with the line of the valve cusps closing, and its length is equal to the initial diameter of the vein, the vein compression will take place only in the direction perpendicular to the line of cusps closing, and the efficient relative length of the cusps will not be subjected to any changes.

Such a necessity to switch over to the elliptic section of corrector is caused by a requirement to improve the venous blood flow. This is conditioned by the below calculations.

It is known that in order to achieve a complete correction of the incompetent valve i.e. to ensure the closing of the valve leaves (Fig. 2), the valvular zone of the vein should be compressed to approximately one fourth of its initial diameter. If the circular corrector is used for this operation, the vein cross-section after correction area becomes

$$S_1 = \frac{\pi}{4} D_1^2 = \frac{\pi}{4} \left(\frac{3}{4} D_O\right)^2, \tag{3}$$

where D_0 is the initial vein diameter and D_1 is the vein diameter after correction by one fourth of the initial diameter (Fig. 2). Taking into account that the vein cross-section area before correction was:

$$S_O = \frac{\pi}{4} D_O^2, \tag{4}$$

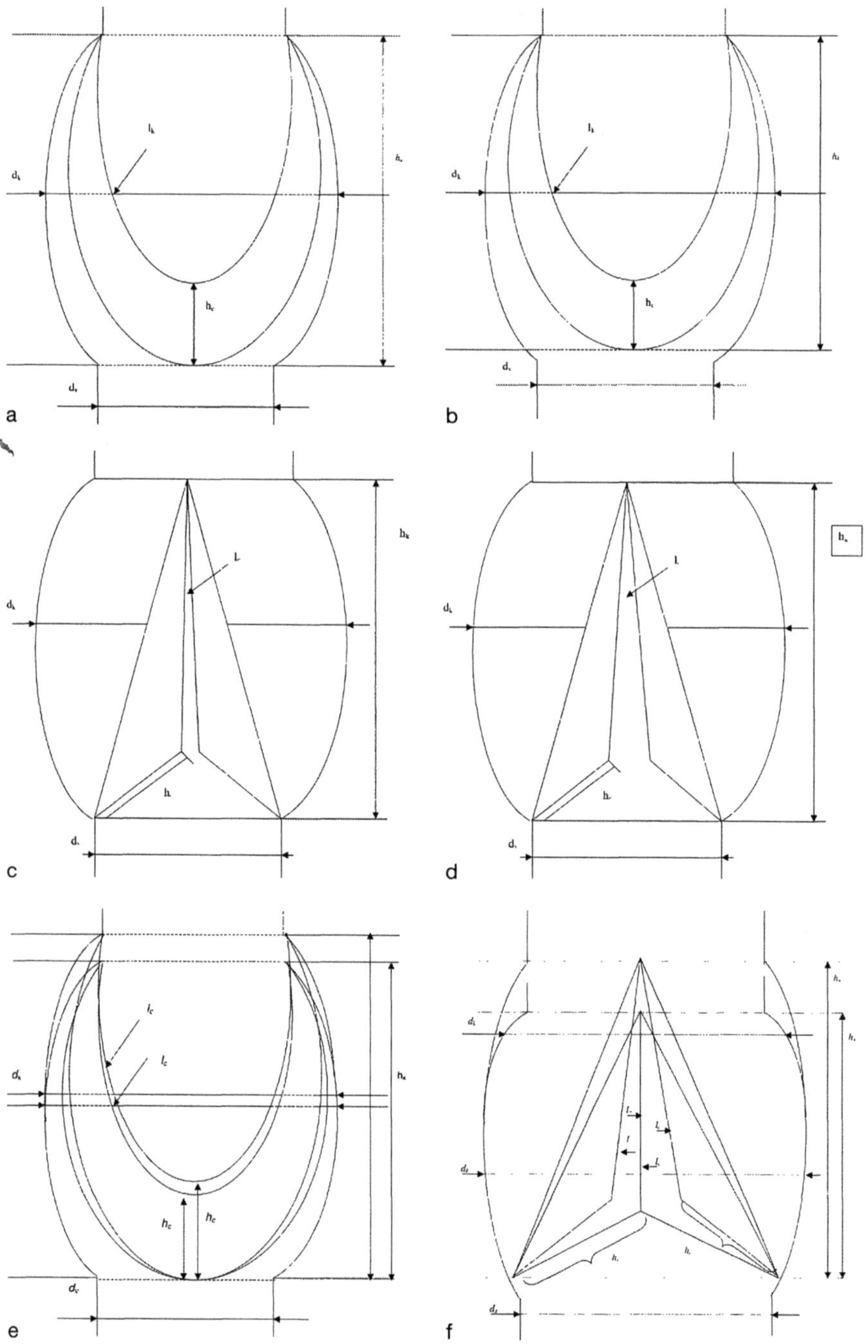

Fig. 1a–f. Scale cartogram of sufficient (**a, b**) and insufficient (**c, d**) valves and their combinations (**e, f**). *1*, the valve cusps; *2*, the circular corrector; *3*, the elliptic corrector; *4*, the joining line of valve cusps. D_0 the initial vein diameter; D_1 the vein diameter after its correction

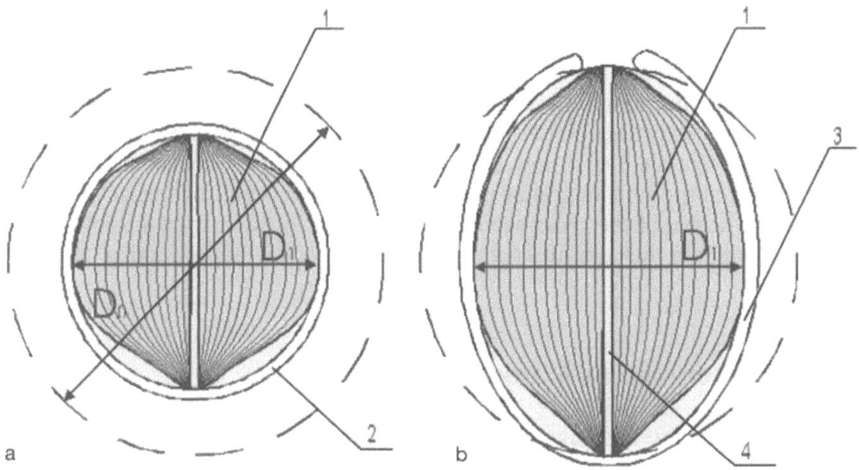

Fig. 2a, b. A schematic representation of a change in a venous cross-section due to contraction by the circular (**a**) and ellipsoidal spiral (**b**)

The reduction of the cross-section area after correction with circular corrector (Fig. 2) is

$$\delta_1 = \frac{S_O - S_1}{S_O} \text{ x } 100\% = \frac{D_O^2 - (0.75Do)^2}{D_O^2} = 43\%, \tag{5}$$

where S_0 is the initial vein cross-section area and S_1 is the vein cross-section area after correction by the circular corrector.

Moreover, because the radial compression of the valvular zone of the vein by means of circular correctors causes a uniform reduction of the vein diameter, while the length of the valve leaves is still unchanged, this correction leads to shrinking of the valve leaves, thus affecting the ultimate result. When the claimed elliptic corrector is applied, the vein cross-section area after correction becomes:

$$S_2 = \frac{\pi}{4} D_O D_1 = \frac{\pi}{4} \left(\frac{3}{4}\right) D_O^2, \tag{6}$$

where D_O is the initial diameter $D1 = \frac{3}{4}$ DO, where D_1 is the short diameter of the elliptic cross-section (Fig. 2).

The reduction of the cross-section area after correction with the elliptic corrector is

$$\delta_2 = \frac{S_O - S_2}{S_O} \text{ x } 100\% = \left(1 - \frac{0.75D_O^2}{D_O^2}\right) \text{ x } 100\% = 25\%. \tag{7}$$

Therefore, the gain in the vein cross-section in case of an elliptic corrector in comparison to the circular one is:

$$\delta = \frac{S_2 - S_1}{S_1} \text{ x } 100\% = \left(\frac{\frac{3}{4} D_O^2}{(\frac{3}{4} DO)^2} - 1\right) \text{ x } 100\% = 33\%. \tag{8}$$

Hence, when the vein compression is performed with an elliptic corrector, the free edges of which are parallel to the free edges of the valve leaves so that the joining line of valvular leaves (Fig. 2) divides in half the gap between free edges of the corrector, the valvular leaves (Fig. 2) move closer to each other, thus restoring the normal functioning of the valve.

In this case, the shrinking of the leaves along their free edges does not take place, and the area of the total vein cross-section is reduced by 25% thus ensuring a 33% higher venous blood flow than in the case of a circular corrector. When compression of the vein should be performed to one third of the initial diameter, the total reduction of the cross-section reaches 50% in the case of a round spiral corrector, while in the case of an elliptic spiral corrector, this reduction does not exceed 33%.

3 Shape-Memory Nitinol Extravasal Correctors

Results of anatomic investigations and calculations described in the preceding section show the necessity for an application of modified spiral in the form of elliptic cylinder. Manufacturing such spirals made of materials being used for this purpose earlier leads to several difficulties connected to their winding deformation based on residual elasticity of the material.

This problem was solved by means of using shape memory alloy "nitinol" for this spiral manufacturing (Fig. 3). The ratio of the ellipse's axes varies between 0.7 and 0.9 and the distance between spires varies between 1.5 mm and 2.5 mm. Such inter-spire distance allows the rapid revascularization of the vein wall, while preventing its ingrowth between spires. These values are based on theoretical and experimental results.

Finally, the spiral tips are made in the shape of closed loops lying on the surface of the corrector. Such construction of the loops ensures the absence of trauma during corrector implantation and service.

The device (extravasal corrector) pinched on a specially manufactured mandrel is subjected to a special thermomechanical treatment conferring the memory and superelastic effects to the material in order to ensure complete shape restoration of the corrector after its implantation in the organism as well as necessary stiffness during service life. In such conditions, the corrector is oriented on the vein so as to provide the vein diameter reduction in the direction perpendicular to the joining line of the valvular leaf free edges.

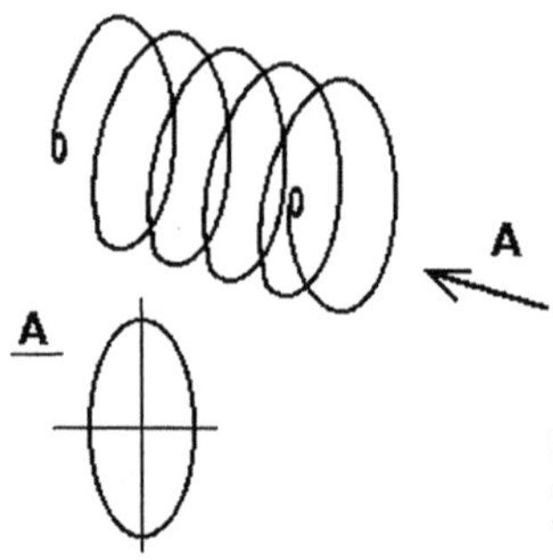

Fig. 3. An extravasal valve corrector made in the form of a hollow elliptic cylinder closed along its forming line and formed by a wire spiral

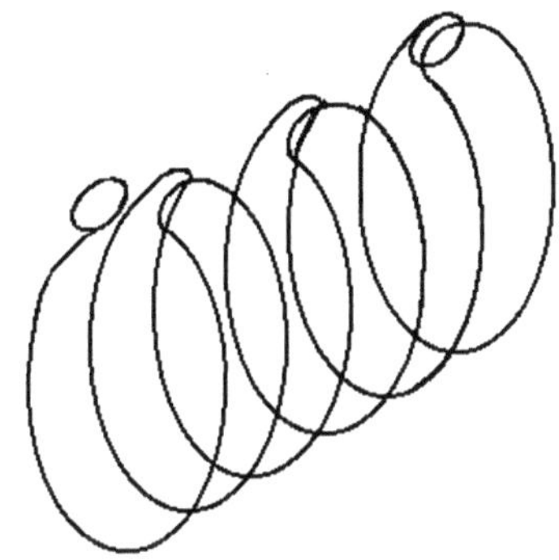

Fig. 4. An extravasal valve corrector made of wire in the form of a quasi-sinusoid rolled around the longitudinal axis and forming the surface of unclosed elliptic cylinder

The ellipse-like spiral device was applied in surgical treatment for three patients of age from 4 years to 30 years old suffering from varicose disease of low extremities. The patients have been operated at the Russian State Medical University (Pediatric Faculty Department of Gennneral Surgery) based on the Munucipal Clinical Hospital No. 4, Moscow. Pre-operative observations showed a pathological venous reflux of the second or third degree. Control retrograde phlebography in post-operative period showed adequate correction of valvular functioning.

Further experimental work on the corrector improvement led to the elaboration of a new device, allowing to simplify significantly the corrector installation procedure [53–55]. The developed corrector represents a framework made of a nickel-titanium wire in the form of quasi-sinusoid wire (diameter of 0, 3-gauge, 6 mm; Fig. 4), rolled around its longitudinal axis along the surface of an elliptic cylinder, such as that the sinusoid peaks form a gap on the cylinder surface along its forming line. The ratio of ellipse axes being from 0.7 to 0.9 and the distance between spires being of 1.5–2.5 mm. If the corrector is installed in such a way that the vein diameter is decreased only in the direction perpendicular to the joining line of free ends of the valve cusps, it allows the improvement of the vein blood flow with a relatively reduced decrease of the vein section than in the case of circular cross-section correctors [53–55]. The gap in the cylinder surface allows vein walls to better adapt to the internal surface of the corrector and prevent the contortion of vein walls.

Another modification of this corrector may be a framework made of a perforated plate with the same ellipse's axes ratio (Fig. 5). To avoid possible venous traumas in the case of wire corrector, sharp spiral tips form closed loops located on the framework surface. The same result in the case of the perforated plate framework is reached by the plate edges rounding.

The plate must be sufficiently openwork, i.e. the ratio of the plate area occupied by perforations to the whole plate area must vary between 0.3 to 0.7. The framework wire diameter and the plate thickness chosen provide the framework stability required to sustain the intravenous pressure.

A wire of a diameter smaller than 0.3 mm does not provide the framework stability required to sustain the intravenous pressure; furthermore, the application of a wire with a diameter greater than 0.6 mm is not recommended either

Fig. 5a, b. An extravasal valve corrector made in the form of a hollow elliptic cylinder with a surface unclosed along its forming line, formed by the plate with round perforations (**a**) and longitudinal perforations (**b**)

because of complications during the correctors' manufacturing. If the wire diameter exceeds 0.6 mm, corrector flattening prior to installation can provoke its plastic deformation and thus an incomplete shape restoration after installation.

An inter-spire distance chosen between 1.5–2.5 mm and an appropriate ratio of the perforated area versus the total area of the corrector (0.3–0.7) ensures the normal external vein the rapid revascularization of the vein wall, while preventing its ingrowth between spires. This corrector's functionality is based on the shape memory effect. The section of the vein with incompetent valve is uncovered and the vein in the incompetent valve domain is mobilized. Prior to the installation on the vein, the corrector is quasi flattened at a temperature lower than the implantation one (from –10°C to +20°C; Fig. 6). In such conditions, the corrector is transported to the injured area of the vein and orientated on the vein so as to provide the vein diameter reducing in the direction perpendicular to the joining line of valvular leaf edges. Upon contact of the corrector with the human body, it heats up and "recalls" its initial ellipsoidal shape, surrounding the vein, and remains fixed on it correcting its cross-section in the appropriate place.

Thus, the described extravasal corrector allows to achieve adequate correction of a function of vein valves, to improve venous blood flow with a relatively lower decrease of the vein mobilization at its long extension in the case of use of the said corrector. It lowers trauma possibility and in some cases, when the direction of free edges of valve cusps is sagittal, a partial mobilization of the vein is possible (without its back wall). Moreover, operation time is radically shortened: the procedure of the corrector placing takes several seconds.

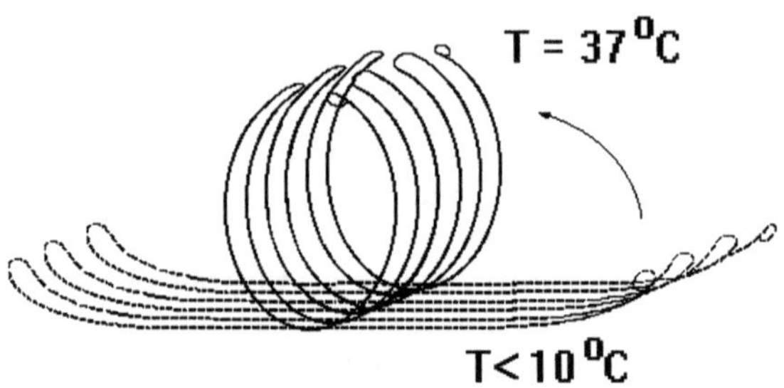

Fig. 6. The schematic representation of the corrector deformation while cooling and further shape restoration while being brought to the vein

4
Clinical Approving and Effectivity of the Nitinol Shape-Memory Extravasal Correctors

For the estimation of the efficacy of extravasal correction by the elaborated shape memory corrector having elliptic cross-section, a comparative clinical investigation has been carried out. 246 patients suffering of a varicose disease were investigated and operated at the General Surgery of the Russian State Medical University (Moscow) between 1992–1997. 180 patients (73.2%) were of age 30 years to 59 years, i.e. of the most able-bodied age. Note a distinct prevalence of women that corresponding to the data of other investigators [5]. The overwhelming majority of patients (73%) were of more than 10-year-long disease. The character of clinical manifestations of the varicose disease and its complications are shown in Table 4.

The patients subjected to extravasal correction were distributed in two clinical groups: (1) the main group in which the correction was performed using the new nitinol corrector (since 1995) and (2) the control group in which the correction was performed using traditional lavsan spiral of circular cross-section. The majority of operations in the main group of patients had been performed using the correctors having the shape of a rolled sinusoid (excluding the first three cases where the nitinol spiral of elliptic cross-section had been used). Comparative distribution of patients by sex and age is shown in Table 5.

Table 4. Localization of damage and its character in varicose patients

Number	Clinical manifestations of varicose disease and its complications	Number of cases	%
1	Varicose expanding of veins in both LSV system	207	84.2
2	Varicose expanding of veins in SSV system	8	3.3
3	Varicose expanding of veins in both LSV and SSV system	22	8.9
4	Varicose veins of lateral surface	8	3.3
5	Perforans insufficiency	235	95.9
6	Hyperpigmentation of skin and lipoderma to sclerosis	97	39.4
7	Allergic dermatitis, weeping eczema	41	16.7
8	Chronic ulcers of the leg	67	27.2
9	Acute trombosis of varicose veins	33	13.4
10	Bleeding from the varicose veins	6	2.4

LSV, long saphenous vein; SSV, short saphenous vein

Table 5. Distribution of varicose disease patients operated using extravasal correctors by sex and age

Patient's age (years)	Main group (nitinol corrector)		Control group (lavsan spiral)		All
	Men	Women	Men	Women	
30–39	1	5	5	12	23
40–49	3	7	3	20	33
50–59	5	8	5	12	30
60–69	2	7	2	6	17
>70	–	3	–	1	4
All	11	30	15	51	107
	41			66	107

Table 6. Character of surgical operations

Number	Kind of operation	Operated patients Number	%
1	Striping of LSV and epifascial perforator ligation	68	24.5
2	Striping of LSV and SSV	7	2.5
3	Striping of LSV and Linton procedure and resection of varicose vein tibialis postoperatively	12	4.3
4	Striping of LSV and autovenous occlusion of varicose vein tibialis postoperatively	45	16.3
5	Striping of LSV and EVC by lavsan spiral	71	25.6
6	Striping of LSV and EVC by nitinol corrector	41	16.7
7	Linton procedure and skin grating	19	6.9
8	Crossectomy	4	1.4
9	Crossectomy and thrombectomy	5	1.8
10	Valve transplantation	2	0.8
11	EVC of popliteal vein by lavsan spiral	1	0.4
All		277	100

EVC, extravasal valve correction; LSV, long saphenous vein; SSV, short saphenous vein

Periods of the varicose disease in main and control groups are comparable to each other and exceed 10 years for the majority of patients (69%). Complications of the varicose disease in the form of dermatitis, lipodermatosclerosis, pigmentation in the gaiter area, chronic varicose uncers, throbophlebits, etc. are noticed for the main and control groups with a comparative frequency. Therefore, as seen from the above, both patient groups are comparable and representative.

Solving of diagnostic problems, distribution of patients in groups, and study of treatment results were performed on the basic of anamnesis, clinical examination and experimental study of regional venous blood flow. The following methods of study were used: radionuclide phleboscintigraphy, and X-ray contrast phlebography investigation were performed on all the patients.

All 246 patients were operated. The character of the surgical operations is shown in Table 6.

The number of operations exceeds the number of patients because for several patients underwent a two-stage surgical treatment. During the 1-month post-operation period, a control phlebography investigation of patients has been performed.

In the main group of 41 patients treated with the new extravasal nitinol corrector, the retrograde phlebography showed a perfect valve correction for 35 patients (85.4% of patients). Four patients showed a reduced blood reflux and two patients show unchanged reflux.

In the control group of patients treated with lavsan spiral, 46 patients were investigated. 37 patients (80.4%) showed valve sufficiency and nine patients (19.6%) showed remaining pathological reflux. In this group, ten patients (21.7%) demonstrated a dependent edema in the ankle region. It is probably connected with to an excessive vein contraction which impeded blood flow. In the main group of patients, dependent edema in the ankle region was noted for two patients (4.9%).

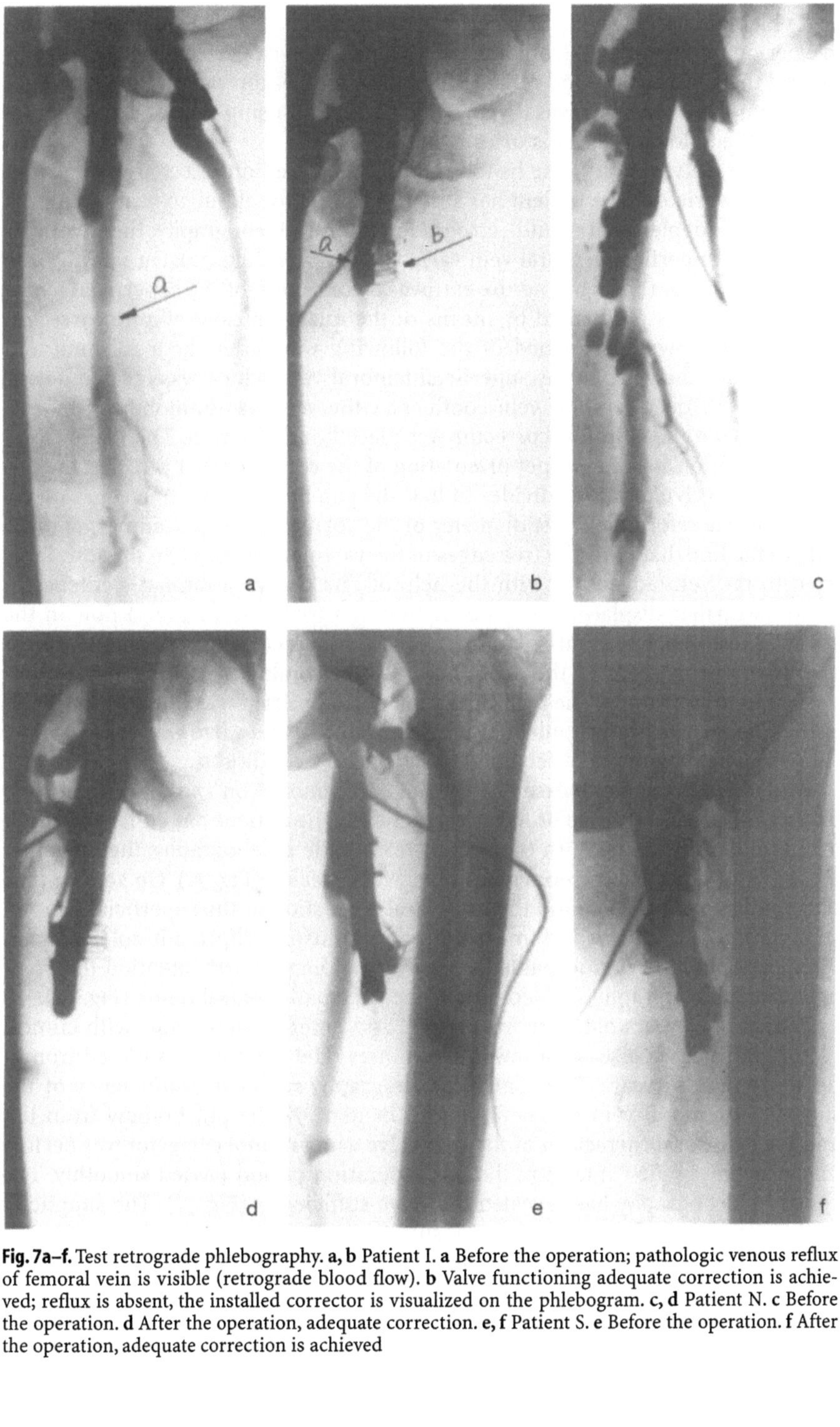

Fig. 7a–f. Test retrograde phlebography. **a, b** Patient I. **a** Before the operation; pathologic venous reflux of femoral vein is visible (retrograde blood flow). **b** Valve functioning adequate correction is achieved; reflux is absent, the installed corrector is visualized on the phlebogram. **c, d** Patient N. **c** Before the operation. **d** After the operation, adequate correction. **e, f** Patient S. **e** Before the operation. **f** After the operation, adequate correction is achieved

A phlebography investigation performed after 6 months to 3 years post-operative period of ten patients from the main group having a perfect valve correction showed the absence of blood reflux and maintaining of the valve sufficiency. Specific examples of the extravasal vein valve correction using nitinol can be illustrated by the following case histories.

Patient I, 62 years old, case history No. 2474, entered on 07.02.96 with a right lower limb varicose. The patient has suffered from it for about 20 years. With the help of the dopplerometry and retrograde femoral phlebography, the incompetence of the superficial femoral vein valve was revealed (Fig. 7a). On 14.02.96, the phlebotomy from the right and the extravasal correction of the superficial femoral vein valve was performed by means of the quasi-sinusoid elliptic corrector. The correction was performed in the following way: after distinguishing and visualizing of the valve in the superficial femoral vein, which was located downstream of the deep femoral vein confluence, the vein was immobilized by two turnstiles, and the elliptical corrector was placed onto the vein. The major point of the operation was the proper orientation of the corrector in a way that the joining line of valvular leaves divides in half the gap between the free edges of the corrector. Therefore the short diameter of the corrector was placed perpendicularly to the line that joins the free edges of the valvular leaves. Then the spiral was fixed to the vein adventitia with the help of one or two atraumatic stitches to avoid its further displacement. After removing turnstiles, the blood flow in the vein was restored. It was sufficient in this case to correct in such a manner at least one valve downstream of the deep femoral vein confluence, which was accomplished in this example. The post-operation period passed without any particularities. The post-operative phlebography showed that the valve was working well (Fig. 7b). The patient was discharged in satisfactory conditions.

Patient N, 57 years old, case history No. 4985, entered on 17.03.97 with clinical picture of varicose disease of lower extremities. The patient has suffered from it for more than 10 years. With the help of retrograde phlebography, the incompetence of the superficial femoral vein valve was revealed (Fig. 7c). On 20.03.97, the phlebotomy from the left and the extravasal correction of the superficial femoral vein valve was performed by means of quasi-sinusoid elliptic nitinol corrector. The post-operation period passed smoothly. Healing by first intention the post-operation phlebography showed the absence of a pathological reflux (Fig. 7d).

Patient S, 39 years old, case history No. 2117, entered on 03.02.97 with clinical picture of varicose disease of lower extremities. The patient has suffered from it for more than 10 years. The control phlebography revealed insufficiency of the superficial femoral vein valves (Fig. 7e). On 16.02.97, the phlebotomy from left and the extravasal correction of the vein valve using nitinol corrector was performed. Healing by first intention the post-operation period passed smoothly. The control phlebography has revealed the valve sufficiency (Fig. 7f). The simplicity of the correction of the function of insufficient vein valves, low trauma possibility, and the achievement of the adequate correction at lower decrease of the venous cross-section area, make a great perspective for a more extensive application of this method in clinical practice.

References

1. Kuzin MI, Anitchkov MN, Zolotarev VY, et al. (1984) Improvement of surgical treatment in case of the varicose disease. Surgery 5:152–153
2. Lisitzin KM, Revskoi AK, Juraev TJ (1986) Epidemic vein diseases of the lower limbs under different climatic and geographic conditions. Surg News 137:71–72
3. Pokrovski AV (1979) Clinical angiology. Medicine, Moscow
4. Savelev VS (1984) Pathogenesis and pathogenetic therapy of the varicose disease of inferior limbs. Surgery 5:152–155
5. Oshiro T (1986) Varicose veins of the leg in Japan. In: Ninth World Congress of Phlebology. Kyoto, p 64
6. Hachen HJ, Lorenz P (1982) Klinische und photopletysmographishe Doppelblind-Storungen der Mikrozirkulation. Angiology 33:480–486
7. O'Donnel TF jr, McEnroe CS, Heggerick P (1990) Chronic venous insufficiency. Surg Clin North Am 70:159–180
8. Winkler M (1990) Social medicine importance and treatment strategy of chronic venous disease. Z Gesamte Inn Med 45:144–145
9. Goldman MP, Mitchel P (1990) Solerotherapy treatment for varicose and telangiectatic veins in the United States: past, present and future. J Dermatol Surg Oncol 16:606–607
10. Lukitch GI, Baikova ZZ, Lipnitzki EM (1977) Treatment of trophic ulcers of the venous etiology by combutek. Mod Med 3:97–99
11. Baker SR, Stacey MC, Jopp-McKay AG, et al. (1991) Epidemiology of chronic venous ulcers. Br J Surg 76:864–867
12. Castro SM (1995) Chronic venous insufficiency of the lower limbs and its socio-economic significance. In: Negus D, et al. (eds) Phlebology '95. Proceedings of the 12th World Congress. 3–8 September, London. pp 23–24
13. Vasutkov VY (1986) Choice of the treatment for trophic ulcers of knee among the patients having chronic venous insufficiency of the lower limbs. Surgery 10:103–108
14. Massi S, Ruggieri C, Botta G, et al. (1995) Leg ulcers. Casuistry from January 1985 to December 1994. In: Negus D, et al. (eds) Phlebology '95. Proceedings of the 12th World Congress. 3–8 September, London. pp 761–763
15. Smith PC, Sarin S, Hasty J, Sourr J (1990) Sequential gradient pneumatic compression enhances venous ulcer healing: a rando-mixed trial. Surgery 108:871–875
16. Konstantinova GD, Kartashev VB (1986) Long-term results of radical venous ectomy with correction of the valve insufficiency of the deep veins. Surgery 12:51–54
17. Bemmelen PS, Bedford G, Beach K, Strandness DE (1991) Status of the valves in the superficial and deep venous system in chronic venous disease. Surgery 109:730–734
18. Konstantinova GD, Bogdanov AE (1990) The modern aspects of treatment for the chronic vein diseases. Ter Arkh 10:125–128
19. Vedenski AN (1983) Varicose disease. Medicine, Leningrad
20. Jarikov VI (1984) Extravasal correction of valves of deep veins and dissection of perforal veins in surgical treatment of the varicose disease. Thesis, Gorki University, Gorki
21. Savelev VS, Dumpe EP, Yablokov EG (1972) Diseases of the main veins. Medicine, Moscow
22. Johnson WG, Burnham S (1986) Vascular surgery. Moore, Miami pp 1093–1138
23. Bradbury AW, Ruckely CV (1993) Foot volumetry can predict recurrent ulceration after subfascial ligation of perforators and saphenous ligation. J Vasc Surg 18:789–795
24. Nachbur B (1984) Surgical treatment of venous leg ulcers. Ther Umsch 41:873–877
25. Hopkins NFG, Wolfe JHN (1992) Deep venous insufficiency and occlusion. BMJ 304:107–110
26. Zelenin RP, Kurakov NP (1979) The question of surgical correction of the valvular insufficiency. Surgery 9:27–29
27. Zhang K, Liang F (1993) Autovenous sleeve around superficial femoral venous valve to treat primary deep venous incompetence. In: 35th World Congress of International Society of Surgery. Hong Kong
28. Askerchanov RG (1984) Choice of the treatment for the first venous varicose of the lower limbs. Surg News 6:40–43
29. Raju S, Jackson S (1991) Valvuloplasty in chronic venous insufficiency: a worthwhile procedure? Vasa 33(Suppl):42–43
30. Belokonev EV (1983) Extravasal correction of the valvular insufficiency of deep veins for the tretment of varicose of the lower limbs. Thesis. Leningrad University, Leningrad
31. Tzukanov UT (1985) Ectasia correction of the femoral vein. Surgery 6:59–62
32. Zuev NS, Sedlovskaya ON, Zuev LN, Lobanova IG (1986) Long-term results of the surgical treatment for several kinds of veinous insufficiency of the lower limbs. In: Regional Conference of Surgeons. Tumen, pp 40–42

33. Kistner RL (1975) Surgical repair of the incompetent femoral vein valve. Arch Surg 110:1336–1342
34. Moore DJ, Himmel PD, Sumner DS (1986) Distribution of venous valvular incompetence in patients with the postphlebitic syndrome. J Vasc Surg 3:49
35. Raju S (1983) Venous insufficiency of the limb and stasis ulceration. Ann Surg 197:688–697
36. Raju S, Fredericks R (1988) Valve reconstraction procedure for nonobstructive venous insufficiency: rational, technique and results in 107 procedures with two to eight years follow-up. J Vasc Surg 7:301–310
37. Strandness DE Jr, Thiele BL (1981) Selected topics in venous disorders. Futura, New York, p 186
38. Eriksson I, Almgren BA (1986) Influence of the profunda femoris vein on venous hemodynamics of the limb. J Vasc Surg 4:390–395
39. Taheri SA, Pendergast DR, Lazar E, et al. (1985) Vein valve transplantation. Am J Surg 150:201–202
40. Sottiurai VS (1991) Surgical correction of reccurent venous ulcer. J Cardiovasc Surg (Torino) 32:104–109
41. Perrin M, Hiltbrand B, Bayon JM, Calvignac JL (1995) Valve repair in deep veins of the lower limb. Techniques, indications and results. In: Negus D, et al. (eds) Phlebology '95. Proc of the 12th World Congress. 3–8 September, London. pp 983–885
42. Bergan JJ (1985) Overview of potential benefit of direct venous reconstruction. Int Angiol 4:463–466
43. Kistner RL (1990) Surgical technique of external venous valve repair. Straub Found Proc 55:15–16
44. Prokoshkin SD, Ryklina EP, Khmelevskaya IU, Sytchev PA, Chadaev AP, Butkevitch AZ (1995) Shape memory titanium nickelide device for extravasal correction of the function of main vein valves. In: Proceedings of the First Russian-American Seminar and 31st Seminar on Actual Problems of strength 13–17 November, St. Petersburg. pp 55–59
45. Borschberg E (1967) The prelevance of varicose veins in the lower extrimities. Karger, Basel
46. Bernstein EF (1986) Future prospects in the treatment of venous disease. World J Surg 10:959–967
47. Bagev II (1982) Errors and dangers during the operative treatment of varicose of the lower limbs' veins. Surg News 128:141–143
48. Dumpe EP, Ukhov UI, Shvalb UI (1982) Physiology and pathology of venous blood circulation of the lower limbs. Medicine, Moscow
49. Khorochaev VA (1981) Vein morphology in cases of varicose and venous stagnation. Thesis. Moscow University, Moscow
50. Ukhov UI (1979) Mechanism of reactive and pathologic reconstruction of the valves. In: Dumpe EP (ed) Development, morphology and plasticity of the venous canal under normal, pathological and experimental conditions. Medicina, Moscow
51. Tchitchinadze NA (1975) Some particularities of the valves of veins. In: Proceedings of the First Caucase Conference of Morphologists. Tbilisi, pp 254–256
52. Leu HJ (1990) Chronisch-venose insuffizienz heute (eine Standortbestimmung). Vasa 19:195–202
53. Prokoshkin SD, Khmelevskaya IYu, Ryklina EP et al. (1998) Device for the extravasal correction of the function of main vein valves. Russian Federation patent No. 2102016, Russian Patent Agency, Moscow
54. Chadaev AP, Butchevitch AC, Prokoshkin SD et al. (1999) Method of extravasal corection of the function of main veins and device for its realization. Russian Federation patent No. 2128966, Russian Patent Agency, Moscow
55. Prokoshkin SD, Khmelevskaya IYu, Ryklina EP et al. (1997) Extravasal correctors of main vein valves function and experience of their application. Proceedings of Second International Conference on Shape Memory and Superelastic Technologies, SMST-97, Pacific Grove, SMST, p 613–616

Large-Caliber NiTi SMA Stents and Stent Grafts

John D. Pazienza, Willard Hennemann

1 Introduction

To help stabilize conduits within the human body, implantable endoluminal stents and stent-grafts have been developed. These devices serve as scaffolds to the surrounding tissue and/or create new fluid conduits acting in both palliative and curative functions. Stents (Fig. 1) are best known for their use in maintaining vessel patency in the coronary arteries following balloon angioplasty, but are also used throughout the peripheral vasculature (carotid, renal, iliac, femoral, and popliteal arteries) and other non-circulatory conduits (esophagus, ureter, colon, and biliary tract). Stent-grafts (Fig. 2) are a more recent development which have combined the features of endoluminal stents with those of surgically implantable fabric grafts (woven or knitted polyester or PTFE) to provide both a scaffold to the existing tissue and an intact replacement lumen. Stent-grafts are best known for their use in the repair of abdominal aortic aneurysms (AAA), but they are

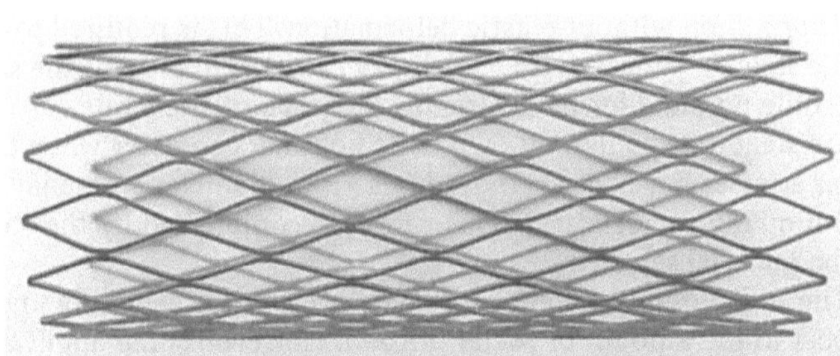

Fig. 1. Large-caliber NiTi stent

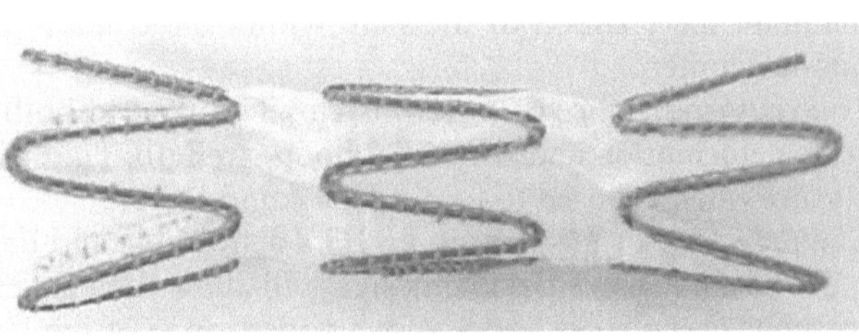

Fig. 2. Large-caliber Large-caliber NiTi stent graft

also used to treat arterial occlusive disease, arterio-venous fistulae, thoracic aortic aneurysms (TAA), peripheral aneurysms, pseudoaneurysms, and dissections [26]. The majority of available stents are stainless steel designs, which require a balloon to inflate from within the stent to plastically deform the metal and transform the stent to its expanded, deployed state. However, large caliber (>6 mm) stents are one type of implant which have effectively utilized the unique properties and features of NiTi shape memory alloys (SMA) to provide self-expanding designs with attractive performance characteristics.

2
Design Constraints

In their efforts to provide a scaffold, self-expanding endoluminal stents are bound by general and application specific design constraints [43].

- As for all metallic implants, the stent must be *biocompatible*, fully functional after *sterilization* (γ irradiation, ethylene oxide), have a low susceptibility for *corrosion* (general, pitting, crevice, galvanic), and be capable of withstanding the anticipated *monotonic and cyclic loading*. Stents are designed for a minimum implant life of ten years (a vascular stent may be exposed to 10 years at 76 beats per minute; approximately 400 million cycles).
- Stents should be at least minimally *radiopaque* to allow the implanting physician to visualize the device during positioning, deployment, and follow-up. Additionally, stents must be constructed of materials, which are not affected by the strong magnetic fields experienced during *magnetic resonance imaging* (MRI). Stents made from ferromagnetic and paramagnetic materials may be susceptible to movement, resistive heating, and create image artifacts when exposed to MRI [37].
- The self-expanding endoluminal stent must be capable of *large scale elastic deformation* without plastic deformation. This is required to radially compress the stent, in concert with its delivery catheter, to provide the smallest geometric profile possible for navigating to the implant site while subsequently allowing the stent to expand on it own to its nominal geometry when the external catheter stresses have been removed (a 24 mm diameter stent may be compressed to 6 mm, 18F, for delivery). This can be accomplished by the stent design and/or the material selected.
- The *radial hoop strength* exerted by the stent is generally maximized with respect to the amount of metal utilized. However, some applications may seek to match the radial hoop strength of the stent to the compliance of the conduit being supported, but most applications require high hoop strengths in order to maintain the position of the stent within the conduit and provide the largest possible lumen.
- The *crush resistance* of a stent is closely linked to both the capacity for large scale deformation and the radial hoop strength. The stent should resist lumen closure when exposed to focal radial compression.
- The *flexibility* of the stent in both its deliverable and deployed state should be maximized. A more flexible design will allow the delivery catheter to navigate tortuous vasculature and access a wider range of implant sites while allowing

the stent to conform to a greater degree of angulation and maintain the cross-sectional area of the lumen without kinking.
- Depending upon the clinical application, the stent should possess *column strength* to prohibit geometrical deformation and lumen closure when exposed to longitudinal compressive loading.
- The stent should be designed to minimize *shortening* (stent length change as a function of stent diameter) in order to provide the greatest accuracy during positioning and deployment.
- The *metallic surface area* of a stent, relative to the cylinder that it encompasses in its deployed state, must be balanced with respect to the design features described above and also the biological response. Some applications may require higher percentages of metal to prevent tissue ingrowth (stenting in a cancerous biliary duct) while others will minimize the percent metal to reduce neointimal formation (stenting in arterial vasculature). Typical designs range from 5–30% metal.

3
Review of NiTi SMA Stent Designs

NiTi shape memory alloys have been utilized to address the design constraints of large caliber, self- expanding endoluminal stents [30]. The greatest benefit of NiTi SMA compared to stainless steels or other typical Hookian implant metals is derived from the austenite/martensite transformation which can produce superelasticity and shape memory.

Both superelasticity and the shape memory effect allow large diameter stents to be compressed to relatively small diameters for endoluminal delivery via a catheter. Some stent designs exploit the inherent superelasticity, which is present at temperatures near the austenitic transformation, by inducing large strains to compress the stent for delivery (forming stress induced martensite) and then removing those strains during deployment with no permanent deformation. Other stent designs exploit the shape memory effect by presenting the stent in a deformed (martensitic) state for delivery and allowing the stent to transform to its "programmed" (austenitic) shape after exposure to the implant environment (greater than the austenitic transformation temperature, A_f).

NiTi stents also present the possibility of exhibiting a low modulus (martensite) flexible behavior during navigation to the implant site and subsequently a high modulus (austenite) stiff behavior after deployment. The superelastic properties of NiTi make it especially suitable for stents designed to resist crushing (carotid artery and superficial femoral artery, SFA). Large deformations can be absorbed with little or no permanent deformation upon removal of the load due to the formation of stress induced martensite. NiTi stents are typically implanted in an oversized condition to provide diametrical interference, which helps to maintain the position of the stent and maximize the radial force applied to the vessel wall. The stent will continue to exert a load against the surrounding tissues until it has returned to its original "programmed" shape. The self-expanding nature of NiTi stents allows them to change shape with changes in the implant site morphology.

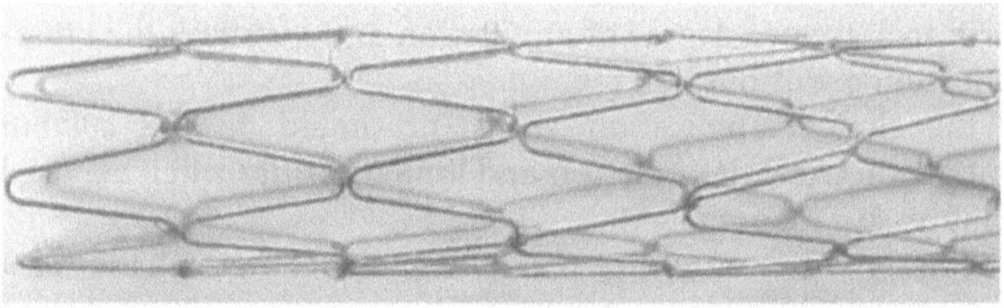

Fig. 3. Large-caliber Typical bent-wire stent design

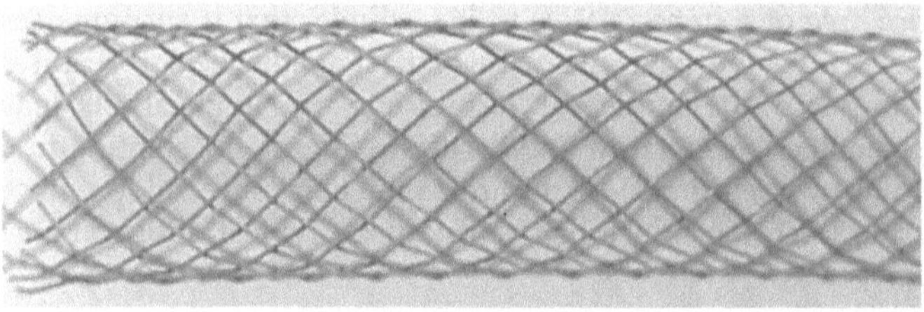

Fig. 4. Large-caliber Typical braided-wire stent design

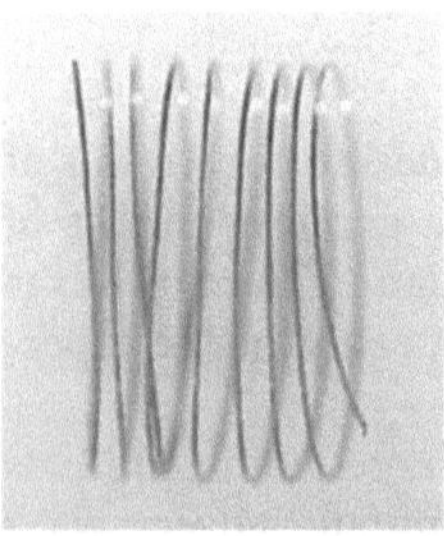

Fig. 5. Large-caliber Typical coiled-wire stent design

Five basic NiTi SMA stent designs have been developed as depicted in Figures 3–7: bent wire, braided wire, coiled wire, laser cut tube, and laser cut sheet. NiTi is most readily available in wire form, but tubing is also available in diameters up to 6 mm and hot or cold rolled sheet is becoming more prevalent. Of the three forms of NiTi, the mechanical properties of wire are the most homogeneous and best documented. Tubing will exhibit anisotropy with respect to the drawing axis, but it can be beneficial in encouraging the stent to maintain its cylindrical shape when radially compressed. Sheet is anisotropic with respect to the rolling direction and may require additional processing to provide a suitable surface finish.

- *Bent wire* stents are made by forming round or flat wire into a repeating sinusoidal pattern that is subsequently formed into a tubular shape by wrapping into a helix, suturing together, crimping together, or welding together. The sinusoidal geometry, the number of repeating sinusoidal units, the wire geometry, and the connectivity of the successive rings can be altered to create stents with different properties. These stents typically provide moderate flexibility, moderate radial hoop strength, and minimal foreshortening.
- *Braided wire* stents are made by braiding multiple round or flat wires over a core. The wires form a helix about the core with the clockwise oriented wires

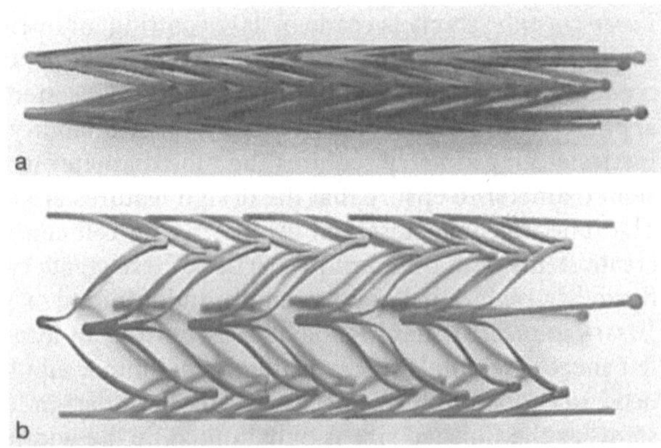

Fig. 6a,b. Large-caliber Typical laser-cut tube stent design. **a** Compressed. **b** Expanded

intertwined with the counter-clockwise oriented wires. The number of wires, wire geometry, braid pattern, and braid angle can be altered to create stents with different properties. These stents typically provide excellent flexibility, but also feature reduced radial hoop strength and extensive foreshortening.

- *Coiled wire* stents are made by winding single or multiple round or flat wires around a core in a helical pattern. The number of wires, wire geometry, and pitch of the helix can be altered to create stents with different properties. These stents typically provide the best flexibility, may have high radial hoop strength, but are susceptible to collapse of the coil and thus the lumen under certain load conditions, and extensive foreshortening.

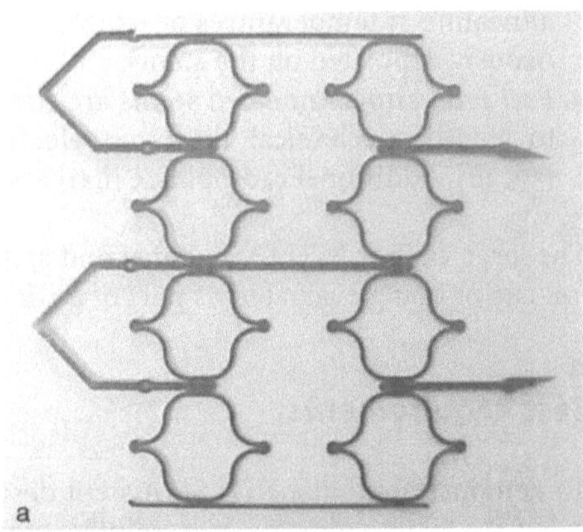

Fig. 7a,b. Large-caliber Typical laser-cut sheet stent design. **a** Flat. **b** Tubular

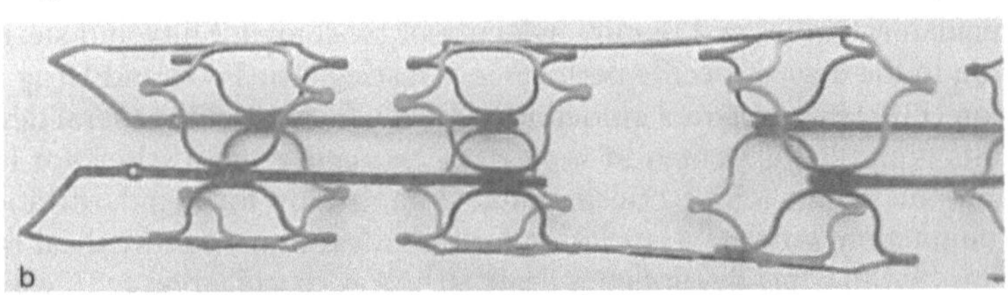

- *Laser-cut tube* stents is made by laser cutting an intricate array of "cells" into a small diameter tube. The cellular array is then expanded to the desired geometry and annealed to "program" the shape. This stent design is limited in its application to large caliber stents by the availability of large diameter tubes. Intricate cell geometry requires the tube diameter to closely match the desired stent diameter to ensure that the design features are not lost during expansion. The tube wall thickness, cell geometry, and cell connectivity can be altered to create stents with different properties. These stents typically provide moderate flexibility, high radial hoop strength, and minimal or no foreshortening.
- *Laser-cut sheet* stents is made by laser cutting an intricate array of "cells" into a flat sheet. The cellular array is then rolled into a tubular configuration and welded or crimped to form the stent. This stent design lends itself to large caliber stents as the finished size is only limited by the width of the rolled sheet. Intricate cell geometry can be cut from the sheet and retained as the sheet is formed into a stent without any required deformation. The sheet thickness, cell geometry, and cell connectivity can be altered to create stents with different properties. These stents typically provide moderate flexibility, high radial hoop strength, and minimal or no foreshortening.

Manufacturing of NiTi SMA stents typically occurs with four general steps:

1. *Material preparation.* The NiTi material is prepared for subsequent operations by such processes as grinding, pickling, ultrasonic cleaning, and annealing.
2. *Stent forming.* Stents are created using such processes as wire forming, braiding, laser cutting, deburring, grit blasting, annealing, laser welding, expanding, suturing, and ultrasonic cleaning.
3. *Final annealing.* The raw stents are "programmed" to their final shape by annealing at temperatures of ~500°C (a specific oxide coating may be simultaneously deposited on the stent).
4. *Post processing.* Annealed stents are completed using such processes as cutting to length, mechanical tumbling, electropolishing, and chemical passivation [39, 40]. Additional radiopaque markers may be added at this time.

The large caliber NiTi SMA stents and stent-grafts, which are currently available for sale or being evaluated as part of a clinical study, are described in Table 1.

4 Test Requirements

To demonstrate that a NiTi SMA stent design is suitable for clinical applications, extensive testing must be successfully completed prior to regulatory approval. All regulatory bodies will require evidence of biocompatibility and sterility in addition to the device specific performance testing. The Food and Drug Administration (FDA) has prepared guidelines [15, 16] which provide general descriptions of tests typically performed in support of regulatory submissions for intravascular and biliary stents. The International Organization for Standardization (ISO), in conjunction with the Association for the Advancement of Medical Instrumentation (AAMI), has developed a draft technical specification [22], which provides

Table 1. Large-caliber NiTi SMA stents and stent grafts

Stent/ stent graft	Manufacturer	Material	Design	Application	Relaxed diameter	Deliverable diameter	References
AneuRx	Medtronic	Tube	Laser-cut tube; diamond pattern	AAA	12–16 mm 20–28 mm	16F 21F	[9] [46]
Excluder	Gore and Associates	Round wire	Bent wire; wire formed into zig-zag rings; rings connected with strands of ePTFE	AAA	23 mm	18F	NA
				TAA			
Expander	MediCorp	.006–.007" round wire	Braided wire; 24 wires; one-over-one braid pattern; braid angle <90°	Peripheral	6–12 mm	7–8F	NA
Hemobahn	Gore and Associates	Round wire	Bent wire; wire formed into zig-zag rings	Peripheral	6–13 mm	8–12F	[2]
			Rings connected with strands of ePTFE				[11]
Integra	Boston Scientific	Tube	Laser-cut tube	Carotid	NA	7.5F	NA
Memotherm	Bard	Tube; polished	Laser-cut tube; diamond lattice	Aortic	14–20 mm	10F	[29]
				Peripheral Biliary TIPS	4–12 mm 7–10 mm 10–14 mm	7–8F 7F 10F	[38]
Passager MIBS	Boston Scientific	.009–.015" round wire (shape-memory effect)	Bent wire; wire formed into zig-zag rings	Peripheral	10–12 mm	10F	[8]
			Rings connected using polypropylene sutures				
Sinus-Stent	Optimed	Tube; polished	Laser-cut tube; sine-wave pattern	NA	NA	NA	NA
SMART	Cordis	Tube (shape-memory effect; superelasticity)	Laser-cut tube; segmented	Carotid, biliary	6–10 mm	NA	NA

Table 1. Continued

Stent/ stent graft	Manufacturer	Material	Design	Application	Relaxed diameter	Deliverable diameter	References
Stenway	Stenford	Round wire	Bent wire; wire formed into zig-zag rings; rings connected using polypropylene sutures	Peripheral, biliary	6–12 mm;	NA	NA
				AAA, TAA	22–32 mm		
Symphony	Boston Scientific	.010–.015" round wire (superelasticity)	Bent wire; wire bent and welded to form hexagonal cells	Peripheral	6–14 mm	7F	NA
Talent	Medtronic	Round wire; oxide coating for improved radiopacity, corrosion resistance, and biocompatibility	Bent wire; wire springs in a serpentine pattern with wire bars connecting rings to prevent shortening and add column strength; loose wire ends crimped together	AAA,	18–36 mm;	18F or 24 F	[13, 19, 36, 41]
				TAA	As large as 46 mm	27F	
Vanguard II EAG	Boston Scientific	.009–.015" round wire (shape-memory effect)	Bent wire; wire formed into Bzig-zag rings; rings connected using polypropylene sutures	AAA	22–30 mm	18F	[1, 12, 18, 27, 33, 42]
IntraCoil	Intratherapeutics	.010–.013" round wire (superelasticity)	Coiled wire; helical coils with terminal balls	Peripheral	4–8 mm	7–9F	[10]

NA, not available

Table 2. NiTi self-expanding stent-test requirements

Test description	FDA (intra-vascular)	FDA (biliary)	ISO/ AAMI (vascular)	CEN (arterial)	Design/ application specific	Standardized test method
% Metal	X		X	X		
Chemical composition/ materials	X	X				ASTM (draft)
Compression force/ local compression		X	X			
Corrosion/oxidation	X	X	X	X		ASTM (draft)
Crush resistance			X	X		
Dimensional	X	X	X	X		
Durability			X			
Fatigue FEA/ stress–strain analysis	X		X	X		
Fatigue in vitro	X		X	X		
Magnetic resonance imaging compatibility	X		X	X		
Migration resistance			X	X		
Radial strength/radial force/hoop strength/ expansion force	X	X	X	X		
Radiopacity			X	X		
Surface-scanning electron microscopy/ visual	X	X		X		
Shortening	X		X	X		
Sizing study	X					
Tensile strength/ elongation	X	X	X	X		
Uniformity after deploy-ment/conformability	X	X	X	X		
Usage/deployment		X	X	X		
Metallurgy					X	
Thermal analysis (M_s, M_f, A_s, A_f)					X	ASTM (draft)
Flexibility					X	
Column strength					X	
Joint strength (welds, crimps)					X	
Kink angle					X	
Torsional resistance					X	
Erosion/wear					X	

general descriptions of the tests typically performed to demonstrate safety of vascular stents. The European Committee for Standardization (CEN) has developed one general standard (EN 14630) and has drafted two more specific standards (prEN12006-3 [31], prENxxxxx [32]) which provide general descriptions of the tests typically performed to demonstrate safety of arterial stents. The American Society for Testing and Materials (ASTM) has drafted standards and specifications which address the corrosion resistance of implantable medical devices [4], the chemical composition, thermal analysis, and terminology of NiTi alloys [5–7] and are considering the development of additional test standards for evaluating the properties of coronary stents. Table 2 provides a cross-reference of the tests typically specified, as interpreted for NiTi SMA stents. As it is relatively early in the life cycle of stents, the various standards organizations have not yet developed standardized test methods for the tests that they have described. Each stent manufacturer is responsible for developing specific test methods to address the test requirements, which have been outlined.

5 Clinical Applications

Currently, stenting is used to treat two main disease conditions in the body:

- *Occlusive disease* in which an atherosclerotic plaque (vascular disease) or a tumor (biliary and esophageal applications) is impinging upon the lumen of a conduit and restricting the flow of its contents, often with serious clinical consequences. Clinically significant occlusive disease is most common in the coronary, renal, carotid, and iliac arteries.
- *Aneurysmal disease* in which a diseased vessel wall has weakened to the point that the wall is no longer able to resist the pressure within and begins to gradually expand outwards (like a balloon) often until it ultimately ruptures. Although aneurysmal disease can be found in virtually any arterial system, the most common clinically significant aneurysmal disease is found in the abdominal aorta (AAA) and thoracic aorta (TAA).

Management of occlusive disease in which the stent is used to physically scaffold the vessel opens and allows unrestricted flow within the lumen is by far the most common clinical application for stenting. Clinical data are available for certain clinical indications including coronary arteries [17, 23, 35], coronary vein grafts [34], carotid arteries [3, 45], and renal arteries [21] indicating that stents reduce the risk of vessel closure and improve clinical outcomes acutely and/or long-term.

Management of aneurysmal disease is projected to become the fastest growing indication for stenting. In the case of aneurysmal disease, a stent graft provides a new conduit for the contents of the vessel (blood) to flow, bypassing the dilated aneurysmal section of the wall, and avoiding the risk of rupture. In order for the risk of rupture to be eliminated, the stent graft must completely exclude the aneurysm and prevent the transmission of pressure to the aneurysmal portion of the vessel wall.

Although plastically deformable, balloon-expandable stents currently dominate certain clinical applications, including coronary and renal applications, the deliverability and compression-resistance of self-expanding designs offer significant clinical advantages in many indications including AAA, carotid, and SFA stenting. In addition, the ability to cover long arterial segments with one stent (as opposed to deploying sequential overlapping stents) has made self-expanding designs popular for iliac artery and biliary applications [20]. The earliest self-expanding designs (and still the dominant design in most peripheral indications except AAA) were constructed of Elgiloy or stainless steel wires in a bent wire or braided wire configuration. More recently, NiTi SMA self-expanding designs have become available to threaten the dominance of the braided wire stents.

In the treatment of occlusive disease, all self-expanding designs offer resistance to permanent deformation/compression. This compression resistance may be critical in preventing vessel obstruction and occlusion in areas of the body where physical compression of plastically deformable stents has been reported (principally the carotid arteries and the SFA) [24]. In addition to compression-resistance, the key clinical advantages offered by most of the NiTi SMA designs are enhanced radial force and minimal or no shortening upon deployment. These benefits are significant whether managing occlusive or aneurysmal disease.

In the case of occlusive disease, enhanced radial force may be important in preventing plaque from encroaching upon the lumen either acutely or long-term (restenosis). The ability of metallic stents to reduce the incidence of restenosis and enhance long-term vessel patency thereby improving clinical outcomes has been clearly demonstrated in randomized clinical trials, especially in the coronary arterial system [17, 23, 35]. Virtually all of these clinical trials have been performed using balloon-expandable, plastically deformable stents. Whether the increased radial force and continuous expansion characteristics of NiTi SMA stents offers additional protection against restenosis is currently being investigated.

In the case of aneurysmal disease, enhanced radial force increases the forces required to dislodge the stent-graft (theoretically preventing it from migrating). Migration of the stent-graft can allow blood to flow into the aneurysmal sac with potentially catastrophic consequences. Balloon-expandable stent-graft designs (AnCure, Guidant) rely on hooks to imbed themselves into the vessel wall [28], thereby preventing the stent-graft from becoming detached from the vessel wall and moving downstream. Most NiTi SMA stent-grafts (AneuRx and Talent, Medtronic; Hemobahn and Excluder, Gore; Vanguard, Boston Scientific) do not utilize hooks or barbs, but rely exclusively upon the force between the NiTi SMA stent and the vessel to hold the stent graft in place.

The lack of significant shortening during expansion from the constrained state to the "programmed" diameter allows increased precision in placing the stent at the site of the disease (whether occlusive or aneurysmal). Braided stents shorten significantly upon deployment, especially in larger vessels, making accuracy during deployment difficult. As such, occasionally more than one stent must be deployed to fully cover the diseased area. Some braided elgiloy stents (Magic Wallstent, Boston Scientific) allow the stent to be recaptured after partial (up to 50%) deployment to help circumvent this shortcoming. The non-shortening characteristics of some of the newer NiTi SMA designs (SMART, Cordis) may encourage penetration of NiTi SMA stent designs into applications currently dominated by balloon-expandable designs such as coronary and renal artery stenting where precise positioning is critical.

Although the efficacy of stenting in the management of occlusive disease has been demonstrated in a variety of clinical applications as discussed above, the long-term value of stent-grafts in the treatment or palliation of aneurysmal disease is more controversial. Whereas the acute results of stent-grafts in bypassing aneurysms in the thoracic and abdominal aorta, as well as the iliac, femoral and popliteal arteries have been encouraging, the long-term patency and durability of the results have yet to be clinically proven. Acute and chronic endoleaks, migration of the stent-graft, kinking of the stent and dislodgement of one or more components (modular systems) have occurred with balloon-expandable and self-expanding systems alike [18, 25, 44, 46].

Two stent-graft systems, the balloon-expandable EVT/Guidant AnCure device (unibody/non-modular), and the NiTi SMA Medtronic AneuRx system (modular), have recently been recommended for market approval in the United States by the FDA Circulatory System Devices Advisory Panel. This recommendation was based on superior one-year clinical outcomes versus conventional surgery for the repair of AAA disease. Both companies will be required to do extensive post-market surveillance to answer the serious questions of long-term clinical durability versus surgery.

6
Future Development

Stent development will continue to focus on designs, which allow access to a greater range of implant locations through smaller and more tortuous anatomy. These stents will be more flexible and present a smaller profile when compressed for delivery. Stents will be developed to address specific implant site requirements for accessibility, radial strength, and flexibility. Other stents, or modular stent systems, will be designed to treat complex anatomy such as bifurcations and side-branches. Existing applications will be enhanced to provide greater durability, bio-active coatings, and reduced manufacturing costs.

References

1. Alimi YS, Chakfe N, Rivoal E, Slimane KK, Valerio N, Riepe G, Kretz J-G, Juhan C (1998) Rupture of an abdominal aortic aneurysm after endovascular graft placement and aneurysm size reduction. J Vasc Surg 28:178–183
2. Allen BT, Hovsepian DM, Reilly JM, Rubin BG, Malden E, Keller CA, Picus DD, Sicard GA (1998) Endovascular stent grafts for aneurysmal and occlusive vascular disease. Am J Surg 176:574–580
3. Al-Mubarak N, Roubin GS, Gomez CR, Liu MW, Terry J, Lyer SS, Vitek JJ (1999) Carotid artery stenting in patients with high neurologic risks. Am J Cardiol 83:1411–1413
4. American Society for Testing and Materials (1998) ASTM Z7421Z standard test method for conducting cyclic potentiodynamic polarization measurements for corrosion susceptibility of implantable medical devices. ASTM, West Conshohocken
5. American Society for Testing and Materials (1999) ASTM Z7783Z specification for wrought nickel titanium (nitinol) shape memory alloys for medical devices. ASTM, West Conshohocken
6. American Society for Testing and Materials (1999) ASTM Z7784Z standard test method for transformation temperature by thermal analysis. ASTM, West Conshohocken

7. American Society for Testing and Materials (1999) ASTM Z7785Z standard terminology for nickel titanium (nitinol) shape memory alloys. ASTM, West Conshohocken
8. Beregi JP, Prat A, Willoteaux S, Vasseur MA, Boularand V, Desmoucelle F (1999) Covered stents in the treatment of peripheral arterial aneurysms: procedural results and midterm follow-up. Cardiovasc Intervent Radiol 22:13–19
9. Biasi GM, Piglionica MR, Meregaglia D, Ferrari SA, Cao PG, Barzi F, Verzini F, Coppi G, Pacchioni R, Gennari S, Moll FL, Nolthenius RP, Van der Berg JC, Stancanelli V, Piccinini E, White R, Allen R (1998) European multicentre experience with modular device (Medtronic Aneurx) for the endoluminal repair of infrarenal abdominal aortic aneurysms. J Mal Vasc 23:374–380
10. Bosanac Z, Jain M, Monsour MJ (1998) Treatment of malignant ureteric stricture with use of a Vascucoil spring stent. J Vasc Interv Radiol 9:855–856
11. Bürger T, Meyer F, Tautenhahn J, Halloul Z, Fahlke J (1998) Initial experiences with percutaneous endovascular repair of popliteal artery lesions using a new PTFE stent graft. J Endovasc Surg 5:365–372
12. Coppi G, Pacchioni R, Moratto R, Gennai S, Farello GA, Bergamaschi G, Rabbia C, Rossato D, Ponzio F, Stancanelli V, Piccinini E (1998) Experience with the stentor endograft at four Italian centers. J Endovasc Surg 5:206–215
13. Criado FJ, Fry PD, Machan LS, Twena M, Patten P (1998) The talent endoluminal AAA stent-graft system: report of the phase-I USA trial and summary of worldwide experience. J Mal Vasc 23:371–373
14. EN/ISO 14630 (1998) Non-active surgical implants: general requirements
15. FDA/CDRH/ODE/DCRND (1994) Guidance for the submission of research and marketing applications for interventional cardiology devices: ptca catheters, atherectomy catheters, lasers, intravascular stents
16. FDA/CDRH/ODE/DRAERD (1998) Guidance for the content of premarket notifications for metal expandable biliary stents
17. George CJ, Baim DS, Brinker JA, Fischman DL, Goldberg S, Holubkov R, Kennard ED, Veltri L, Detre KM (1998) One-year follow-up of the stent restenosis (STRESS I) study. Am J Cardiol 81:860–865
18. Harris P, Brennan J, Martin J, Gould D, Bakran A, Gilling-Smith G, Buth J, Gevers E, White D (1999) Longitudinal aneurysm shrinkage following endovascular aortic aneurysm repair: a source of intermediate and late complications. J Endovasc Surg 6:11–16
19. Hausegger KA, Mendel H, Tiessenhausen K, Kaucky M, Aman W, Tauss J, Koch G (1999) Endoluminal treatment of infrarenal aortic aneurysms: clinical experience with the talent stent-graft system. J Vasc Interv Radiol 10:267–274
20. Huibregtse K, Carr-Locke DL, Cremer M, Domschke W, Fockens P, Foerster E, Hagenmüller F, Hatfield AR, Lefebvre JF, Liquory CL, et al. (1992) Biliary stent occlusion: a problem solved with self-expanding metal stents? (European Wallstent Study Group). Endoscopy 24:391–394
21. Isles CG, Robertson S, Hill D (1999) Management of renovascular disease: a review of renal artery stenting in ten studies. QJM 92:159–167
22. ISO/CD (TS) 15539 (1999) Cardiovascular implants-endovascular devices
23. Macaya C, Serruys PW, Ruygrok P, Suryapranata H, Mast G, Klugman S, Urban P, Den Heijer P, Koch K, Simon R, Morice MC, Crean P, Bonnier H, Wijns W, Danchin N, Bourdonnec C, Morel MA (1996) Continued benefit of coronary stenting versus balloon angioplasty: one-year clinical follow-up of Benestent trial. J Am Coll Cardiol 27:255–261
24. Mathur A, Dorros G, Iyer SS, Vitek JJ, Yadav SS, Roubin GS (1997) Palmaz stent compression in patients following carotid artery stenting. Cathet Cardiovasc Diagn 41:137–140
25. May J, White GH, Yu W, Waugh R, Stephen MS, Sieunarine K, Chaufour X, Harris JP (1997) Endoluminal repair of abdominal aortic aneurysms: strengths and weaknesses of various prostheses observed in a 4.5-year experience. J Endovasc Surg 4:147–151
26. Medical Data International (1998) Stent-grafts: the tortuous road ahead. MedPro Month 8:122–128
27. Mialhe C, Amicabile C, Becquemin JP (1997) Endovascular treatment of infrarenal abdominal aortic aneurysms by the Stentor system: preliminary results of 79 cases. J Vasc Surg 26:199–209
28. Moore WS, Rutherford RB (1996) Transfemoral endovascular repair of abdominal aortic aneurysm: results of the North American EVT phase-1 trial (EVT Investigators). J Vasc Surg 23:543–553
29. Nawaz S, Cleveland T, Gaines P, Beard J, Chan P (1999) Aortoiliac stenting: determinants of clinical outcome. Eur J Vasc Endovasc Surg 17:351–359
30. Ohki T, Veith FJ, Sanchez LA, Marin ML, Cynamon J, Parodi JC (1997) Varying strategies and devices for endovascular repair of abdominal aortic aneurysms. Semin Vasc Surg 10:242–256
31. (1998) prEN 12006-3.Non active surgical implants-particular requirements for cardiac and vascular implants. Part 3: endovascular devices

32. (1999) prENxxxxx (WI 00285 071). Non-active surgical implants: particular requirements for cardiac and vascular implants-specific requirements for arterial stents
33. Riepe G, Heilberger P, Umscheid T, Chakfe N, Raithel D, Stelter W, Morlock M, Kretz JG, Schroder A, Imig H (1999) Frame dislocation of body middle rings in endovascular stent tube grafts. Eur J Vasc Endovasc Surg 17:28–34
34. Savage MP, Douglas JS Jr, Fischman DL, Pepine CJ, King SB 3rd, Werner JA, Bailey SR, Overlie PA, Fenton SH, Brinker JA, Leon MB, Goldberg S (1997) Stent placement compared with balloon angioplasty for obstructed coronary bypass grafts. Saphenous Vein De Novo Trial Investigators. N Engl J Med 337:740–747
35. Serruys S, de Jaegere P, Kiemeneij F, Macaya C, Rutsch W, Heyndrickx G, Emanuelsson H, Marco J, Legrand V, Materne P, et al. (1994) A comparison of balloon-expandable-stent implantation with balloon angioplasty in patients with coronary artery disease: Benestent Study Group. N Engl J Med 331:489–495
36. Stelter W, Umscheid T, Ziegler P (1997) Three-year experience with modular stent-graft devices for endovascular AAA treatment. J Endovasc Surg 4:362–369
37. Taal BG, Muller SH, Boot H, Koops W (1997) Potential risks and artifacts of magnetic resonance imaging of self-expandable esophageal stents. Gastrointest Endosc 46:424–429
38. Tesdal IK, Adamus R, Poeckler C, Koepke J, Jaschke W, Georgi M (1997) Therapy for biliary stenoses and occlusions with use of three different metallic stents: single-center experince. J Vasc Interv Radiol 8:869–879
39. Trepanier C, Tabrizian M, Yahia L'H, Bilodeau L, Piron DL (1998) Effect of modification of oxide layer on NiTi stent corrosion resistance. J Bio Mater Res 43:433–440
40. Trepanier C, Leung TK, Tabrizian M, Yahia L'H, Bienvenu J-G, Tanguay J-F, Piron DL, Bilodeau L (1999) Preliminary investigation of the effects of surface treatments on biological response to shape memory NiTi stents. J Bio Mater Res 48:165–171
41. Uflacker R, Robison JG, Brothers TE, Pereira AH, Sanvitto PC (1998) Abdominal aortic aneurysm treatment: preliminary results with the talent stent-graft system. J Vasc Interv Radiol 9:51–60
42. Umscheid T, Stelter WJ (1999) Time-related alterations in shape, position, and structure of self-expanding, modular aortic stent grafts: a 4-year single-center follow-up. J Endovasc Surg 6:17–32
43. Veith FJ, Abbott WM, Yao JST, Goldstone J, White RA, Abel D, Dake MD, Ernst CB, Fogarty TJ, Johnston KW, Moore WS, Van Breda A, Sopko G, Didisheim P, Rutherford RB, Katzen BT, Miller DC (1995) Guidelines for development and use of transluminally placed endovascular prosthetic grafts in the arterial system. J Vasc Surg 21:670–685
44. White GH, May J, Waugh R, Harris JP, Chaufour X, Yu W, Stephen MS (1999) Shortening of endografts during deployment in endovascular AAA repair. J Endovasc Surg 6:4–10
45. Wholey MH, Wholey M, Bergeron P, Diethrich EB, Henry M, Laborde JC, Mathias K, Myla S, Roubin GS, Shawl F, Theron JG, Yadav JS, Dorros G, Guimaraens J, Higashida R, Kumar V, Leon M, Lim M, Londero H, Mesa J, Ramee S, Rodriguez A, Rosenfield K, Teitelbaum G, Vozzi C (1998) Current global status of carotid artery stent placement. Cathet Cardiovasc Diagn 44:1–6
46. Zarins CK, White RA, Schwarten D, Kinney E, Diethrich EB, Hodgson KJ, Fogarty TJ (1999) AneuRx stent graft versus open surgical repair of abdominal aortic aneurysms: multicenter prospective clinical trial. J Vasc Surg 29:292–305

Shape-Memory Alloy for Interventional Stenting in View of its Development in China

Mi Xujun, Zhu Ming, Guo Jinfang, Yuan Guansen

1 Introduction

Interventional stenting is the clinical practice of interventional delivery of stent or stent graft (prosthesis) into human vascular or non-vascular vessel lumen either as a scaffolding device to treat luminar stricture, stenosis or restenosis following local pathologies or anatomical complications or for the purpose of fluid redistribution as in case of vascular aneurysm or dissection. As a significant branch of interventional medicine (IVM) and a complementary approach to balloon dilatation surgery, this technology of minimally invasive surgery (MIS) has been extensively applied in different sectors of the medical circle in recent decades to treat urethra-prostatic [1, 2], esophageal [3], tracheal [4], biliary [5, 6] and also vascular constriction [7, 8] and aneurysmal pathologies [9–11], thanks to fruitful experimental and clinic explorations, latest progresses in monitoring technologies such as roentgenography, ultrasonography, computer tomography (CT), magnetic resonance imaging (MRI), digital subtraction imaging (DSI) as well as structural and material innovations with the stent-graft itself.

2 Structural and Material Considerations for the Stent Design

In the past few decades, there have been incessant efforts to modify the stent structure to achieve the best biomechanical compatibility and hence, to increase clinic success rate, both intra-operatively and post-operatively. As a result, many types of stent have been introduced into different sectors of interventional medicine. There are two kinds of stent shapes, as seen from the basic configuration, i.e., the spirals and the mesh stents. Classified from the mechanism of supporting force exertion, the prevalent mesh stents can be described as balloon-expandable (Palmaz-schatz stent, Gianturco-Roubin stent, Strecker stent), self-expandable (Wallstent) or thermoexpandable (Nitinol stents) [12]. These stents can be either naked or coated (covered) with a film of a particular polymer material. Competitive as they are in terms of different manufacturers seeking for a better market share, this diversity also benefits the medical profession in view of the inter-complementary nature of different configurations of stents as one application field (urologic, gastrointestinal, vascular) or combination of case indications may favor the choice of one particular kind of stent while another design may be preferable if the clinic details are changed. Theoretically, there should be clear-cut niches each type of stent occupies or, at least, different indications of the stents. There have been endeavors for their clarification as seen from multi-center comparative studies [13]. Unfortunately insufficient work to this purpose and lack of consensus among different authors have made a pervasive conclusion largely improbable at present.

Biomechanical compatibility is the principal consideration for the structural design of a stent. Three main factors are to be borne in mind for the evaluation of the stent design: functional effectiveness, biochemical reactions and delivery. A stent should have proper combination of rigidity and flexibility for achievement of sufficient radial strength to guarantee effective scaffolding and to prevent stent migration without significantly affecting tubular curvature and physiological peristalsis of the supported lumen, or causing traumatic pressure to the duct wall. It should be thin-walled and low-profiled, and overlapping parts or protruding nodes should be avoided for ease of endothelialization and minimization of thrombus formation. To accommodate the use of small caliber catheter system for easy delivery, a stent must also have high expandability, i.e., a high ratio between the expanded and compressed (constrained) diameters. In addition to these considerations, some stents are constructed with hooks for better fixation and certain designs, such as the Strecker stent, can be easily retrieved through pulling off at one end of the wire while the location of some other stent can be readjusted through delicate design of the stent and also of the delivery system.

The success of stenting surgery depends not so much on mere structural designs as on ideal choice of the constituent biomaterial itself. A biomaterial is defined as " a non-viable material used in a medical device intended to interact with biological systems, in that they possess a combination of properties (physical, chemical, mechanical and biological) that render them safe, effective and reliable for use within a physiological environment, that is both sensitive to and unforgiving of irritating bodies" [14]. The quest for an ideal stent-graft biomaterial is a tortuous history, not unlike the trial-and-failure process for selection of prosthetic material for aortic surgery as described interestingly by Dr. Smith [15]. Currently available stents generally fall into two categories of polymeric and metallic materials, and a great amount of work has been done on evaluation and comparison of different materials for stenting in different sectors of interventional medicine. Sean and Michael [16] compared the long-term struvite and hydroxyapatite encrustation of ureteral stents fabricated with silicone, polyurethane, hydrogel-coated polyurethane (HPU), siliteck and percuflex and concluded that their effectiveness "is still hindered by the formation of encrusting deposits which may cause obstruction and blockage of the device", while the work of Steven [13], Joseph [2] and Holmes [17] seems to establish the effectiveness of stainless steel, tantalum, titanium and gold-plated metallic urethra-prostatic stents as compared with results for polymeric stents. For this application, as well as for other sectors such as vascular, esophageal and biliary stents, stainless steel has come to dominate the stent fabrication before the advent of the shape memory biomaterial. In addition to comparative studies aimed at evaluating the viability of the bulk materials, there has also been attention focused on surface preparation and modifications in consideration of the effect of factors such as surface coating, smoothness, hydrophilicity and surface net charges. These factors will be discussed in some detail later in this paper in relation with their effect on Nitinol stents.

3
SMA Stent and Its Application in China

The feasibility study of TiNi shape memory alloy used as a biomaterial started in the early 1970s when Williams [18] and other investigators reported the histological embedding reaction, corrosion resistance and toxigenecity tests with the TiNi material. These in vivo and in vitro studies led to a positive attitude toward its biocompatibility and aroused worldwide interest in its potential as a biomaterial.

Researches on shape memory alloy did not start in China until the late 1970s, followed immediately by investigations on its medical significance including physiological solution immersion [19], histological observations [20], toxigenic and carcinogenic tests [20, 21] and trace nickel element analysis [22]. Encouraging results from these early investigations triggered active response on the part of the materialists as well as from the medical circle. Typical applications in this period include orthodontic wires [21] and orthopaedic devices such as the scoliosis correction rod [23] and bone fixation staples [24, 25]. The experimental and clinical explorations of nitinol stenting started in the early 1990's and this new technology quickly spread to involve different sectors of interventionology such as digestive, respiratory and vascular clinics.

3.1
Non-Vascular Applications

The earliest application of Nitinol spiral stents in China was reported by Qiu and his collaborators [26] in Beijing General Research Institute for Nonferrous Metals (GRINM), the herald of shape memory alloy researches in China. Among the 15 cases of interventional stenting for treatment of prostatic hyperplasia, 14 succeeded with uroflow rate reaching 7–12 ml/s, contrasting with the less than 2 ml/s before stenting. Follow-up of the successful cases showed that all symptoms of urine retention and urethral stricture had disappeared. No deposition or stone formation was found on the stent surface. This is in sharp contrast with reports on polymeric stents [13]. The same authors further reported their experience with 39 nitinol stents for treatment of benign prostatic hyperplasia (BPH) in 1993 [27]. The stents were again spiral shaped, though the design was modified a little. The success rate was 89.7% and a 3- to 26-month follow-up showed no encrustation or lithogenesis on the stent wire, an encouraging, if not conclusive, observation. Since then, many other doctors throughout China have followed this practice and the spiral stent has been gradually substituted by the mesh stent, except for treatment of anterior uritery strictures. Wang [28], for example, reported their satisfactory results with the urethra-prostatic mesh stent as evaluated from such objective criteria as international prostate symptom score

(IPSS), peak uroflow rate (PUFR) and retentive urine volume (RUV). Again, the stent wire was found to be overgrown with urinary mucosa and no lithogenesis was detected for the 21 cases followed for over 6 months. It is interesting – and testifying to the relative maturity of urethra-prostatic stent development – that the authors tend to lay store on operation discretion and meticulous catering to clinical details such as stricture length measurement, stent size selection and correct stent location for achievement of higher success rate rather than on desire for stent modification. Now the application of shape memory urethra-prostatic stent has become a nearly routine procedure and, up to this day thousands of nitinol stents have been employed for this purpose with some in service for over 5 years.

The first application of shape memory spiral esophageal stent was reported by Dong [29]. Among the eight cases of benign esophageal strictures, five were alleviated through nitinol stenting, while the other three who underwent chemical burning had restenosis only one month after operation. The author concluded with generally positive attitude that "the stent could be used for the short segment of esophageal stricture safely, simply and effectively". This practice has since been adopted throughout China [30–32] for treatment of digestive diseases including benign and malignant esophageal stricture, duodenal or pyloric stenosis and stenotic gastroenterostomical carcinoma with thousands of clinic cases favoring this interventional technology. At the same time, studies aimed at overcoming its drawbacks, such as reflux esophagitis and recurrent stenosis caused by mucosa or tumor ingrowth into the stent and food retention, have resulted in improvement of the stent design such as membrane covering and development of an anti-reflux esophageal stent [33, 34]. These structural and material modifications to the stent-graft as well as the accumulation of clinical experience in terms of indication analysis and operation skills have remarkably increased the safety and success rate of esophageal stricture management and made this practice one of the most successful and popular field of interventional application of shape memory alloy.

The development of nitinol tracheal and biliary stents followed similar course as that for esophageal stents. Clinical application of the tracheal stent initiated in 1993 following a successful applied research by Liu [35] in which the stenting experience with seven experimental mongrels and two clinical cases revealed that the NiTi tracheal stent could "maintain long-term patency without severe infection or obstruction by granulation or secretion", thus paving the way for nitinol tracheal stenting in China. The widespread application of this technology also resulted in some modifications of the stent structure from the original spiral design to double-end flared mesh stent [36, 37], with generally favorable results recommending popularization of this innovation at least as a palliative method. The appearance of nitinol biliary stent occurred at roughly the same time as the application of esophageal and tracheal stents. On successful experimental researches, Gu [38] treated 13 patients with malignant biliary strictures with nitinol stent since June 1991, and a 1- to 16-month follow-up suggested that "nitinol alloy endoprosthesis could maintain long term patency without severe infection or obstruction due to biliary sludge" [39]. Jiang [40] conducted a comparative study between different materials, i.e., plastics, stainless steel and nitinol for biliary stenting, and found that there is no conspicuous difference between their clinical performance in terms of decrease in serum total bilirubin level and early complications although the metallic stents were easier to manipulate due to their high expandability and better conductivity of biliary flow. The authors concluded that "percutaneous placement of biliary endoprostheses was effective and safe for palliation of malignant biliary obstruction".

3.2 Vascular Applications

Vascular interventionology represents the most challenging innovation and also the most promising field for shape memory stents. Application in this field puts strict biological and biomechanical demands on the stent material and structure design. At the same time, the large number of cardiovascular patients who are in desperate need of interventional treatment also holds great expectations for the technology of vascular stenting and intensifies researches in this aspect.

China is a heavily populated country with a large number of patients afflicted with coronary strictures and aortic aneurysmal diseases, and many of them are subject to coronary interventions and stenting treatments. According to a study released by Section of Interventional Cardiology, Chinese society of Cardiology [41], the 51 hospitals enrolled in the registry had 6213 cases of PTCA treatment and 1735 stents were implanted in 1520 of the patients during the follow-up period from 1984 to 1996. The drastic increase in the number of stenting and its percentage in the total PTCA cases in these recent years (Table 1) demonstrate the rapid growth of stenting application in China and also underscores the importance of development and clinical research of intravascular stents.

While a material which is compatible with blood is also tissue compatible [14], biocompatibility test carried out in other medical fields do not necessarily suffice for the biological requirements of the material in the hematological environment. For this reason, preliminary investigations involving the hematological compatibility have been carried out in recent years. Tong [42] implanted nitinol spiral stents into the iliac arteries of 8 dogs and observed growth of smooth neointima on the inside surface of the stent after one month. The neointima grew to a 229.8 ± 98.1-μm thickness after 3 months and the thickness kept steady even after 6 months ($P>0.5$). The nitinol wire displayed no corrosion and caused no inflammatory reactions. Ma [43] carried out similar investigations with 6 normal pigs and found that there was no changes ($P>0.05$) in results of routine blood and urine tests, the function of liver and kidney or the content of titanium and

Table 1. Growth of PTCA and stenting cases in recent years in 51 hospitals across China

Year	PTCA cases	Stenting cases	Stenting percent
1992	343	6	1.7
1993	484	40	8.2
1994	817	106	13.0
1995	1341	442	33.0
1996	1803	926	51.3

nickel element as compared to preoperative results. It was concluded that "intravascular stent of nitinol alloy had good tissue compatibility". On successful animal experimentation with nitinol mesh stents implanted into the precavae, postcavae and abdominal aortae of six healthy dogs, Wang [44] used the stents for the transjugular intrahepatic portosystemic stent shunting (TIPSS) on 12 patients with a history of post-hepatitis cirrhosis and increased the maximum blood flow velocity in the main portal vein from 14.0 ± 4.5 cm/s to 48.0 ± 16.5 cm/s. An average of 4 months follow-up confirmed shunt patency in all patients and no occurrence of rebleeding. The authors concluded that the advantages of nitinol stent used for TIPSS, in addition to the hematological compatibility, lie in its radiopacity which makes exact placement easy, its good longitudinal flexibility which renders effective management of curved shunt tract and tortuous vessels possible and its non-ferromagnetic property which enables follow-up to be performed noninvasively with MRI.

One of the most promising field for intravascular application of nitinol stent is for treatment of aortic aneurysm and, for this purpose, GRINM has coordinated a series of systematic researches with some of the most prestigious hospitals in China. Jiang [45], for example, conducted experimental studies on transluminar implantation of nitinol mesh stents for treatment of pseudoaneurysm of abdominal aorta and concluded that "the mesh stent is feasible for reparation of the leaks of abdominal aorta and recanalization of vessels". The same authors also used the nitinol mesh stents in a clinical application to treat thoracic abdominal pseudoaneurysm and met preliminary success [46]. Although further investigations are necessary, especially in clarification of the effect of stent configuration, mesh hole area, film coverage and material selection. We are nevertheless confident that this new technology will mature in the near future to offer a promising application for shape memory stent.

4 Discussion and Comments

This decade will pass as very important period for the development and applied researches of shape memory alloy interventional stents in China. Up to now there has been widespread investigations of the nitinol stenting technology, and thousands of clinical cases for the esophageal, urethra-prostatic, tracheal and biliary stents for which GRINM has been given market authorization by the state medical administration. At the same time, shape memory vascular stents and other cardiovascular and vena cava devices are being intensively developed to be pending for full-scale applications. These preliminary efforts has aroused a great interest in the stenting technology from the related medical circles and basically established the nitinol shape memory alloy as an ideal biomaterial for interventional devices. This success, however, must not be overestimated in face of the fact that much is still to be desired for maturity of this innovation, especially in aspects of systematic clinical investigations, stent manufacturing and material preparation.

4.1
Systematic Clinical Investigations

Although many large hospitals have been involved in the development of this technology, and some of the researches are organized and coordinated by related administrative departments, few of them, however, can really be termed systematic multi-center investigations. As we can see from the literature listed in this paper, many of these studies are done and reported in a similar manner, and fall short of originality. Due to the lack of a scientific method or manner of clinical follow-up, the data so acquired are incomprehensive and render it difficult to make persuasive conclusions on a sound inductive basis. As a result, most of the critical problems affecting clinical success of this technology are left unsolved or even undelt with. Lack of prolonged follow-up leaves the problem of long-term stenting results hardly accounted for; the mechanism of post-operative luminar restenosis is yet to be clarified, and the effective method of tackling with it has rarely been discussed; insufficient clinical data, especially on comparison of the effects of different configurations and sizes of the stent, make it impossible to tabulate the indications of this technology and different kinds of stents, so that selection of stent is almost done at random as the circumstances suffice. These deficiencies, nevertheless, have been noticed by some of the doctors in this field [47] and both administrative and professional circles are paying attention to these problems.

4.2
Stent Manufacture

Concerns over the effect of stent configuration and size has been an important factor for stent design and manufacturing. Work in this respect, however, was largely on an experimental level due to a lack of systematic manner in dealing with stent characterization in aspects of both geometric considerations and mechanical performances.

4.2.1
Geometric Considerations

Stent parameters which can guide a stent evaluation and offer an objective comparison among different stent designs may include such items as the stent profile (in terms of post implantation metal surface and total cross area of the component) and wall thickness, which may influence alien reaction of the target lumen and the extent of adverse effect of the stent to the transported fluid and also thrombus formation in case of vascular prosthesis, the expandability and the mesh hole area. The last factor is particularly important for vascular stents as this may determine whether neointima can grow into an integrate film to stop blood flow into or out of the stent. In case of aneurysmal separation, this is desirable for prevention of endoleaks which is closely related to aneurysmal saccular rupture and hence determines the failure of endovascular graft treatment [48]. But if the stent is to span over a branch vessel, this should be carefully avoided since this occlusion may lead to fatal complications. Hagen [49] carried out experimental research and found out that the stents with a mesh hole area of 0.84 mm^2 did not separate the aneurysm from the blood flow while those with an area of 0.68 mm^2 succeeded in this function. Ruiz [50] also did similar tests and concluded that one of the advantages of uncovered mesh stent is that it can span over aortic branches without excluding the blood flow and hence affecting the blood supply of important organs. While preliminary results are published, a comprehensive investigation is necessary for clarification of the effect of mesh hole area to provide guide to stent manufacture and guarantee proper stent selection for different clinical details.

4.2.2
Mechanical Performance

Mechanical performance is critical in determining the function of a stent and how the supported lumen will react to the implant. Two main properties are of general concern, i.e., the longitudinal flexibility and the radial strength. While the stent flexibility is reduced to a rough conceptual evaluation and comparison, efforts to characterize the stent force have been made using newly developed methods such as flat-plate compression or wrap-around collar tension test [51, 52]. But none of the published studies are correlated to biomechanical parameters and hence, unable to advise on how strong a stent should be for a certain application. In contrast to this mechanical approach, diameter criteria have been suggested though it may differ among different authors. Gu [39] advised that a biliary stent should oversize the expanded bile duct diameter by 1–2 mm while Jiang [46] recommended a vascular stent 10–15% larger than the aorta diameter for aneurysmal treatment. Undoubtedly, such a merely geometric approach irrespective of the mechanical parameters could not guarantee optimum performance of a stent, not to say that precise determination of luminar diameters is still problematic.

4.3
Stent-Material Preparation

Although we have relatively mature technology to manufacture shape memory alloys, its medical applications have put forward more stringent demands upon its mechanical and biochemical properties. It is necessary to clarify the effect of bulk material production and surface preparation on the properties so that we can make the best of the novel material.

4.3.1
Bulk-Material Production

As elucidated by numerous studies, the composition, processing and heat-treatment of shape memory alloy have a great effect on the properties of the material, but we are not so certain about the potential effects of these factors in a biomedical environment, especially the effect of impurity content, inclusions, lap defects and residual stress etc. on the corrosion and fretting abrasion resistance and biocompatibility of a shape memory product within a particular body fluid. Insufficient researches in this aspect are partly to be blamed on the fact that standardization of nitinol manufacture for medical purpose is still left blank.

4.3.2
Surface Preparation

Potentially important as it is for the biocompatibility of the nitinol stent, this is also a controversial issue which defies any simplified approach. While excessive surface roughness has been shown to promote more rapid clot formation than highly polished surfaces [14], and it is commonly believed that a smooth surface will be conducive to normal blood flow, the study of Trigwell [53] demonstrated that the smoothest surface produced by mechanical polishing to a mirror finish was determined to be more susceptible to corrosion. While ASTM recommends passivation of implant metallic material prior to implant in order to have a proper corrosion resistance to the body fluid [54], the problem is that the thickness of oxide film is not easily known and the possibility of flaking or spalling of the oxide may hazard the stent fatigue resistance. While net electrical charge is very important because most of the metals used have a positive electrical charge when placed in an electrolytic solution contrasted to the fact that all biological intravascular surfaces are negatively charged, this may not be a problem because the charge allows the plasma proteins to cover the surface of the stent with a layer of fibrinogen and this negatively charged layer helps decrease thrombogenicity before the arrival of platelets and white blood cells [12, 47]. Other concerns may include such hot issues as surface hydrophilicity–hydrophobicity (wettability) [14] and various processes for surface modification, e.g., ion-beam-assisted sputtering [55], glow discharge plasma treatmen [56], ion implantation [57] and spark-erosion [58] as applied to other biomedical materials. But again research results in these areas could neither establish definite correlation nor failed to bring remarkable surface improvements and further standardization investigations are needed.

5
Summary

Experiences with shape memory stents in the passing decade has established nitinol alloy as an ideal material for interventional stenting treatment though further comprehensive studies are still necessary for optimization of its properties in the physiological environment. Intensive experimental and clinic researches of the application of shape memory stents in different branches of interventional medicine have met with preliminary success and aroused worldwide interest.

China has been on the forefront of the applied researches on shape memory alloy for interventional stenting and extensive experiences with this innovation have upgraded the nitinol stenting from an investigational application to a popularization stage in fields of esophageal, urethroprostatic, biliary and tracheal stricture management with vascular stenting at the horizon. Stent manufacture standardization and systematic multi-center clinical investigations are urgently needed to achieve full maturity of this novel technology.

References

1. Fabian KM (1980) Der Introprostaticsche "partielle katheter" (urologische spiral). Urologe A 10:236–238
2. Joseph EO (1991) The obstructive prostate and the intraurethral stent. Contemp Urol Jan 1991:61–70
3. Watkinson AF, Ellul J, Entwisle K, Mason RC, Adam A (1995) Esophageal carcinoma: initial results of palliative treatment with covered self-expanding endoprosthesis. Radiology 195:821–827
4. Johnston MR, Loeber N, Hillyer P, Stephenson LW, Edmunds LH Jr (1980) External stent for repair of secondary tracheomalacia. Ann Thorac Surg 30:291–296
5. Lammer J, Klein GE, Kleinert R, Hausegger K, Einspieler R (1990) Obstructive jaundice: use of expandable metal endoprosthesis for biliary drainage. Radiology 177:789–791
6. Gordon RL, Ring EJ, LaBerge JM, Doherty MM (1992) Malignant biliary obstruction: treatment with expandable metallic stents – follow-up of 50 consecutive patients. Radiology 182:697–700
7. Schatz RA, Baim DS, Leon M, Ellis SG, Goldberg S, Hirshfeld JW, Cleman MW, Cabin HS, Walker C, Stagg J, et al. (1991) Clinical experience with the Palmaz-Schatz coronary stent. Circulation 83:148–150
8. LaBerge JM, Ring EJ, Gordon RL, Lake JR, Doherty MM, Somberg KA, Roberts JP, Ascher NL (1993) Creation of transjugular intrahepatic portosystemic shunts with Wallstent endoprosthesis: results in 100 patients. Radiology 187:413–416
9. Blum R, Voshage G, Lammer J, Beyersdorf F, Tollner D, Kretschmer G, Spillner G, Polterauer P, Nagel G, Holzenbein T (1997) Endoluminal stent-grafts for infrarenal abdominal aortic aneurysms. N Engl J Med 336:13–20
10. Ahn SS, Obrand DI (1998) Current status of intraluminal grafts for aortic aneurysms. Vasc Surg 32:215–219
11. Chuter TA, Wendt G, Hopkinson BR, Scott RA, Risberg B, Walker PJ, White G (1995) Transfemoral insertion of a bifurcated endovascular graft for aortic aneurysm repair: the first 22 patients. Cardiovasc Surg 3:121–128
12. Wu KH (1997) Development and future of intravascular stents. In: Chu YY, Otsuka K (eds) Proceedings of the China-Japan Bilateral Symposium on Shape-Memory Alloys, Hangzhou, China. International Academic Publishers, Beijing pp 229–241
13. Steven AK (1993) Can a stent succeed in keeping the prostatic urethra open. Contemporary Urology April 1933:19–33
14. Banerjee R, Nageswari K, Puniyani RR (1997) Hematological aspects of biocompatibility: review article. J Biomater Appl 12:57–76
15. Robert BS (1998) Presidential addresses: the foundations of modern aortic surgery. J Vasc Surg 27:7–15
16. Sean PG, Michael MT (1996) Assessment of encrustation behavior on urinary tract biomaterials. J Biomater Appl 12:136–166
17. Holmes SAV (1992) Encrustation of intraprostatic stents: a comparative study. Br J Urol 69:383–387
18. Williams DF (1982) Biocompatibilities of clinically implemented matererials. CRC, Boca Raton, pp 145–150
19. Xue M, Chen XX, Li YM, et al (1981) Basic study of NiTi shape memory alloy-stimulating corrosion test. Stomatology 1:40–43
20. Xue M, Li YM, Gu GZ, et al (1983) Basic study of NiTi shape memory alloy. Chinese J Biomed Eng 2:28–33
21. Xue M, Jia WT (1986) Application of NiTi shape memory alloy to medicine and dentistry. In: Chu YY, et al. (eds) Proceedings of the International Symposium on Shape Memory Alloys, Guilin, China. pp 411–415
22. Bao YY (1983) A clinical study of orthodontic application of NiTi alloy. Chin J Stomatol 18:15–17
23. Lu SB, et al (1986) Treatment of scoliosis with shape memory alloy rod. Chung Hua Wai Ko Tsa Chih 24:129–132

24. Dai KR (1983) Orthopaedic application of shape memory compression staple. Chung Hua Wai Ko Tsa Chih 21:343–345
25. Xue M, Guo JF, Shen L, et al (1993) Application of NiTi shape memory alloy on maxillofacial surgery. Stomatology 13:131–132
26. Qiu CY (1991) Titanium-nickel alloy stent for urethrostenosis caused by prostatauxe. Chung Hua Wai Ko Tsa Chih 29:369–371
27. Qiu CY (1993) Shape memory alloy spiral for urethrostenosis caused by benign prostatic hyperplasia. Chung Hua Wai Ko Tsa Chih 31:272–274
28. Wang XF, Zhu J, Hou S (1996) Mesh-like tubular stent treatment for chronic urinary retention caused by BPH. Chung Hua Wai Ko Tsa Chih 24:107–109
29. Dong ZJ (1993) The nickel titanium alloy esophageal stent. Chung Hua Wai Ko Tsa Chih 31:264–266
30. Wu X, Ge R, Li PJ, et al (1997) The clinical application of three types of esophageal stent designed by ourselves. Chin J Radiol 31:172–175
31. Mao AW, Gao ZD, Yang RJ, et al (1998) Treatment of duodenal malignant stenosis using stent implantation combined with arterial chemotherapy. Chin J Radiol 32:655–657
32. Li TX, Han XW, Ma WZ, et al (1998) Treatment of benign and malignant gastroduodenal obstruction with self-expanding metal stents. Chin J Radiol 32:658–660
33. Dai DK, Zhai RY, Yu P (1998) Follow up study of esophageal stent placement. Chin J Radiol 32:391–394
34. Wu X, Ge R, Li PJ, et al (1999) Clinical application of the esophageal anti-reflux stent. Chin J Radiol 33:185–187
35. Liu Y, Sun YE, Huang XM, et al (1993) Nitinol alloy endotracheal stent used in treatment of tracheal stenosis, experimental and clinical application. Chung Hua Wai Ko Tsa Chih 31:267–268
36. Wei GZ, Yin XW, Zhong XD, et al (1995) Clinical application of NiTi stent for treatment of tracheal strictures. Jiangsu Med J 21:13–14
37. Feng QX, Li QL, Tan GF, et al (1997) Endoscopic placement of NiTi stent for treatment of tracheal strictures. Chung Hua Chieh Ho Ho Hu Hsi Tsa Chih 20:242–243
38. Gu WQ, Liu YX, Wang YS, et al (1994) Researches on NiTi shape memory alloy biliary stent. Chin J Exp Surg 11:283–284
39. Gu WQ (1993) Shape memory nitinol alloy endoprosthesis for malignant biliary strictures. Chung Hua Wai Ko Tsa Chih 31:260–263
40. Jiang WJ, Yao F, Ren A, et al (1997) Percutaneous placement of endoprostheses for treatment of malignant biliary obstruction: a report of 51 cases. Chin J Radiol 31:729–733
41. Chinese Society of Cardiology (1998) A data analysis of the first national coronary intervention registry. Chin J Cardiovasc Dis 26:25–28
42. Tong J, Su HX, Li GS, et al (1992) A preliminary study of Ti-Ni shape memory alloy intravascular stent. Chin J Thorac Cardiovasc Surg 8:54–55
43. Ma GS, Huang J, Wang JL, et al (1995) Tissue-compatibility of intravascular endoprosthetic stent of nitinol alloy. Chin J Biomed Eng 14:198–201
44. Wang MQ, Zhang JS, Yu M, et al (1994) Nitinol self-expanding stents in TIPSS procedure: animal experimental study and preliminary clinical experience. In: Chu YY, Tu HL (eds) Proceedings of the International Symposium on Shape Memory Materials, Beijing, China. International Academic Publishers, Beijing pp 631–634
45. Jiang WJ, Ren A, Zhang XZ (1997) Experimental study on transluminal implantation of mesh stent for the treatment of pseudoaneurysm of abdominal aorta. Chin J Radiol 31:331–333
46. Jiang WJ, Ren A, Liu P, et al (1995) Application of self-expandable nitinol mesh stent in the treatment of aortic aneurysm. Chin J Radiol 29:444–447
47. Zhang JS, Wang MQ (1997) Some problems in the applications of endostents in China. Chin J Radiol 31:295–297
48. Wain RA, Marin ML, Ohki T, Sanchez LA, Lyon RT, Rozenblit A, Suggs WD, Yuan JG, Veith FJ (1998) Endoleaks after endovascular graft treatment of aortic aneurysms: classification, risk factors and outcome. J Vasc Surg 27:69–80
49. Hagen B, Harnoss BM, Trabhardt S, Ladeburg M, Fuhrmann H, Franck C (1993) Self-expandable macroporous nitinol stents for transfemoral exclusion of aortic aneurysms in dogs: preliminary results. Cardiovasc Intervent Radiol 16:339–342
50. Ruiz CE, Zhang HP, Butt AI, Whittaker P (1997) Percutaneous treatment of abdominal aortic aneurysm in s swine model. Circulation 96:2438–2448
51. Longas JL, Puertolas JA, Rios R, et al (1997) Design characteristics and mechanical properties of a new NiTi stent. In: Pelton AR, Hodgson D, Russell SM, Duerig TW (eds) Proceedings of SMST 1997. Shape Memory and Superelastic Technologies, Pacific Grove, pp 567–572
52. Agrawal CM, Clark HG (1992) Deformation characteristics of a bioabsorbable intravascular stent. Invest Radiol 27:1020–1024

53. Trigwell S, Selvaduray G (1997) Effects of surface finish on the corrosion of NiTi alloy for biomedical applications. In: Pelton AR, Hodgson D, Russell SM, Duerig TW (eds) Proceedings of SMST 1997. Shape Memory and Superelastic Technologies, Pacific Grove, pp 383–388
54. Su YY, Raman V (1997) The quest for nitinol wire surface quality for medical applications. In: Pelton AR, Hodgson D, Russell SM, Duerig TW (eds) Proceedings of SMST 1997. Shape Memory and Superelastic Technologies, Pacific Grove, pp 389–394
55. Pan J, Leygraf C, Thierry D, Ektessabi AM (1997) Corrosion resistance for biomaterial applications of TiO_2 films deposited on titanium and stainless steel by ion-beam-assisted sputtering. J Biomed Mater Res 35:309–318
56. Aronsson BO, Lausmaa J, Kasemo B (1997) Glow discharge plasma treatment for surface cleaning and modification of metallic biomaterials. J Biomed Mater Res 35:49–73
57. Leitao E, Silva RA, Barbosa SA (1997) Electrochemical and surface modifications on N^+-ion-implanted 316L stainless steel. J Mater Sci Mater Med 8:363–368
58. Wennerberg A, Hallgren C, Johansson C, et al (1997) Surface characterization and biological evaluation of spark-eroded surfaces. J Mater Sci Mater Med 8:757–763

Other Medical Applications

An Implantable Drug Delivery System Based on Shape-Memory Alloys

Dominiek Reynaerts, Jan Peirs, Hendrik Van Brussel

1
Introduction

This chapter describes the design of an actively controlled implantable drug delivery device. Its job is to deliver small amounts of drug on a daily basis such that a patient no longer needs to get daily or weekly injections. Besides a reduction of the number of injections, implanted drug delivery systems offer many other advantages to a patient. Implantable drug delivery devices give a more constant drug level in the blood compared to injections. By the use of an active device instead of a passive, the drug level in the blood could be adapted to variations in physical activity, changes in temperature, etc. In chemotherapy and similar treatments, the device can be implanted at the place where the drug is needed such that the overall concentration of the drug in the body is much lower. The proposed device could be useful for hormonal treatments and all other treatments where small amounts of drugs are needed. In a first paragraph, solid drug delivery systems will be discussed. Main part of the chapter will concentrate on the design of a liquid drug delivery system. The proposed design is particular because the shape memory actuated microvalve has only silicone parts in direct contact with the drug. It is also very easy to produce in large volumes and can incorporate sensing functions.

There are three main reasons to take shape memory actuators instead of other types of actuation. First, shape memory actuators have a higher energy density [1] than other driving principles. Also their construction is very easy and they can be driven directly by an electric current using simple resistive heating. Finally, they insure a safe operation even over a long period of time. This last property is due to the fact that shape memory is based on a phase change so that transformation temperatures change very little even after several years. The work of Kao et al. [2] can further illustrate the reliability of shape memory alloys. The most frequently used shape memory alloys are nickel-titanium (NiTi) alloys.

Several attempts have been made to produce shape memory micro-actuators. Johnson et al. presented a tilting mirror [3]. Kuribayashi et al. developed a micron-sized arm using reversible NiTi [4]. Walker et al. made a microspring [5]. Johnson [6] and Kohl [7] have also developed shape memory alloy based microvalves. Some recent applications of shape memory actuators can be found in [8, 9]. In most cases, these micro-actuators were made by sputter deposition of NiTi and subsequent micromachining of the substrate. Many other research groups also try to produce thin films by sputter deposition [10–12]. Other possibilities are laser ablation [13] and melt spinning [14]. In this paper, shape memory thin wires will be used, mainly because of the availability of these materials.

2
Design of a Delivery System for Solid Drugs

2.1
Introduction

Liquid drugs are well known and are basically the same as the ones used with manual injection except that their concentration can be higher. Solid drugs are a new type and consist of crystalline material containing the drug. The drug is implanted in the body, for example just under the skin. The immunosystem of the patient decomposes the material and in that way releases the drug. The decomposition is a very slow process such that the drug can be made highly concentrated. For example, oestradiol (implant 100 mg Organon Laboratories Ltd.) is a solid drug containing female hormones and is supplied in the form of a cylinder 5 mm in diameter and 6 mm long. It is implanted just under the skin, only demanding a small incision, and serves 6 months. It is clear that with larger amounts, the lifetime can be increased. A drug delivery device based on solid drugs offers the possibility to regulate the concentration in the blood or tissue by exposing and covering the drug. If case of a leak or malfunction of the system, the drug stays in contact with the tissue and the concentration will remain high but no dangerous situation will occur.

With liquid drugs, leaks are potentially dangerous especially when the complete reservoir flows into the body at ones. Thus, a system for liquids has to be very reliable and fail-safe. For safety reasons the amount of drug stored and the concentration have to be limited. This forces the patient to refill the reservoir regularly. This is as well a disadvantage as an advantage. By refilling the reservoir, the lifetime of the device can be made higher than for solid drugs where refilling is unlikely. Table 1 gives a summary of the advantages and disadvantages of both types of drug.

Table 1. Solid and liquid drugs compared

	Solid drugs	Liquid drugs
Lifetime	Relatively long	Shorter
Refillable	No	Yes
Safety in case of leaks	Satisfactory	Not safe
Availability	Low (new products)	High

2.2
Design of a Drug-Delivery Device for Solid Drugs

The developed solid drug delivery devices have the form of a shell containing the solid drug. To deliver the drug, the shell opens and brings the drug in contact with the tissue. Figure 1 shows the basic prototype. The body enclosing the drug, elec-

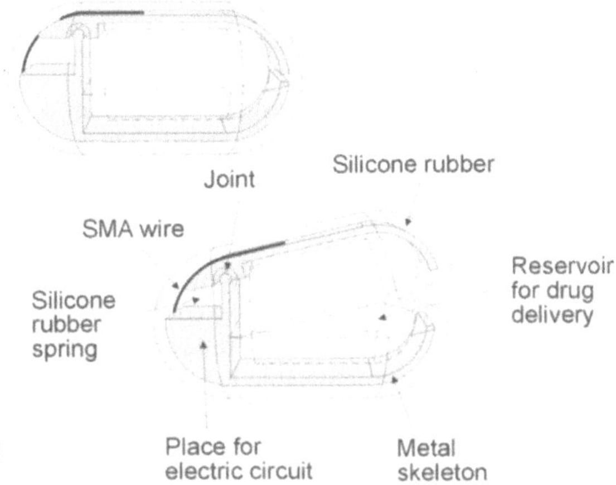

Fig. 1. Basic prototype for solid drug delivery

tronics, and the shape memory actuator is made of silicon rubber both for its biocompatibility and its flexibility. The reservoir opens when the shape memory wire acts on a metal skeleton embedded in the body. Without the metal skeleton the device would be too soft and only local deformations would occur when the wire is heated. Furthermore, the valve would open due to external forces acting on it.

The closing force is delivered by the silicon rubber located between the two surfaces at the left of the joint. This piece of rubber is compressed during opening and acts as a spring to close the device when the shape memory alloy cools down. The joint is a flexible joint based on the deformation of the rubber located in the joint. The use of a structural flexibility to generate the return force for the shape memory actuator greatly simplifies the design. This design feature will also be present in the other prototypes presented in this chapter. The skeleton is produced by electro-discharge machining. The rubber is a two component silicon rubber. Opening gaps in the order of 1 mm are attained. Figure 2 shows a photo of the prototype.

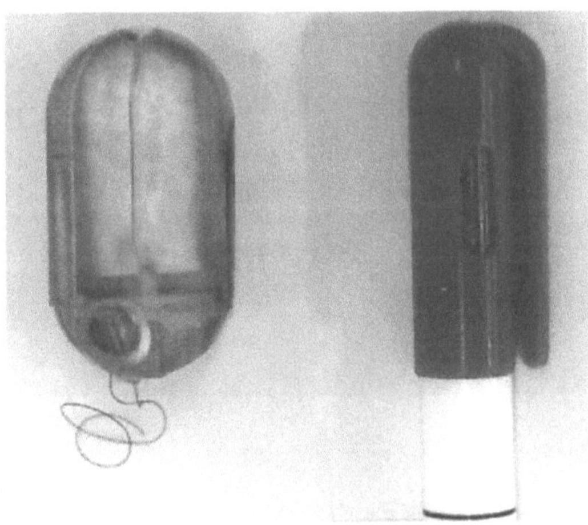

Fig. 2. Photo of the solid drug delivery prototype

2.3 Conclusion on Solid Drug Delivery

These solid drug delivery designs suffer from a number of problems. First, the movement can irritate the surrounding tissue. The body encapsulates all unknown entities in a cocoon of tissue. This can obstruct the flow or prevent the device to open. Moreover, tissue can grow into the opening and prevent the valve to close. Last but not least, long opening times are needed to dissolve the drug. With the oestradiol mentioned above, the valve should stay open several days. In that way, a normal hormonal cycle could be created. However, the proposed valve cannot be powered during several days. Not only would it be very inefficient with energy but also uncomfortable for the patient. These are the main reasons why research was shifted towards liquid drugs, described in the following paragraphs.

3 Design of a System for Delivery of Liquid Drugs

3.1 Introduction

The aim of this design is to manufacture a liquid drug delivery device that is implantable. This limits the weight to a few hundred grams and the size should be in the order of 300 cc. It is the aim that the system stays in the human body for several years. The volume of the single dose is heavily dependent on the type of drug. At the start of this design, a target value of 0.1 ml was aimed at. This also means that a refilling feature should be provided. The device should be remotely powered and controlled by a transcutaneous transformer in order to the need for a battery. Of course all biocompatibility and safety requirements for an implantable eliminate device should be fulfilled. Perhaps less evident is the requirement that the drug to be delivered should be stable over several years and compatible with the materials of the drug delivery device. In this specific design, shape memory alloys will actuate the system. The advantages of this technology were already discussed in Sect. 3.

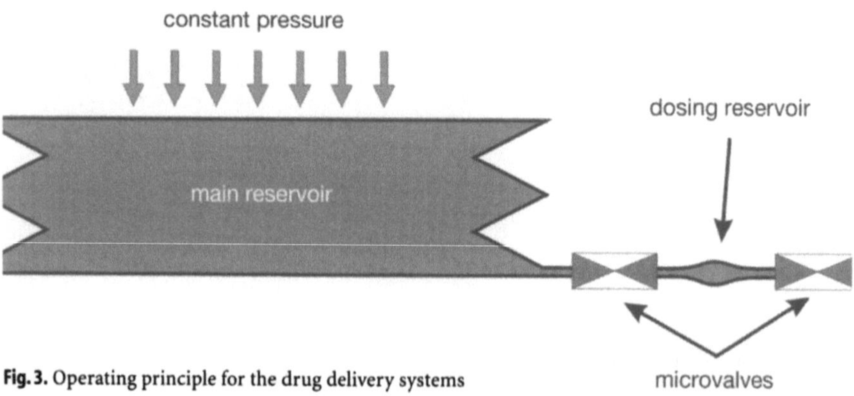

Fig. 3. Operating principle for the drug delivery systems

Figure 3 shows that the drug delivery system is based on a controlled release from a pressurised reservoir. It would be difficult to control the delivered dose exactly with only one valve. The flow will vary due to variations in pressure and variations among different valves and even more important, the cooling time of the shape memory element is uncontrollable. The solution is to use two valves. Two approaches are possible. A first approach is to open the second valve while the first one is closing. By varying the overlap time, the dose changes. This solves only the third problem. Another solution is to put a small reservoir between the two valves as shown on Figure 3. By opening the left valve, the drug flows from the main reservoir into the small one. After closing the left valve the right valve opens and the dose contained in the small reservoir flows into the body. It is clear that the small reservoir and the pressure in the main reservoir determine the dose delivered during one cycle. Because the form of the small reservoir is fixed, only using multiple cycles can change the dose. For this reason the second reservoir must be sufficiently small. This system is chosen for further development. By the use of two valves, the system becomes also safer. When one valve breaks and stays open, there is still a second one to close the tube. To have a fixed dose during one cycle, the pressure in the main reservoir has to be constant. This is an important design aspect for the pressurisation of the main reservoir. It is also important that the two valves are never open at the same time. The adopted dimensions for the whole drug delivery system (with the antenna for remote powering) are a diameter of 50 mm and a height of 15 mm.

3.2 Mechanical Design

3.2.1 *First Prototype*

The general concept of the valve is a pincher placed on an elastic tube. In the normal state the pincher is pressing on the tube such that this is locked. To unlock the tube, the pincher can be opened by actuation of a shape memory element. Different concepts were worked out and several prototypes were built to come to a concept with minimal dimensions and minimal number of parts. The reduction of parts is of extreme importance for both minimisation and production reasons. For this purpose, classic joints were avoided and replaced by elastic joints, screwed clamps were replaced by gluing, melting, and soldering. Following these design rules, a concept was retrieved with only three parts per valve: the body, the shape memory wire and a screw for adjusting the initial wire length. The body contains two elastic joints: one for the opening and closing movement of the valve and another for the pre-tensioning system of the wire. The valve is normally closed and opens when it is elastically deformed by actuation of a shape memory wire.

Figure 4 shows a first prototype of the resulting design made of aluminium. The valve is normally closed and opens when it is elastically deformed by actuation of the shape memory wire. Thereafter, the valve closes again by the elasticity of the joint. The body is produced by electro-discharge machining. Its dimen-

Fig. 4. Aluminium prototype valve

sions are 15 x 10 x 2 mm^3. The valve can be placed on a printed circuit board (PCB) among other electronic components. In this way, the electric connections for the shape memory alloy and eventually build-in sensors can be easily realised. This prototype is not yet optimal. The following sections describe the optimisation of the design.

3.2.2
Tube Characteristics

The first step in the optimisation was the measurement of the characteristics of the tubes. These values are very important to know the forces and displacements needed to open and close the valve. For this reason a dedicated tube-testing instrument was built. The tube is placed between the two tips. One tip is fixed to

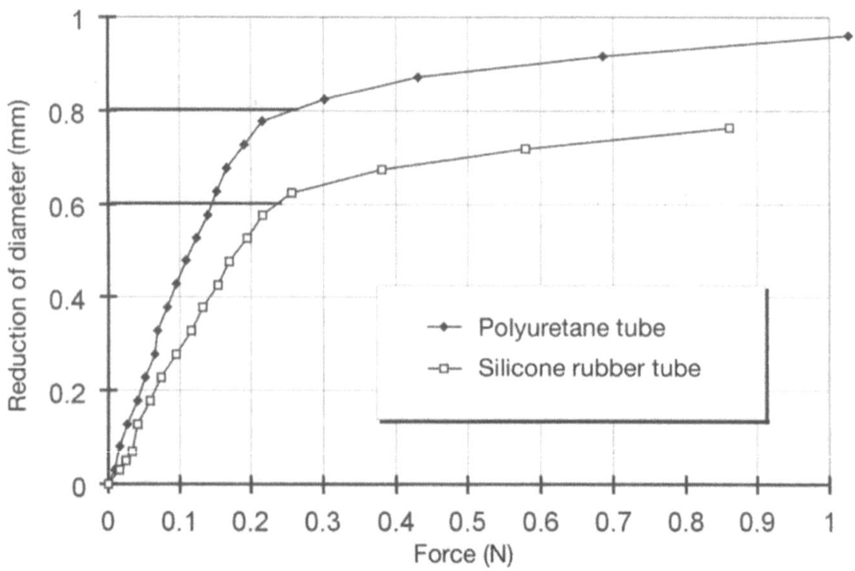

Fig. 5. Elastic characteristics of two types of tubes.

a flexible beam and can be moved by the micrometer that measures at the same time the displacement. The other tip is fixed to a force sensor. The form of the tip is chosen in a way that the contact surface between pincher and tube is small such that the force needed is as low as possible.

Figure 5 shows that two different types of tubes were tested. Tube 1 is a white polyurethane tube with an outer diameter of 1.2 mm and an inner diameter of 0.8 mm. The second one is a silicone rubber tube. It has an inner diameter of 0.6 mm and an outer diameter of 1 mm. For both tubes, the relationship is linear until the inner opening of the tube is closed. This point corresponds with a reduction of diameter equal to the inner diameter, what is to be expected. To avoid leaks, the tubes have to be pressed further. Tests show that an additional 0.1 mm is enough to withstand a pressure of 200 kPa. For both tubes this corresponds to a closing force of 0.5 N.

3.2.3
Valve Finite-Element Model

For the optimisation towards smaller dimensions, a parametric finite element model of the valve was made. The input parameters for the model are displayed in Table 2. The type of material and the values displayed there are the result of an iterative optimisation. Of course the material parameters play an important role. In contrast with the first prototype, plastic was chosen for manufacturing the second prototype. Miniaturisation of the valve means that a smaller mechanism has to offer the same displacement because the tube does not change. This requires higher strains in the pincher, which can be provided by plastics. Also plastics can offer a good electrical isolation of the wire, a good machinability, and the possibility of mass production.

Disadvantages of plastics are creep and the low melt temperature. For this reason, a glass fibre reinforced polyetherimide with a high melting temperature of

Table 2. Design parameters for the optimisation of the pincher

Material	
Composition	Polyetherimide and 20% glass fibre (Ultem 2200)
Heat deflection temperature	210°C
Melting temperature	390°C
Yield stress	150 MPa
Young's modulus	7600 MPa
SMA wire	
Diameter	120 μm
Stress	150 MPa
Strain	3%
Valve	
Thickness	2 mm
Thickness of beam	0.4 mm
Initial gap	0.1 mm
Tube	
Open (thickness, force)	0.6 mm; 0.14 N
Closed (thickness, force)	0.3 mm; 0.5 N

390°C was used. The working temperature of the shape memory alloy (≈100°C) is low enough under the heat deflection temperature of 210°C. Although the wire is not heated above 100°C, there are a number of reasons to work with a large safety margin. Thermoplastic materials are sensitive to creep. Creep becomes more important at higher temperatures. The glass fibres in this plastic have a positive effect in this case. Also the device should resist overheating which could occur and thermoplastic materials become weaker with higher temperatures. The high strength and modulus of a glass fibre reinforced polyetherimide assure that high forces can be generated in a small volume. As will be explained later, this plastic also gives the possibility to put electrical connections on the surface (like on a printed circuit board).

The shape memory element is a wire with diameter 120 μm and is used at its maximum stress and strain respectively 150 MPa and 3%. Parameters related to the valve itself are: the thickness of the valve, the thickness of the bent beam, and the gap between the valve tips. The thickness of the valve is 2 mm which is a bit more than the thickness of the tube. The stiffness of the valve is a third power of the thickness of the bent beam. Thus this value has a very strong influence. The smaller the initial gap between the two tips of the valve, the less stiff and the smaller the valve can be. This value is limited by the production technology of the valve and is chosen to be 0.1 mm. Figure 6 shows the form and final dimensions of the valve. The tips have the same form and dimension as the ones in the tube-measuring instrument.

After choosing the optimal parameters, a two-dimensional finite element model was made to check the stress concentrations at the top and bottom of the beam and to calculate the stresses in the flexible joint right down. In this model, a

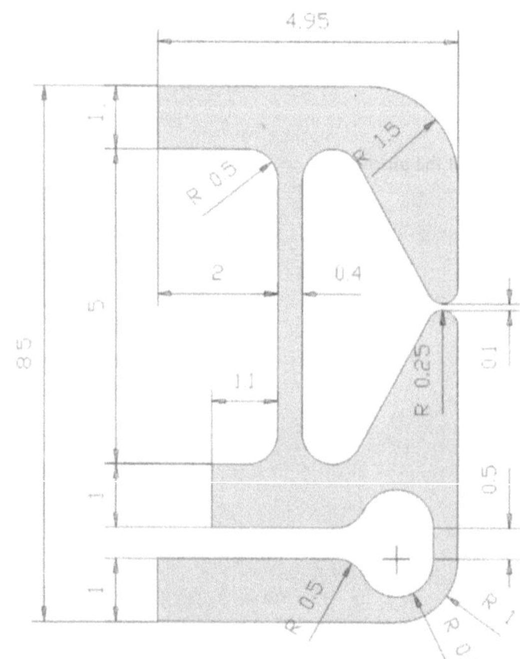

Fig. 6. Valve dimensions as a result of the optimisation

Table 3. Results of FEM-model compared to algebraic

	Opening	Maximum stress
Algebraic model	0.5 mm	62 MPa
Finite-element model	0.52 mm	73 MPa

constant force of 1.7 N, corresponding to 150 MPa simulates the shape-memory wire. Table 3 shows the results of the finite element analysis in comparison with the results of the algebraic model. The resulting deformation is the same and the stress is only slightly higher due to stress concentrations. The maximum stress on the deflected beam and is 73 MPa, still offering a safety factor of two in comparison to the yield stress of the material.

3.3 First Prototype Building

3.3.1 *Introduction*

A complete drug delivery system based on shape-memory actuated microvalves, including the technology for wireless powering and control, has been developed. The dimensions for the whole liquid drug-delivery system (with the antenna) are a diameter of 50 mm and a height of 15 mm. An electronic powering and control circuitry compatible with this system has also been developed. The plastic valve was made by a modified form of injection moulding. The mould is made by a combination of milling and electro-discharge machining. An important production aspect not yet dealt with is the fixation of the shape memory wire to the valve. In macroscopic designs, the wire is normally clamped between two plates using a screw. It is clear that this method is not appropriate for micro-scaled

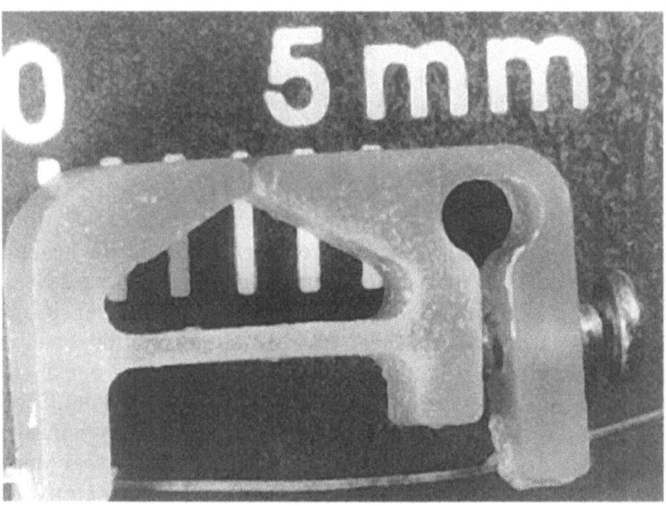

Fig. 7. Photograph of the pincher-type valve (5x8.5x2 mm^3)

devices. Other methods like soldering are impossible on plastic. Only gluing and ultrasonic welding are left. Ultrasonic welding is a technique where two parts are pressed on each other at high frequencies. The friction between the two parts produces the heat to melt the two parts together. This technique requires special equipment such that it was replaced by melting the wire into the valve using a current of 1 A. Because the initial length of the wire can not be controlled during assembly, the valve has a pre-tensioning system. The wire is pre-stressed by adjusting the length between the two "legs" of the pincher using a screw with 1-mm diameter. The threaded hole for this screw is made by classic production methods. Figure 7 shows a photograph of the resulting valve.

3.3.2
Electrical Characteristics

The energy needed to actuate the valve depends on the resistance of the wire, the load and the thermal characteristics of the wire. Values for these characteristics are presented in Table 4. The values about the thermal characteristics are provided by the shape memory supplier (Memry Europe). For the transformation enthalpy, literature gives 3200–12000 J/kg, which is quite different from the values below. The wire has to be heated from body temperature (37°C) to A_f. As indicated by the stress rate, A_f rises approximately 1°C per 5 MPa load. With a load of 150 MPa, this gives an A_f of 106°C such that the temperature difference to be bridged is 69°C. The shape memory wire has an electrical resistance of 0.039 ohm/mm. The length of the wire between the clamps is 6.5 mm but to realise the electrical connection, the wire is longer, so that the total length is approximately 15 mm, resulting in a resistance of 0.59 ohm. The power consumed is equal to the sum of the energy needed to heat the thermal mass, the transformation enthalpy and the heat losses as expressed by the following equation.

$$P_{in}=m.c'.\frac{dT}{dt} + w.S.T$$

with

$$c' = c + \frac{\Delta h}{(A_f - A_s)} \quad \text{if} \quad A \leq T \leq A_f$$

$$c' = c \quad \text{else}$$

T represents temperature, t time and S the surface of the wire. Other symbols are explained in Table 4. The term c' represents the specific heat plus the transformation enthalpy. The transformation enthalpy is only present during transformation and is assumed to be equally spread over the transformation interval. The last term represents the heat loss due to dissipation to the environment. The faster the wire is heated the lower the losses. The above equation was solved assuming that the wire is heated to A_f using a constant power level. Figure 8 shows the results for different power levels. Low power means slow heating and thus more dissipation to the environment. On the other side, the current limits the fastest heating time. For example, for a response time of 0.66 s, the current

Table 4. Characteristics of SMA wire

Mechanical characteristics	
Diameter	0.12 mm
Length	15 mm
Density	6500 kg/m^3
Load	150 MPa
Electrical characteristic	
Resistance/length	0.039 ohm/mm
Thermal characteristics	
Start austenitic transformation (A_s)	65°C
End austenitic transformation (A_f)	76°C
Start martensitic transformation (M_s)	58°C
End martensitic transformation (M_f)	51°C
Stress rate	5 MPa/K
Thermal conductivity austenite	18 W/(mxK)
Thermal conductivity martensite	8.6 W/(mxK)
Specific heat (c)	490 J/(kgxK)
Transformation enthalpy (Dh)	28000 J/kg
Heat transmission coefficient (w)	5.8 W/(m^{2}xK)

needed is 0.4 A. This is much for a miniature electronic system. From the figure below, one can also conclude that heat losses are negligible when the current is higher than 0.25 A such that the current can be chosen between 0.25 A. and the maximum current of the source.

Table 5 compares the results of the model with values measured on the valve and shows that the model gives quite a good prediction. Important for the integration with the electronics is the resistance of the source. The internal resistance of the source has to be significant below the resistance of the wire. If not, the source will be heated instead of the wire.

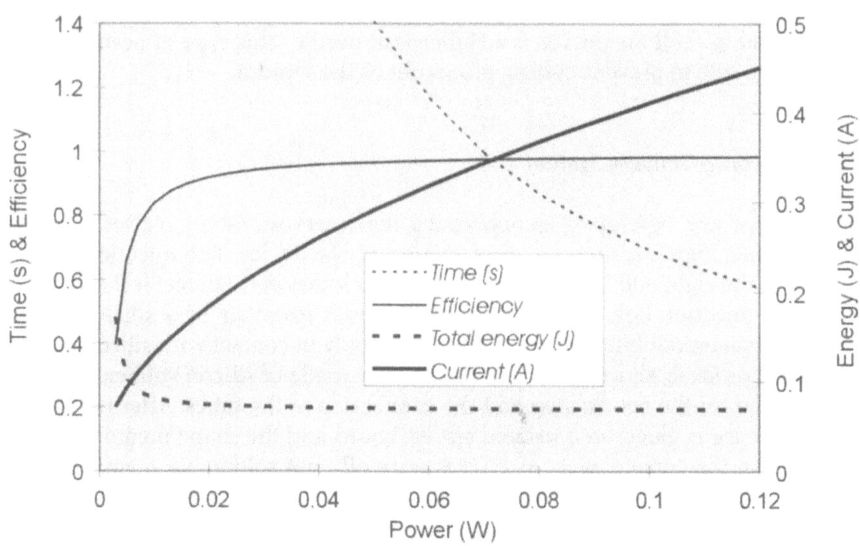

Fig. 8. Influence of power on response time, energy consumption and efficiency

Table 5. Electrical characteristics calculated and measured

	Calculated	Measured
Current	0.42 A	0.4 A
Voltage	0.25 V	0.3 V
Resistance	0.59 ohm	0.75 ohm
Response time	0.66 s	0.66 s
Power	0.11 W	0.12 W
Energy	69 mJ	79 mJ
Efficiency	98%	

3.3.3
Design of the Reservoirs and Refill Port

The main reservoir has a bellow-like form enabling easy deformation of the reservoir. The content of the reservoir is estimated to be approximately 1 ml. It is pressurised by a two-phase liquid with boiling temperature lightly above body temperature. This technique is already in use in commercial drug delivery devices and assures a constant pressure independent of the volume left in the reservoir. The expansion reservoir is made of two elastic membranes. When applying a pressure to this reservoir, the membranes deform and the reservoir fills with drug. When applying no pressure, the internal volume is zero. As the volume contained in the reservoir depends on the pressure, the pressure in the main reservoir should be constant to have a constant dose. The refill port contains a 10 mm thick silicon rubber septum. A needle can penetrate this septum to refill the reservoir. After retraction of the needle the septum closes again. Silicon rubber septa are generally used drug delivery devices like Port-a-Cath and Infusaid. The septum of the Port-a-Cath system can be used up to 2000 times. The stainless steel stop prevents the needle from piercing the bottom of the refill port. The needle used to refill the device, is a Huberpunt needle. This type of needle has its hole on the side to prevent cutting pieces out of the septum.

3.3.4
Prototype Drug Delivery System

A prototype was developed incorporating the reservoir, tubing, valves, housing and antenna. Figure 9 shows a cross-section of the device. The injection port is located at the right and is a little bit elevated such that you can feel it through the skin. The injection hole is connected to the main reservoir by a silicon rubber tube. For biocompatibility reasons, the drug is only in contact with silicon rubber and stainless steel. All reservoirs and tubing are made of silicon rubber. Stainless steel is used for the needle stop and the connection of the tubes. After tuning the valves, one leg is glued on a printed circuit board and the shape memory wire is soldered on the copper pads. As NiTi is normally not solderable, a copper layer was first deposited in an electrochemical way. The wire and the valve are attached to the printed circuit board in a way the movement is not disturbed. The housing is a thin metal shell of stainless steel or titanium. As the housing is a metal part,

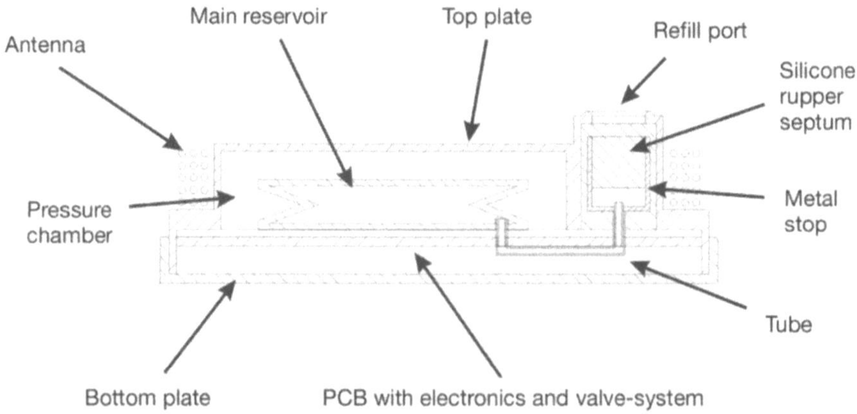

Fig. 9. Cross-section of the prototype

the antenna has to be placed on the outside to avoid electromagnetic shielding. This maximises also its internal surface. The antenna is wound in a form that fits exactly around the housing and has to be covered with some biocompatible resin or plastic. Silicon rubber can not be used because it does not stick on the metal. The two shells of the housing can be welded together by laser, which is a precise, reproducible process with minimal temperature influence.

In order to be able to build a working prototype, some modifications were made. A first one affects the pressurising of the main reservoir. Some pressurising liquids were tested like ether and pentane but these are incompatible with the silicon rubber of the reservoir. The silicon rubber swells and is transparent for these products. Or another liquid has to be found or a metal reservoir has to be used. Metal reservoirs are already in use in commercial drug delivery systems. In the prototype, this type of reservoir is replaced by an elastic one. The pressure is now offered by the elasticity of the reservoir. A second difference is the aluminium that is used for the housing of the device. Normally stainless steel has to be

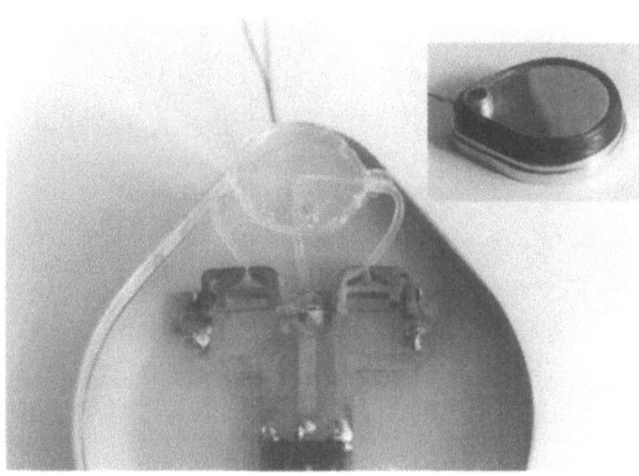

Fig. 10. Drug delivery system prototype

used but this is difficult to mill. In production, this part can be stamped out of a plate. Also biocompatible plastics or ceramics could be used. Figure 10 shows an internal view of the final drug delivery prototype. The inset shows the completely closed device with a mock-up of the antenna.

4 Operational Tests

4.1 System Components

Several operational tests were performed on the components of the drug delivery system. The tubes were tested with air to determine the maximum pressure they resist. This is 200 kPa (always relative pressure). This is not very high in comparison with the pressure a person can generate in a syringe (500 kPa). The valves were tested with air to determine at what pressure they leak. Pressure was raised up to the maximum strength of the tubes, without any leak. Also fatigue tests were done with water as a test liquid. The applied pressure was 100 kPa and current peaks of 0.4 A during 0.66 s were applied to the valve. The desired lifetime is 10,000 cycles, corresponding to three injections a day, 9 years long, or five injections a day, 5.5 years long. The tests showed that the smoothness of the tips pressing on the tubes is of extreme importance: with a rough surface, the tube was perforated already after a few hundred cycles. With a smooth surface, 10,000 cycles were reached without any problem. No leaks occurred and the valve had not to be adjusted. Figure 11 gives an impression of the smoothness of the tips of the valve that performed 10,000 cycles. Only problem were impurities in the water partially obstructing the passage such that the delivered quantity per cycle decreased with time. However, this valve design exhibits no dead corners and is in fact "self-flushing" for impurities contained in the liquid.

Another important aspect is of course the power consumption of the system. The resistance of the wire is 0.75 W. Together with current peaks of 0.4 A during 0.66 s, this results in a power consumption of 120 mW or 80 mJ per cycle and per

Fig. 11. View on the tips of the valve used for the second test

valve. It was experimentally verified that this amount of energy could be transmitted by a system equivalent to a transcutaneous transformer.

4.2 Total System

The total system with refill port, main reservoir, small reservoir and valves was tested as a whole. Distillated water was injected through the silicon rubber stop of the refill system. The stop did not leak after retraction of the needle, even after several refills. After opening the first valve, the expansion reservoir swells. After closing the first valve and opening the second, approximately one drop leaves the drug delivery device.

The main reservoir of the proposed prototype has a capacity of about 1 ml. Depending on the internal pressure, the dose leaving the system each cycle is 5–25 μl, allowing 40–200 doses. Further miniaturisation and integration of the peripheral components, not necessarily of the valves, allows to include a reservoir of 15 ml (50% of the volume of the total device) so that the total number of doses could be 3000.

5 Possible Improvements

Compared to the demonstrated prototype, a number of improvements can still be suggested. These are especially important for further development of the prototype towards a commercial product. To increase the resistance of the shape memory element, thin ribbons [15] can be used instead of wires. As safety is an important issue, some sensors could be built in to detect leaks or fractures in the valve. The most critical place for leaks is between the tips of the pincher. If the drug is conducting, a leak can be detected by covering the two tips of the valve with a conductive material. If a leak occurs, there will be electrical contact between the two tips. To detect fracture, conducting strips can be printed on the critical places. If the device breaks, the electrical connection is broken. As there are two valves, still one is closing the tube. In both cases, the device has to report this to the external controller after which the device has to be removed as soon as possible.

The electrical connections for the sensors and the shape memory alloy can be put on the plastic body of the valve using MID-technology (Mould Interconnect Device). The Ultem 2200 is compatible with this technology. Figure 12 shows how the connections could be realised. Leak detection is done via pads one and two, fracture detection via pads two and three, power supply to the shape memory wire via pads four and five. Because of the small dimensions, soldering is not done on metal pins but via pads on the side of the valve. In this way the valve has become a surface mounted device (SMD). The shape memory wire can be made solderable by electrolytic deposition of copper on the surface. Omitting the adjustment screw can give a further optimisation. This can be realised by pretensioning the shape memory wire during assembly. In this case, a cold assembly method is required.

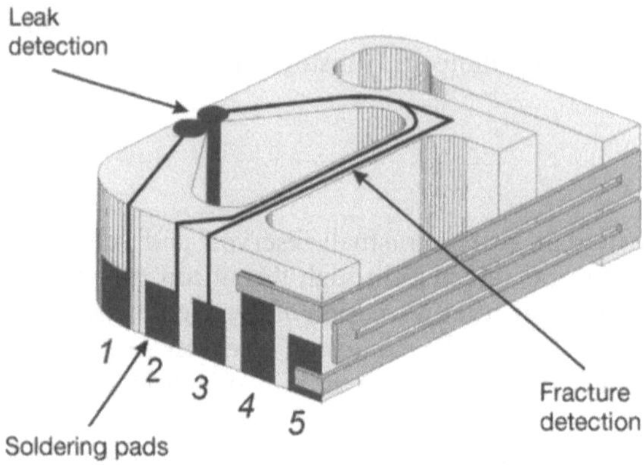

Fig. 12. Electrical connections realised using the MID-technology

6 Conclusion

This paper mainly presented an implantable system for delivery of liquid drugs. The proposed operating principle is based on a precisely controlled, discontinuous release from a pressurised reservoir using a shape memory alloy actuated microvalve system. The system is remotely powered and controlled using a transcutaneous transformer. A refilling possibility based on transcutaneous injections is provided. The prototype liquid drug delivery system contains besides the valve system also a main reservoir, a refill port, an antenna, and a housing. The valve design is optimised towards dimensions, energy consumption, and production aspects. The resulting valve measures 8.5x5x2 mm^3 and consists of only three parts. The small dimensions allow to integrate the valve on a PCB with the electronics, simplifying the electrical connections. Tests showed no leakage and a lifetime of at least 10,000 cycles. The volume of one dose can be controlled with accuracy up to 5 μm. Although not suitable for implantation, the current design clearly shows the potential of this technology for this category of medical applications.

Acknowledgements. This research has been sponsored by the Brite-Euram programme of the European Union, project number BE-7596–93, contract number BRE2-CT93-0579 and by the Belgian programme on Interuniversity Poles (IUAP4–24) of attraction initiated by the Belgian State, Prime Minister's Office, Science Policy Programming. The authors assume the scientific responsibility of this paper. D. Reynaerts is a Postdoctoral Fellow of the Fund for Scientific Research – Flanders (Belgium; F.W.O.).

References

1. Reynaerts D, Van Brussel H (1994) Shape memory alloy based electrical actuation for robotic applications. In: Pelton AR, Hodgson D, Duerig TW (eds) Proceedings of SMST 1994. Shape Memory and Superelastic Technologies, Pacific Grove, pp 271–276
2. Kao M, Schitz D, Thoma P, Klaus P, Angst D (1996) Shape memory alloy ribbon actuator. In: Proceedings of the Fifth International Conference on New Actuators, 26–28 June 1996. Axon Technologie Consult, Bremen, pp 370–374
3. Johnson AD, Busch JD, Ray CA, Sloan C (1992) Fabrication of silicon-based shape memory alloy micro-actuators. In: Jardine AP, Johnson GC (eds) Smart materials fabrication and materials for micro-electro-mechanical systems. Materials Research Society, Pittsburgh, pp 151–169
4. Kuribayashi K, Taniguchi T, Yositake M, Ogawa S (1992) Micron sized arm using reversible TiNi alloy thin film actuators. In: Jardine AP, Johnson GC (eds) Smart materials fabrication and materials for micro-electro-mechanical systems. Materials Research Society, Pittsburgh, pp 167–175
5. Walker JA, Gabriel KJ (1990) Thin-film processing of TiNi shape memory alloy. Sensors Actuators A 21:243–246
6. Johnson AD, Busch JD (1994) Recent progress in the application of thin film shape memory alloys. In: Pelton AR, Hodgson D, Duerig TW (eds) Proceedings of SMST 1994. Shape Memory and Superelastic Technologies, Pacific Grove, pp 299–304
7. Kohl M, Skrobanek KD, Quandt E, Schlossmacher P, Schuessler A, Allen DM (1995) Development of microactuators based on the shape memory effect. In: Proceedings of the International Conference on Martensitic Transformations, 20–25 Aug 1995. Les éditions de physique, Ulis, pp 1187–1192
8. Kohl M, Just E, Strojek A, Skrobanek KD, Pfleging W, Miyazaki S (1998) Shape memory microvalves for high pressure applications. In: Proceedings of the Sixth International Conference on New Actuators, 17–19 June 1998. Axon Technologie Consult, Bremen, pp 473–477
9. Bellouard Y, Clavel R, Gotthardt R, Bidaus JE, Sidler T (1998) A new concept of monolithic shape memory alloy micro-devices used in micro-robotics. In: Proceedings of the Sixth International Conference on New Actuators, 17–19 June 1998. Axon Technologie Consult, Bremen, pp 499–502
10. Miyazaki S, Ishida A, Takei A (1992) Development and characterization of Ti-Ni shape memory thin films. In: Proceedings of the Second International Symposium on Measurement and Control, Nov 1992. Society of Instrument Contract Engineers of Japan, Tokyo, pp 495–500
11. Ikuta K, Fujita H, Ikeda M, Yamashita S (1990) Crystallographic analysis of TiNi shape memory alloy thin film for micro actuator. In: Proceedings of the IEEE Micro Electro Mechanical Systems, 11–14 February 1990. Institute of Electrical and Electronics Engineers, Piscataway, pp 38–39
12. Kohl M, Quandt E, Schuessler A, Trapp R, Allen DM (1994) Characterization of NiTi shape memory microdevices produced by microstructuring of etched sheets or sputter deposited films. In: Proceedings of the Fourth International Conference on New Actuators, 15–17 Jun 1994. Axon Technologie Consult, Bremen, pp 317–320
13. Ikuta K, Hayashi M, Matsuura T (1990) Shape memory alloy thin film by laser ablation. In: Proceedings of IEEE Micro Electro Mechanical Systems, 25–29 January 1994. Institute of Electrical and Electronics Engineers, Piscataway, pp 355–360
14. Donner P, Eucken S (1990) The shape memory effect in meltspun ribbons. Mater Sci Forum, 56:723–728
15. Reynaerts D, Peirs J, Van Brussel H (1995) Production of shape memory alloys for microactuation. J Micromech Microeng 5:1–3

References

Subject Index

MIX
Papier aus verantwortungsvollen Quellen
Paper from responsible sources
FSC® C105338

If you have any concerns about our products,
you can contact us on
ProductSafety@springernature.com

In case Publisher is established outside the EU,
the EU authorized representative is:
Springer Nature Customer Service Center GmbH
Europaplatz 3, 69115 Heidelberg, Germany

Printed by Libri Plureos GmbH
in Hamburg, Germany